SPY PLANES, INTRUDERS, AND WILD WEASELS

SPY PLANES, INTRUDERS, AND WILD WEASELS

ELECTRONIC WARFARE IN THE SKIES OVER VIETNAM

THOMAS WILDENBERG

Naval Institute Press
Annapolis, Maryland

Naval Institute Press
291 Wood Road
Annapolis, MD 21402

© 2025 by Thomas Wildenberg
All rights reserved. No part of this book may be reproduced or utilized in any form or by any means, electronic or mechanical, including photocopying and recording, or by any information storage and retrieval system, without permission in writing from the publisher.

Library of Congress Cataloging-in-Publication Data
Names: Wildenberg, Thomas, author
Title: Spy planes, intruders, and wild weasels : electronic warfare in the skies over Vietnam / Thomas Wildenberg.
Other titles: Electronic warfare in the skies over Vietnam
Description: Annapolis, Maryland : Naval Institute Press, [2025] | Includes bibliographical references and index.
Identifiers: LCCN 2025017332 (print) | LCCN 2025017333 (ebook) | ISBN 9781682476697 hardback | ISBN 9781682476772 ebook
Subjects: LCSH: Vietnam War, 1961–1975—Aerial operations, American | Vietnam War, 1961–1975—Technology | Electronic warfare aircraft—United States—History—20th century | Avionics—United States—History—20th century
Classification: LCC DS558.8 .W56 2025 (print) | LCC DS558.8 (ebook)
LC record available at https://lccn.loc.gov/2025017332
LC ebook record available at https://lccn.loc.gov/2025017333

♾ Print editions meet the requirements of ANSI/NISO z39.48–1992 (Permanence of Paper).
Printed in the United States of America.

9 8 7 6 5 4 3 2 1

To
Bob Breault
Better Known as "The Lieutenant"
Fighter Pilot, Scientist, Entrepreneur, and Humanitarian

And
The other American Aviators who risked their lives flying in the skies over Vietnam

CONTENTS

List of Tables and Figures ... ix
Preface ... xi
Acknowledgments ... xv
Abbreviations and Acronyms ... xvii
Electronic Equipment Numbering System ... xix
Aircraft Type Numbering System ... xxi
Soviet Radar and Missile System Names ... xxii

Introduction. The Air War in Vietnam: A Historical Overview ... 1
1. First Strikes ... 3
2. Operation Rolling Thunder ... 11
3. Electric Spads, Voodoos, Skyknights, and Electric Intruders ... 24
4. The SAM Threat ... 36
5. ECM Pods, RHAWs, and Chaff Dispensers ... 47
6. Suppressing SAM Sites ... 60
7. Black Boxes and Wild Weasels ... 69
8. AGM-45 Shrike: Potential Game Changer ... 83
9. Wild Weasels II and III ... 92
10. AGM-78 Standard Antiradiation Missile ... 103
11. Dealing with the MiG Threat ... 111

12. Big Safari, College Eye, and Combat Apple ... 125
13. Combat Martin and an AGM-45 Friendly Fire Incident ... 134
14. Automating the Tactical Air Control System:
    Combat Lightning, Red Crown, and the NTDS ... 138
15. Proud Deep Alpha, RGM-8H Talos, and
    Operation Linebacker I ... 148
16. Teaball, NTDS, and Red Crown ... 159
17. Combat Tree Phantoms, Prowlers, and Wild Weasel IVs ... 168
18. Linebacker II: The Preliminaries ... 176
19. Linebacker II: Into the Dragon's Teeth ... 187
20. Linebacker II: Losses Lead to New Tactics ... 196
21. Looking Back ... 208

Appendix I. Military and ECM Radar Bands ... 215
Appendix II. Soviet Equipment Deployed by the North Vietnamese ... 217
Appendix III. AGM-45 Shrike Missiles ... 218
Appendix IV. F-4 Phantom EW Upgrades ... 220
Notes ... 223
Bibliography ... 263
Index ... 293

TABLES AND FIGURES

TABLES

2.1	RB-66C Electronic Warfare Suite	21
4.1	ECM Equipment Installation	46
5.1	RHAW Gear	59
8.1	AGM-45 Variants	91
10.1	AGM-78B Bomb Damage Assessment Logic Display	109
21.1	Linebacker II: Total Sorties Conducted	209

FIGURES

1.1	Contemporary map of Operation Pierce Arrow	4
4.1	Fan Song radar	38
5.1	Showing the effects of synchronized noise jamming	49
5.2	F-105D Pod Formation	51
5.3	The CRT display supplied with the APR-25	56
8.1	Angle gating operation	89
8.2	F-100F Super Sabre Wild Weasel I	90
9.1	Route Packages were operating zones within North Vietnam	96
10.1	F-105G Wild Weasel III	110
11.1	The Ironhorse system diagram	113

11.2	North Vietnamese early warning radar coverage	117
11.3	EC-121D mission tracks	120
15.1	North Vietnamese jet aircraft order of battle	149
15.2	Map showing details of first AGM-8H combat firing	153
16.1	The Bull's Eye map used to identify air and land threats to U.S. aircraft	160
16.2	Teaball elements and communication system	162
18.1	B-52 cell formation	178
18.2	B-52G flight deck	179

PREFACE

Technology has been the primary source of military innovation throughout history. It drives changes in warfare more than any other factor.[1]
—ALEX ROLAND, *WAR AND TECHNOLOGY*

The Vietnam War has frequently been termed the first high-tech air war. It was the first major conflict to employ air-to-air missiles and smart ground attack weapons on a large scale. It was also the first in which electronic warfare (EW) played a major role in the air—as a means of defeating the surface-to-air missile—but also as a means of improving the offensive and defensive capabilities of U.S. fighters in air-to-air combat. It was the first aerial campaign in which electronic intelligence (ELINT) and signal intelligence (SIGINT) aircraft were used extensively to collect information on the enemy's electronic order of battle, the characteristics of its radars, identification friend or foe (IFF) equipment, missile control signals, and communication systems. The Vietnam War was also the first war in which strike aircraft were fitted with electronic countermeasure (ECM) pods and radar homing and warning (RHAW) equipment. It was also the first air war in which airborne jamming was used on a wide scale, and it saw the

---

[1] As expressed in the introduction to Alex Roland's *War and Technology: A Very Short Introduction* (New York: Oxford University Press, 2016), 1–2.

introduction of the first aircraft designed specifically for electronic warfare, the EA-6B. It was also the first conflict in which the suppression of enemy air defenses (SEAD) and the aircraft dedicated for this purpose, called Wild Weasels, became a major component of operational planning and use.

Although dozens of books and numerous articles have been written on the air war in Vietnam, the number of books on EW can be counted on one hand, none of which provides a comprehensive history of this unique aspect of the war. There are several reasons for this. First, the Vietnam War was a long and complicated conflict that involved a number of air campaigns involving strike, bombing, interdiction, and air-to-air combat. Historians, authors, and aviation enthusiasts have concentrated on the various air campaigns or the specific aircraft involved. Second is the level of secrecy surrounding EW. Detailed information is hard to come by since operational records concerning the use of EW were classified at the time they were created. Over time, as these records became obsolete and were no longer considered of value, they were destroyed. This particular problem was raised by Phil Hays in his discussion of his attempts to obtain information on the only launch of an RGM-8H Talos antiradiation missile (ARM) during the Vietnam War, which took place five decades ago. "The ARM shots were classified Secret," he wrote in a footnote attached to his internet post titled "Details of the First RGM-8H Anti-Radiation Missile Combat Firing," "and any record of the *OK City's* ARM shot would have been in the classified Commanding Officer's Quarterly Narrative Report (CONAR) attachment to the 1972 Ship's History. I have tried Freedom of Information Act searches from the Navy History and Heritage Center and searches in the National Archives with no success. The Navy tells me that all CONARS were destroyed!"[2]

There are other reasons too why writers have shunned the subject of EW. There is little glory attached to its use and it is not the instrument that determines the result of great battles or memorable actions. The technical nature of EW also discourages those who are not trained in engineering

---

[2] Philip Hays, "Details of the First RGM-8H Anti-Radiation Missile Combat Firing." Okieboat, June 30, 2020, https://www.okieboat.com/Talos%20antiradiation%20shot.html.

or the physical sciences. Neither of these impediments dissuaded me from tackling this heretofore neglected aspect of military history.

Because of the secrecy surrounding documents related to EW, other avenues of research are needed to obtain a full picture of the equipment and tactics utilized in EW. Fortunately, a significant number of individuals have provided recollections of their involvement with EW. Some of these are in the form of oral histories, others are contained in interviews or blogs. Getting to such material has been greatly facilitated by the use of the research capabilities inherent in the internet, especially via the use of Google Books.

The internet and the World Wide Web have also been a useful source for obtaining a number of declassified histories produced by the Central Intelligence Agency and the National Security Agency. These have proved to be of immense value. Also accessible online are the many CHECO (contemporary historical examination of current operations) reports that were prepared by the Air Force.

Several aspects of the Vietnam War make writing an easily readable, comprehensive history of the electronic war in the skies over Vietnam difficult. As in other published works, the history of the air war over Vietnam frequently focuses on one or more of the major air campaigns that took place between 1965 and 1972. Putting all of this information together is an arduous task. Complicating the story is the dual command structure between the Air Force and the Navy and the involvement of National Security Agency and the CIA. As a result, many of the books and articles on the Vietnam War focus on the activities of each service or agency to the exclusion—for the most part—of the other. Likewise is the need to explain the political actions and reasoning that drove the command decisions originating with the Joint Chiefs of Staff. Lastly, the large number and different variants of the aircraft and the various pieces of ECM gear deployed by each of the services adds another layer of complexity.

As readers proceed through the book's contents, they will note the extensive endnotes that are incorporated in the text. These are provided for two reasons. First, they are a means of documenting where certain material came from and verifying its veracity. Second, they are a potential source for use in further research by scholars.

It is my hope that this work provides a useful vehicle for both scholars and current practitioners of the art of electronic warfare to discover and understand the evolution and application of airborne electronic warfare during the Vietnam War.

ACKNOWLEDGMENTS

Several individuals were of great help in preparing this work. I am particularly grateful to my friend Bill Trimble, an experienced editor, scholar, and talented author in his own right, who was kind enough to go through the entire manuscript and provide much-needed editorial corrections and clarifications to the text. I am also indebted to Bob Breault, one of the original Wild Weasel pilots, for his insightful comments and suggestions based on his firsthand knowledge of Wild Weasel combat in Vietnam. I want to thank Shawn Finn, the interlibrary loan manager for the Pima County Library, for his assistance in obtaining much-needed material via interlibrary loan. Thanks are also in order for Martin Streetly and Phil Hays, both of whom were kind enough to allow me to reproduce their excellent illustrations.

ABBREVIATIONS AND ACRONYMS

AAA	Anti-Aircraft Artillery
ACRP	Airborne Communication Reconnaissance Program
AIC	Air Intercept Controller
BDA	Bomb Damage Assessment
CAP	Combat Air Patrol
CBU	Cluster Bomb Unit
CIC	Combat Information Center
CINCPAC	Commander-in-Chief, Pacific Command
CPU	Computer Processing Unit
CRT	Cathode Ray Tube
dbm	Decibel per Milliwatt
D/F	Direction Finding
DMZ	Demilitarized Zone
ECM	Electronic Countermeasures
ELINT	Electronic Intelligence
EWO	Electronic Warfare Operator
GCI	Ground Control Intercept
HRF	High Radio Frequency
IFF	Identification Friend or Foe
ILS	Instrument Landing System
JCS	Joint Chiefs of Staff
MACV	Military Advisory Command

NATO	North Atlantic Treaty Organization
NSG	Naval Security Group
NVA	North Vietnamese Army
PRF	Pulse Repetition Frequency
POW	Prisoner of War
RF	Radio Frequency
RTAFB	Royal Thai Air Force Base
SAC	Strategic Air Command
SAR	Search and Rescue
SIGINT	Signals Intelligence
SIOP	Single Integrated Operational Plan
SIR	Systems Integrated Receiver
TFW	Tactical Fighter Wing
TFS	Tactical Fighter Squadron
TJS	Tactical Jamming System
UHF	Ultra-High Frequency
VF	Fighter Squadron
VHF	Very High Frequency
VNAF	Vietnamese Air Force (South Vietnam)
VPAF	Vietnamese People's Air Force (North Vietnam)
WSO	Weapons System Officer

ELECTRONIC EQUIPMENT NUMBERING SYSTEM

"AN" NUMBERS

Since February 1943 communications and electronics systems in the U.S. military have been designated using the Joint Army-Navy Nomenclature System, which is also known as the Joint Communications–Electronics Nomenclature System or short "AN System" for the letters that prefaced each name. The initial emphasis was on airborne radio and radar equipment, but the system was designed to be extendable and soon included other types of equipment. When the Air Force separated from the Army in 1947, it continued to use the system for its electronic equipment. In 1957 the system was formalized in Military Standard 196 "Joint Electronics Type Designation System" (JETDS).

All designations are prefixed by *AN*. Originally, this stood for "Army-Navy," but this interpretation is no longer valid. Nowadays, *AN* is simply an indicator for the JETDS. In nonofficial references to electronic equipment, the *AN* prefix is often omitted, which will be the practice followed in this work.

In the *AN* system, each piece of equipment is identified by an alphanumeric designator that begins with the letters *AN* followed by a forward slash and three letters. The first letter indicates the installation location of the equipment (e.g., *A* for aircraft). The second letter indicates the type of equipment (e.g., *P* for radar). The third letter defines the purpose of the equipment (e.g., *R* for receiving or passive detection). The final element is the model number sequence. Thus, AN/APR-5 defines the fifth airborne radar receiver produced.

QRC NUMBERS

QRC stands for quick reaction capability for ECM, a program introduced by the Air Force in 1952 for rapidly obtaining high-priority electronic equipment by circumventing the traditional bureaucratic procurement methods for standard electronic gear. The ECM Branch of the Aircraft Radiation Laboratory at Wright Air Development Center, Wright-Patterson Air Force Base, Dayton, Ohio, was assigned technical responsibility for QRC equipment, and the Air Force Supply Depot at Gentile Air Force Base, Kettering, Ohio, was given procurement responsibility.

AIRCRAFT TYPE NUMBERING SYSTEM

Prior to September 18, 1962, when the Department of Defense's Tri-Service Aircraft Description System Aircraft took effect, U.S. Navy aircraft designations were based on the Navy's mission-manufacturer-number system. The dual numbering system that applied to aircraft designed prior to that date sometimes creates confusion with respect to referencing the correct version of certain aircraft described in the text. For historical accuracy, I have chosen to use the Navy system to describe all aircraft that were in use before September 18, 1962, and the Tri-Service system for aircraft in use thereafter.

SOVIET RADAR AND MISSILE SYSTEM NAMES

Wherever possible, upon first use, I will provide the reader with the Russian model number and name, followed in parentheses by the code name assigned by NATO used to identify specific pieces of equipment. NATO code names, which are commonly used in most publications, will be used thereafter. A list of the equipment mentioned in this book showing the Russian model numbers, Russian name (where available), along with the NATO code assigned can be found in appendix II.

INTRODUCTION
THE AIR WAR IN VIETNAM

A HISTORICAL OVERVIEW

I have tried, for the most part, to discuss the development and application of electromagnetic warfare in Vietnam in chronological order. Nevertheless, it was sometimes necessary to focus on the development of a specific aircraft type, piece of equipment, or subject as it originated and or evolved over time. Material of this nature sometimes covers an extended time period that may or may not overlap the discussion involving specific air campaigns. For those readers not familiar with the Vietnam War, the overview that follows will assist the reader in placing the discussions of electronic warfare tactics and equipment covered herein in their historical context.

The air war in Vietnam can be divided into five phases: the Maddox incident/Operation Pierce Arrow, Operation Rolling Thunder, the Bombing Interlude, Linebacker I, and Linebacker II. A brief summary of each of these phases is outlined below.

The *Maddox* Incident/Operation Pierce Arrow: August 2–5, 1964
The *Maddox* was a U.S. destroyer operating in international waters that was thought to have been attacked by North Vietnamese torpedo boats during the night of August 3, 1964. Before the events of the attack could be confirmed, President Lyndon B. Johnson, in a televised address to the nation, announced the U.S. ships had been attacked twice, and he called for retaliation. This led to the Gulf of Tonkin Resolution, which was the legal basis for the subsequent air attacks by the United States on North Vietnam. The first air strikes were conducted on August 5, 1965, under Operation Pierce Arrow.

Operation Rolling Thunder: March 1965 to October 1968
Operation Rolling Thunder was a bombing campaign authorized by President Johnson that was intended to put increasing pressure on the North Vietnamese to stop the infiltration of personnel and equipment into South Vietnam. Although the first air strike took place on March 2, 1965, it was preceded by several air strikes conducted under Operation Flaming Dart that took place a few weeks earlier in February. The campaign, which was originally intended to last eight weeks, continued for three and a half years.

The Bombing Interlude: November 1968 to March 1972
President Johnson terminated all bombing of North Vietnam on November 1, 1968, in an effort to start serious negotiations with the North to end the war. The bombing ceased until the negotiations ended and the North Vietnamese invaded South Vietnam on March 30, 1972.

Linebacker I: April to October 1972
President Nixon responded to the invasion with Operation Freedom Train, which involved air strikes throughout North Vietnam above the 20th parallel. Operation Freedom Train was renamed Linebacker. Linebacker ended on October 23, 1972, when the U.S ended all tactical air sorties into North Vietnam above 20th parallel as a gesture of good will designed to help the peace negotiation being held in Paris.

Linebacker II: December 16–29, 1972
When the peace negotiations stalled in November 1972, Nixon ordered a massive bombing campaign with the intention that its psychological damage would cause the North Vietnamese to negotiate a meaningful peace agreement. Starting on December 16 and for eleven days with an interruption for the Christmas holiday, B-52s struck targets in Hanoi and Haiphong, dropping thousands of tons of bombs. The campaign was successful in bringing the North Vietnamese back to the bargaining table, but did not end the war.

CHAPTER 1

FIRST STRIKES

The air war against North Vietnam began on August 5, 1964, when sixty-four aircraft from the USS *Ticonderoga* (CVA 14) and the USS *Constellation* (CVA 64) attacked three North Vietnamese patrol boat bases at Qunk Khe, Loc Chao, Hon Gai (Ha Long) and the petroleum storage facility supporting these bases at Vinh. The airmen claimed to have sunk eight torpedo boats, damaged another twenty-one, and destroyed 90 percent of the oil tanks at Vinh. The air strikes, conducted under Operation Pierce Arrow (see figure 1.1 next page), were ordered by President Lyndon B. Johnson. Although Pierce Arrow was successful in accomplishing its mission, it was not achieved without cost. The aircraft attacking the base at Hon Gai encountered moderate to heavy anti-aircraft fire that downed the A-4 Skyhawk flown by Lt. (jg) Everett Alvarez and the A-1 Skyraider piloted by Lt. (jg) Richard C. Sather. Alvarez ejected and was captured by the North Vietnamese, becoming the first of 179 Navy flyers to become prisoners of war. He spent the next eight years in captivity. Sather was killed when his aircraft crashed into the water.[1]

President Johnson ordered the air strikes in retaliation for what was thought to be two unprovoked attacks on the U.S. destroyer *Maddox* (DD 731). On August 2, *Maddox*, which was equipped with a communications van and seventeen SIGINT specialists assigned to the Naval Security Group, was conducting a SIGINT gathering mission (dubbed a Desoto Patrol) in international waters off the North Vietnamese coast when she detected three North Vietnamese patrol boats approaching her position from the west. The ship had already received a message, based on a SIGINT intelligence warning, that an attack on the ship "was possibly imminent." Capt. John Herrick, the ship's commander, aware of the threat to his vessel, ordered her gun crews to open fire if the patrol boats closed to 10,000 yards. At 3:05

FIG. 1.1

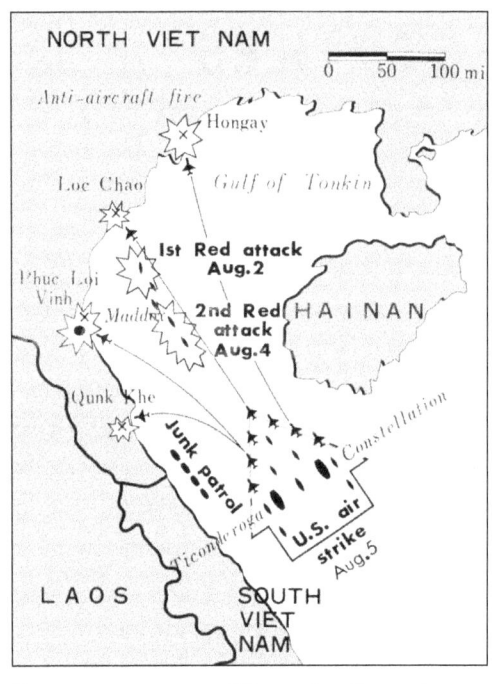

Contemporary map of Operation Pierce Arrow, August 5, 1964, from the USS *Ticonderoga* (CVA 14) 1964 cruise book *United States Navy*

p.m. local time he ordered the 5-inch gun crews to fire warning shots across the bow of the closest boat, which then launched a torpedo and veered away. *Maddox* opened fire on the enemy boats and was supported by four F-8 Crusaders that the *Maddox* had called earlier from the *Ticonderoga*. The F-8s, led by Cdr. James B. Stockdale, the commanding officer of VF-51, made multiple firing runs on the enemy vessels, severely damaging two of the boats and leaving a third dead in the water on fire.[2]

The next day *Maddox*, joined by the *Turner Joy* (DD 951), resumed her Desoto Patrol in the Gulf of Tonkin. At 6:15 p.m. local time, the Naval Security Group detachment on board the *Maddox* received an intelligence report from the Marine contingent at Phu Bai, charged with the collection, processing, and reporting of North Vietnamese naval communications, stating that the North Vietnamese appeared to be preparing some sort of naval activity. Although the report did not specify the nature of the military operations, the Marines appear to have concluded that it was an attack against the Desoto Patrol. This information was passed to Captain Herrick. A short time later, the *Maddox* reported the first of numerous radar contacts thought to be North Vietnamese torpedo boats closing the ship.[3]

At 9:34 p.m., what appeared to be a single boat suddenly appeared on the *Maddox*'s radar screen east of the two destroyers at 9,800 yards closing at nearly forty knots. Six minutes later, Captain Herrick radioed the

Pacific Fleet command that he had commenced firing on the patrol boats. During the next two hours, radar targets coming from all directions suddenly appeared a few miles from the destroyers, showed on the radar screens for a while, and then disappeared. One officer on the *Turner Joy* later told of getting blotches on the radar screen, "but nothing real firm, so we were whacking away at general areas with proximity fuzes, hoping to get something."[4]

Commander Stockdale, who was later awarded the Medal of Honor for his actions as a POW, flying off the *Ticonderoga*, arrived overhead to support the two destroyers. Stockdale spent the next ninety minutes at low altitude seeking in vain the enemy vessels that were attacking the *Maddox*. "I had the best seat in the house," he later reported, "to watch that event and our destroyers were just shooting at phantom targets there were no PT boats there . . . there was nothing there but black water and American firepower."[5]

"The two destroyers," according to cryptologic historian Robert J. Hanyok, "gyrated wildly in the dark waters of the Gulf of Tonkin, the *Turner Joy* firing over 300 rounds madly at swarms of attacking North Vietnamese boats—maybe as many as thirteen—and dodging over two dozen torpedoes. Another twenty-four star shells had been fired to illuminate the area and four or five depth charges had been dropped to ward off the pursuing boats and the torpedoes. The *Maddox* vectored overhead aircraft to the surface contacts, but time and again the aircraft reached the designated point, dropped flares, and reported they could not find any boats."[6]

Information on the attack—both SIGINT and action messages from the two ships—were closely monitored in Washington, D.C., which was twelve hours behind the time in Vietnam. "Calls between the Joint Chiefs of Staff," noted the author Pat Paterson, "the National Military Command Center, headquarters of the Commander in Chief, Pacific (CINCPAC); and Secretary of defense Robert McNamara were frequently exchanged during the phantom battle." Within an hour after the end of the engagement, Herrick had relayed his doubts about the attack in an after-action report suggesting that no torpedo boats had been positively identified. But another SIGINT report, which seemed to be an intercept of a North Vietnamese after-action report, led Secretary of Defense Robert McNamara and Adm. Ulysses G. Sharp, the Commander-in-Chief of the Pacific Fleet, to believe that an attack had occurred and so advised the president. At 2:00 p.m. EST, three hours after the attack had been reported, President Johnson approved a retaliatory strike against the North Vietnamese. In a televised speech to the nation that night starting at 11:36 p.m. EST, Johnson told the audience that U.S. ships had been attacked twice in international waters in the Gulf of Tonkin near North Vietnam and he would retaliate against the North Vietnamese. "Repeated acts of violence against the armed forces of the United States," he stated, "must be met not only with alert defense, but with positive reply. By then the aircraft assigned to Operation Pierce Arrow were already on the way to their targets.[7]

In response to the incident in the Gulf of Tonkin, Congress, at Johnson's request, passed the Gulf of Tonkin Resolution on August 7, 1964. The joint resolution granted the president authority "to take all necessary measures to repel any armed attack against the forces of the United States and to prevent further aggression." As historian and Vietnam War expert Ronald B. Frankum Jr. noted, "In effect, it gave the president a blank check to deploy whatever force he deemed necessary to defend American forces and installations in Vietnam against attack and to assist the South Vietnamese in their war against North Vietnam."[8]

According to William Cahill, a retired Air Force intelligence officer, "Reconnaissance forces were now under pressure to produce more intelligence on which to base additional U.S. reaction strikes." On August 14, 1964, an RB-47H Stratojet assigned to the 55th Strategic Reconnaissance Wing (SRW) deployed to Kadena Air Base, Japan, to conduct an ELINT survey of North Vietnam. The 55th SRW, which was headquartered at Forbes Air Force Base in Topeka, Kansas, was the Strategic Air Command's (SAC) top-secret strategic reconnaissance force. Its aircraft were deployed around the world to conduct daily probes of the defenses of communist countries. Their missions were so secret that they were conducted under strict radio silence and tight security.[9]

The RB-47H was the ELINT version of the B-47, the first swept-wing, multiengine, jet-powered bomber to enter the U.S. inventory. Thirty-two Silver King RB-47s were built to meet the Strategic Air Command's need for electronic reconnaissance. The aircraft cruised at 442–446 knots, had a top speed of 516 knots, and had a combat radius of 1,520 nautical miles in a 6.4-hour flight. Features that differentiated the upgraded Silver King RB-47H from the B-47 were its blunt, rounded nose, the teardrop-shaped antenna pod attached to an offset pylon mounted under the belly, and the pylon-style antenna under each wing outboard of the engine. A pressurized compartment in the area formerly occupied by the B-47's bomb bay housed the specialized electronics equipment gear and the three crew stations for the Electronic Warfare Operators (EWOs), called "ravens" or "crows," who operated the equipment. The electronic receivers and direction finding gear housed in the bomb bay consisted of an APD-4 automatic receiver/analyzer, supplemented by manually operated APR-17 receivers, an ALA-6 direction finding (D/F) system, and ALA-74 pulse analyzers. On a typical mission the EWOs spent many long hours working in the cramped, poorly insulated, windowless compartment.[10]

The Gulf of Tonkin Resolution and the retaliatory air strikes against the North did not deter the communist insurgents in the South—the Viet Cong—from attacking U.S. military bases. On November 1, 1964, Viet Cong troops conducted a mortar attack on the Bien Hoa Air Base northeast of Saigon, lobbing sixty to eighty rounds onto parked aircraft and troop billets. The barrage killed four Americans, wounded thirty others, and destroyed five B-57 Canberra bombers. Then, on December 24, Christmas Eve, the Viet Cong set off a car bomb at the Brinks Hotel in downtown Saigon that served as a bachelor officers' quarters for American military officers serving in the Saigon area. Two Americans were killed and fifty-eight wounded. Although the Joint Chiefs of Staff (JCS) urged President Johnson to strike targets in North Vietnam in retaliation for the Viet Cong attacks against U.S. personnel in the South, Johnson refused to grant permission until the attack on the military complex at Pleiku on February 7, 1965.[11]

The attacks at Pleiku began around 2:00 a.m., when a small band of Viet Cong breached the barbed wire protecting the small detachment of the U.S. Military Advisory Command Vietnam (MACV) compound three miles north of Pleiku in the central highlands of South Vietnam. The attackers set off numerous demolition charges against the main building and the mess office while spraying the building with automatic weapons fire. During the attack, which lasted ten or fifteen minutes, one American soldier was killed and twenty-four wounded.[12]

As this attack was taking place, two five- to six-man Viet Cong assault teams entered the runway and aircraft parking area at nearby Camp Holloway, which served as headquarters for the U.S. Army's 52d Aviation Battalion, and placed demolition charges under the fuselage of several aircraft and helicopters. As the charges exploded, the Viet Cong fired 81-mm mortars at nearby billets. Seven American soldiers were killed and 104 were wounded. The Viet Cong destroyed five Army UH-1B helicopters and damaged eleven UH-1Bs, two CV-2 transports, three O-1F forward air control aircraft, and one Vietnamese air force O-1F that was temporarily stationed at the airfield.

When news of the attacks on the MACV compound and Camp Holloway reached the White House on the afternoon of February 6, 1965, President Johnson ordered a meeting of the National Security Council (NSC) for 7:45 that evening. During the ensuing meeting, Gen. Earle G. Wheeler, chairman of the Joint Chiefs of Staff, urged a quick reprisal air strike on the North. The other NSC members and attendees agreed that this was necessary. Based on their recommendations, Johnson selected four targets from a list previously prepared by the U.S. military in the southern part of North Vietnam and directed U.S. aircraft to hit three and the South Vietnamese Air Force (VNAF) to strike one of the objectives.

Air strikes by the Air Force and Vietnamese aircraft against the North Vietnamese Army (NVA) barracks at Chap Le and Vu Cong, scheduled to begin at 3:00 p.m. local time on February 8, had to be scrubbed due to bad weather, as was the Navy's scheduled attack on the Vit Thu Lu barracks. But the attack on the barracks at Dong Hoi, lying slightly north of the demilitarized zone, known as Operation Flaming Dart, was conducted as scheduled. Twenty-nine A-4 Skyhawks from the carriers *Coral Sea* (CVA 43) and *Hancock* (CVA 19) supported by twenty other aircraft including four F-8E crusaders to suppress anti-aircraft fire struck the administration building and barracks that housed the 15,500 troops of the NVA's 352 Division. "The Flaming Dart strike," wrote Air Force historian Jacob Van

Staaveren, "was not without penalty: apparently on the alert for the air strike, the North Vietnamese downed one A–4E and damaged seven other aircraft with anti-aircraft and small arms fire."[13]

Although the JCS and the service commanders in Honolulu and Saigon wished to continue the air strikes, the consensus within the NSC was "no." President Johnson agreed with their recommendation in the belief that a second strike by U.S. planes might give Hanoi and Moscow the impression that the United States had begun a sustained air offensive against the North. President Johnson and the members of the NSC knew that Alexei Kosygin, the Soviet premier, had arrived in Hanoi on February 6 to inspire the North Vietnamese in their "struggle against American interventionists and their puppets." The presence of Chief Air Marshal Kostantin A. Vershinin, the deputy defense minister who was commander of the Soviet Air Force, indicated to analysts specializing in relations between the Soviet and Asian communist parties that Moscow was prepared to provide Hanoi with more aid, including jet planes, advanced anti-aircraft guns, and guided missiles. The Central Intelligence Agency (CIA) had already reached this conclusion, noting in their February 1 report on the near-term communist military capabilities, that in order to rebuild influence in Hanoi and deter the United States from expanding the scope of hostilities, "the Soviets will probably increase their military and economic assistance to North Vietnam to include additional air defense equipment and perhaps jet fighters and surface-to-air missiles." The missiles, which would be supplied to the North Vietnamese in due course, would radically change the air war over Vietnam and was the harbinger of a revolution in military affairs that would transform the nature of air warfare.[14]

On February 10, the Viet Cong struck the American Army enlisted men's barracks at Qui Nhon, killing twenty-three and wounding twenty-one American servicemen. President Johnson summoned the NSC once again, and after listening to the arguments—some of which suggested that the strikes be postponed until Kosygin left Hanoi—decided that another strike was necessary. The U.S. Navy's Pacific Command (CINCPAC), which was directed to execute the air strike, nicknamed it Flaming Dart II. The operation, which was not launched until late on February 11 Saigon time, involved ninety-seven aircraft launched from the *Coral Sea*, *Hancock*, and *Ranger* (CVA 61). Seventy-one A-1H Skyraiders and A-4C Skyhawks struck the North Vietnamese Army (NVA) barracks at in Chann Hoa. About two hours later, twenty-eight Vietnamese Air Force A-1H Skyraiders, supported by thirty-two USAF aircraft, mostly F-100 Super Sabers for flak suppression, carried out a second strike on Chap Le barracks. Both attacks were costly. The Navy lost three aircraft (two A-4s and one F-8) and while no Vietnamese Air Force aircraft were lost, eighteen were damaged.[15]

The two Flaming Dart II strikes ended a protracted debate within the Johnson administration concerning the political and military risks of bombing North Vietnam, a country officially known as the Democratic Republic of Vietnam. On February 19, explained one Air Force historian, President Johnson "decided to begin a campaign of regular bombing of the North, though he chose not to announce it publicly for the time being, apparently to protect his domestic programs, and to shield himself from criticism from military 'hawks' in Congress and among the public, and an immediate sharp response from China, the Soviet Union, or both." The campaign to bomb the North would be named Rolling Thunder.[16]

CHAPTER 2
OPERATION ROLLING THUNDER

Prior to initiation of air operations over NVN, the Tactical Air Commands were practically devoid of EW equipment and personnel. . . . This situation was fostered by lack of emphasis on EW equipment, manpower, and organization at the command level.

—GEN. JOSEPH J. NAZZARO,
COMMANDER-IN-CHIEF, PACIFIC AIR FORCE[1]

Rolling Thunder was an eight-week campaign proposed by the Joint Chiefs of Staff to put increasing pressure on the North Vietnamese to stop the infiltration of personnel and equipment into South Vietnam. The program called for a series of U.S. and Vietnamese air strikes ascending in intensity and risk, initially on targets lying along Route 7 and south of the 19th parallel. The air strikes were designed to cut the supply lines between Hanoi and Vinh. The targets included railways, bridges, and water transport along with radar and communication stations. The program was delayed by weather and a reorganization of the South Vietnamese government. The latter affected the South Vietnamese Air Force's ability to supplement U.S. commitments for the needed aircraft.[2]

The first Rolling Thunder strike against the Xom Bang ammunition depot ten miles above the Demilitarized Zone (DMZ) took place on March 2, 1965. It was the first attack on the North conducted solely by aircraft assigned to the Tactical Air Command (TAC) and was composed of a mix of aircraft that included a large force of F-105 Thunderchiefs and B-57

Canberras supported by F-100s for flak suppression and combat air patrol (CAP), along with two RF-101 Voodoos for reconnaissance. No MiGs were encountered and the 120 tons of bombs dropped by the air strike destroyed 75 percent of the facility according to the follow-on bomb damage assessment. Almost simultaneously, twenty South Vietnamese A-1 Skyraiders, supported by about sixty F-105s, F-100s, and RF-101s from the 2d Air Division, hit the naval base at Quang Khe. As with the air strike on Xom Bang, bomb damage assessment disclosed that 75 to 80 percent of the target area, consisting mainly of buildings, was destroyed or damaged. But as Van Staaveren noted, "The cost of the air strikes was high: the Air Force lost three F-105 Thunderchiefs in the assault on Xom Bang and two F-100s over Quang Khe." All the losses occurred during the attempts to suppress the enemy's anti-aircraft positions and were brought down by conventional anti-aircraft artillery.

The ground fire experienced by the aircrews attacking the targets in North Vietnam was more formidable than expected. Capt. Robert V. "Boris" Baird recalled his unexpected encounter with the North Vietnamese anti-aircraft fire in an *Aerospace Historian* article written in 1975. Baird, who was assigned to the 67th Tactical Fighter Squadron flying from Korat Royal Thai Air Force Base in Thailand, was one of the F-105 pilots shot down over Xom Bang. During the squadron briefing that preceded the mission, Baird and his fellow pilots were told they could expect no significant ground fire in the target area. Baird, in the number two position in the lead squadron, described what happened during the attack: "When we came in over the target area, making a straight and level high speed pass to deliver our CBUs, the sky was sort of black and greasy. We knew then we had hit that 'unexpected ground fire.' In fact, the sky was full of it." Baird's F-105 was fifty feet off the ground flying straight and level when it was hit with 23-mm fire. He remembered having smoke in the cockpit and holes in the wings when he ejected from the aircraft. He was rescued by an HH-43 Huskie helicopter and continued flying missions over North Vietnam.[3]

One intelligence report produced in early 1965 estimated that the NVA possessed 1,039 anti-aircraft guns (322 14.5-mm, 23-mm, and 37-mm, 709 57-mm, and eight 85-mm) positioned in 161 active anti-aircraft sites. Eleven aircraft had been shot down and fifty-five damaged while the first three air strikes were conducted during Rolling Thunder. Most were hit by small arms fire when the aircraft were flying at low level below 1,000 feet. To some pilots it seemed that every soldier in the target area with an automatic weapon was firing at them.[4]

Secretary of Defense McNamara and the leadership within the Air Force and the Navy were deeply concerned about the aircraft attrition inflicted by enemy ground fire. McNamara requested the JCS to analyze the loss rates with those in World War II and the Korean War and to recommend ways of reducing the loss of aircraft. Secretary of the Air Force Eugene M. Zuckert foresaw a "credibility" problem arising from the loss of tactical aircraft to "unsophisticated" weapons. A change in tactics was imperative: fewer aircraft—no more than sixteen—should be sent against a defended target complex in a single attack, using cluster bombs dropped from no lower than 1,000 feet whenever possible. The Air Force and the Navy quickly initiated their own studies. The Navy study, conducted by the CINCPAC staff, concluded that "bad weather seldom permitted dive bombing. The staff concluded that pilots could expect cloud layers at 3,500 feet more than 50 percent of the time and usually did not have good strike weather until the middle of the afternoon, a circumstance not lost on the North Vietnamese who knew when to be on the alert for most of the attacks."[5]

In separate briefings and reports for the Air Force, Lt. Gen. Gordon M. Graham, TAC's head of operations, stressed the need for pilots to receive more realistic training. A follow-on study by operations analysts in the Tactical Air Command emphasized the need for more practice in gunnery and weapons delivery. These conclusions showed the pernicious culture within the Air Force against the need to provide a nonnuclear strike and interdiction capability stemming from the Strategic Air Command's (SAC) domination during the previous two decades.

The Strategic Air Command, established in 1946, was founded on the operational concepts developed during the U.S. long-range bombing campaigns of World War II. It was the first intercontinental strategic bombing force. When Lt. Gen. Curtis E. LeMay took command in October 1948,

SAC's mission was to deliver a devastating nuclear attack on the Soviet Union. One SAC historian described LeMay as "an energetic, get-things-done type who had never met a challenge he couldn't overcome." General LeMay believed that all wars could be won by using strategic bombing to devastate a country's infrastructure and industry. His attitude toward the use of tactical air power was embodied by the adage (attributed to Gen. James H. Doolittle) that "Tactical bombing is breaking the milk bottle; strategic bombing is killing the cow." LeMay was so adamant about the primary importance of SAC that he considered the other Air Force commands secondary. At one point he even put forward a proposal to absorb TAC into a SAC-dominated "Air Offensive Command." Fiscal subordination to SAC during the years preceding the outbreak of hostilities in Korea left the Tactical Air Command without sufficient assets to wage a conventional war. Both the Air Force and the Navy were forced to rely heavily on World War II fighters for close air support and Air National Guard and Reserve Navy pilots because jet aircraft, pilots, and ground crew were limited in supply.[6]

Although the Korean War had shown the shortcomings of U.S. tactical air forces and underscored the need for more funding, domestic policies continued to work in SAC's favor. As Craig Hannah explained, "nuclear bombers and intercontinental ballistic missiles with nuclear warheads maintained their prominent position in the air force, and SAC continued to receive a disproportionate share of the defense budget." Interservice competition between the Tactical and Strategic Commands for funding forced TAC to concentrate on delivering nuclear weapons, which fit into the most likely scenario for a nuclear war with the Soviet Union. As TAC became increasingly specialized for the nuclear strike role, conventional air-to-ground attack and air-to-air operations fell into neglect. Training to deliver tactical nuclear weapons was complex, leaving little time and money for much else.[7]

The need for more training in the delivery of conventional munitions became glaringly evident when bomb damage assessments reports after the first two reprisal raids in February were issued. The results were not very encouraging. There had been 267 sorties directed against 491 buildings, but only 47 were destroyed and 22 damaged.[8]

The F-105 Thunderchief was destined to carry the burden of the Air Force's ground attack during the war in Vietnam. It was designed in the early 1950s to deliver tactical nuclear weapons from European bases at supersonic speeds and low altitude; it was also the first fighter to have an internal bomb bay and was the heaviest single-engine fighter ever built. The aircraft was built under the assumption that it would have to face anti-aircraft defenses consisting mostly of large, sophisticated surface-to-air missiles. Since a hit by a large missile warhead was presumed to mean an automatic kill, the F-105 was built with little emphasis on system redundancy and resistance to battle damage. It was designed around an internal bomb bay that would be large enough to carry a nuclear weapon using delivery tactics that emphasized minimizing exposure to defenses and a speedy getaway, not precision. Instead, in Vietnam, as Lt. Col. Donald R. Baucom, USAF, noted, "it was expected to deliver conventional munitions with pinpoint accuracy on the most difficult of targets—bridges, road cuts, camouflaged storage areas—under the worst conditions."[9]

Although the bomb bay never carried a bomb into combat, it provided secure storage for an extra 308-gallon fuel tank that gave the F-105 extra range without an increase in drag. In the ground attack support role, the "Thud," as the F-105 was dubbed, carried six 750-lb. Mark-117 bombs on a centerline-mounted multiple ejection rack and two Mark-117 bombs on outboard pylons. The lack of redundancy in the airplane's design proved to be a pilot's nightmare if the F-105's hydraulic lines were hit. Because the primary and redundant lines were placed side-by-side in the fuselage, one round from an anti-aircraft gun near either hydraulic line would usually destroy both, making the aircraft uncontrollable. There was no way to fly the airplane once the hydraulics were gone. Even a hit from the smallest caliber weapon could destroy an F-105.[10]

To improve the aircraft's survivability, Republic Aviation engineers who built the F-105 developed a modification that locked the horizontal stabilizer in the neutral position before the hydraulic system completely failed. The pilot could then use the electrically powered aileron trim switch to provide a limited amount of control. Theoretically this provided the pilot with just enough control to reach friendly territory before ejecting from the aircraft.

Further design changes to augment controllability in the event of hydraulic failure were included in the Thunderstick II modification program that added a more precise navigation system that improved the F-105D's blind bombing capability. Unfortunately, these modifications were only completed in 1969 and involved a total of just thirty aircraft. One authority claimed that had this modification been available at the start of the war it would have saved one hundred pilots "who are now statistics." The F-105 was not a very good air-to-air fighter either, but this will be discussed later on.[11]

By March 19, Rolling Thunder missions were taking place on a daily basis, targeting choke points in North Vietnam with the objective of limiting the NVA's ability to supply and replenish troops in the South. Three days later the missions began to concentrate on radar installations below the 20th parallel in an effort to eliminate the North Vietnamese early warning system. By then the North Vietnamese had deployed a network of no less than twenty-two early warning radars and at least four SON-4 (Whiff) fire-control radars. The latter, when used with a PUAZO-5 director, increased the effective range of the Soviet supplied 57-mm S-60 anti-aircraft gun from 4,000 meters (1,300 feet) using optical sights to 6,000 meters (19,600 feet). As air attacks on targets in North Vietnam increased, the Soviet Union and the Chinese government began supplying the NVA with more anti-aircraft guns accompanied by more Soviet-supplied fire-control SON-4 and SON-9 (Fire Can) radars for use with the heavier caliber (57-mm, 87-mm, and 100-mm) guns.[12]

When Operation Rolling Thunder got underway, special aircrews assigned to the EC-121Ms of the Navy's Fleet Air Reconnaissance Squadron One (VQ-1)* flying from Don Muang International Airport, Bangkok, Thailand, and Da Nang airfield in South Vietnam, were already conducting secret electronic intelligence (ELINT) collection missions using the Brigand system to locate and accurately position the early warning and ground control interception (GCI) radars in the North. The EC-121M was

---

* For the origins of VQ-1 and their early operations in Southeast Asia, see Thomas Wildenberg, *Fighting in the Electromatic Spectrum: U.S. Navy and Marine Corps Electronic Warfare Aircraft, Operations, and Equipment* (Annapolis, MD: Naval Institute Press, 2023), 46, 86–87.

a specially modified version of the EC-121 Warning Star airborne early warning aircraft equipped to perform the ELINT mission. It was a four-engine propeller-driven aircraft derived from the Lockheed L-1049 Super Constellation. The EC-121M was crewed by six aviators on the flight deck and eleven to twenty-five EWOs in the electronics cabin. It had a top speed of 295 knots and a maximum range of 4,250 miles.[13]

In July 1964, a detachment from VQ-1 based in Atsugi, Japan, started flying EC-121M SIGINT aircraft from Don Muang Airport, Thailand, to locate and accurately position the early warning and ground control interception (GCI) radars in North Vietnam. The squadron, which had been flying missions over the Gulf of Tonkin since 1952, had been tasked to look for evidence of MiGs and SA-2 missile sites, neither of which had previously been known to be in North Vietnam. The single plane detachment, under the command of Lt. Cdr. Norman S. Bull, flew two nine-hour missions over the Gulf of Tonkin on the 21st and 22nd of July before returning to Cubi Point Naval Air Station in the Philippines.

Although the exact ELINT equipment of the EC-121M flown by Lieutenant Commander Bull is not known for sure (it changed over time), it seems likely that the electronic equipment carried by the aircraft included APR-9 radar receivers, an APR-13 panoramic radar receiver, and an ALQ-28 ELINT receiver, along with the standard APS-20 radar. Bull's aircraft probably carried an early version of the highly sensitive, top-secret ELINT receiver invented by Charles Christman that was just entering service dubbed "Brigand" by Cdr. Charles McMakin, a Senior Electronic Warfare Officer in VQ-1. The name was an acronym for Bistatic Radar Intelligence Generations New Development. Brigand used an AS-400 signal processor interfaced with the EC-121M's APR-6 panoramic radar receiver to locate enemy radars. Alfred Price, in volume II of *The History of US Electronic Warfare*, revealed that "it could measure the location of an E band surveillance radar to within about 350 yards—a level of accuracy sufficient to enable a reconnaissance plane to photograph the site on a single-run pass." To achieve this accuracy the Brigand operator had only to tune the equipment to the radar's frequency and photograph the signal's display through a full rotation of the radar antenna, about ten seconds. An aircraft equipped with

Brigand did not need to fly a special pattern that might betray its mission and could locate search radars at a distance of 250 miles. Brigand only worked against a search radar whose antenna rotated through 360 degrees. The system did not work against intercept radars, height finders, and most types of missile and gun control radars.[14]

When Air Force aircraft began flying strike missions as part of Operation Rolling Thunder, there were no tactical electronic warfare assets in the Pacific Air Force, the lack of which severely hampered its operations in Rolling Thunder. Although the Tactical Air Command had fifty-eight RB-66 Destroyer reconnaissance aircraft ready for combat, only a handful had been modified for electronic warfare. Despite opposition from the U.S. Air Force in Europe, which was opposed to withdrawing any electronic warfare assets from Europe while the Soviet Union and the Warsaw Pact continued to improve and expand their forces in Central Europe, TAC sent six Big Sail RB-66Cs to Southeast Asia. The first two RB-66Cs arrived at Tan Son Nhut Air Base in South Vietnam in April 1965. The following month two Big Sail RB-66Cs were moved to Takhli Royal Thai Air Force Base, Thailand, where they were joined by four more Big Sail RB-66Cs that had been deployed directly from the United States. This deployment exhausted the pool of RB-66Cs until September, when TAC increased the number at Takhli to nine. The Tactical Air Command also had a specialized jamming version of the Destroyer designated as the "Brown Cradle" B-66B, but it was unable to deploy any of these aircraft to Southeast Asia until October.[15]

The RB-66 was the reconnaissance version of the B-66 light bomber developed from the Navy A-3D Skywarrior. The electronic reconnaissance variant, designated the RB-66C, was the only purpose-built electronic warfare version of the B-66. The first RB-66C entered service with the Tactical Air Command on May 11, 1956, with the 9th Tactical Reconnaissance Squadron (TRS) at Shaw Air Force Base, South Carolina. It had a crew of seven that included the pilot, navigator, four EWOs housed in a pressurized electronics compartment in what had been the bomb bay, and a tail gunner, who was later removed when the tail turret was replaced with a tail cone housing additional electronic gear.[16]

At the time the RB-66C entered service, the heart of its ELINT system was based on the APD-4 radar direction finding system manufactured by ITT. Originally developed for the RB-47, the system consisted of the receiver and thirty-six horn antennas dispersed in clusters in the nose, on each side of the rear fuselage, and in the rear of each engine mounting to give 360-degree coverage. It was an automated system that recorded the received signals on photographic film. In order to provide accurate direction finding capability, the APD-4 had to be carefully calibrated and the electronic tubes matched for amplification. It was not an easy system to use, as the film was difficult to process and required lengthy analysis on the ground once it was developed. But its biggest failure was its susceptibility to being overwhelmed by multiple radar intercepts that made the results virtually useless. As one authority noted, "Electronic intelligence soon reverted back to the tried and proven method of paper and pencil. Each EWO monitored an assigned frequency band. The intercepted signals were graphically displayed on a pulse analyzer, which identified the type of signal. Geographic plots were then made using the airborne direction finder to obtain relative bearings and, by triangulation, determined the approximate position of the intercepted signal. Since range information was not available, these plots were dependent upon knowing the exact location of the RB-66C at the time a bearing was taken, emphasizing the importance of accurate navigation."[17]

The RB-66C also carried APR-4 and APR-9 superheterodyne receivers, an ARR-8 wideband receiver, ALA-6 direction finders, and APA-74 pulse analyzers. The plane was also equipped with an APS-54 warning receiver for self-protection.[18]

In 1959, TAC decided to modify a few of its B-66 bombers to provide jamming for the second series of Weapons Evaluation (WEXVAL)† tests. These were scheduled for October and were aimed at evaluating the effectiveness of U.S. Navy defenses against air attacks supported by electronic countermeasures. The modification program, called "Brown Cradle," involved the conversion of thirteen B-66Bs taken from the 17th

---

† In 1958 the Institute for Defense Analysis formulated a series of elaborate tests, dubbed WEXVAL tests, to evaluate the effectiveness of various parts of the U.S. armed forces.

Bomb Wing. Each airplane's bomb bay was stripped of racks and replaced with a pallet carrying jammers and their power supplies. The rear gun turret was removed and replaced with a cone-shaped fairing. The gunner's position in the cabin became the position for the Electronic Warfare Officer, who had access to large fold-away panels on which he could monitor the performance of the jamming gear. The jammers in the bomb bay were preset on the ground, making it impossible to retune them in the air, as the crew did not have access to the bomb bay. This arrangement required that an RB-66C fly sorties to gather the frequencies that were to be jammed and upon its return pass this information on to the B-66B Brown Cradle.[19]

During the WEXVAL exercise, the Brown Cradle B-66Bs, equipped with batteries of Hallicrafters QRC‡ noise jammers tuned to frequencies of the Navy's shipborne radars, flew jamming missions in support of F-100s that carried out mock attacks on a Navy task force off the coast. The RB-66Cs assigned to the exercise were used to keep track of the radars the ships were using. One observer reported that the jamming was very effective and prevented several interceptions by Navy planes. After the exercise ended, the Hallicrafters sets were removed and the Brown Cradle B-66Bs reconfigured to carry ALT-B slow-sweep jammers tuned to Soviet radar frequencies. These planes were sent to Europe where they joined the 42nd TRS.

Before the RB-66Cs were sent to Southeast Asia, the leadership within TAC deemed it imperative that their ELINT and electronic countermeasures (ECM) capabilities be enhanced prior to their deployment. To accomplish this task, Project Big Sail was undertaken to provide the RB-66C with the ability to collect ELINT and provide ECM against all known radars employed by the North Vietnamese. The jamming capability of the RB-66C was so limited that it was not even able to protect the aircraft from SA-2 missile guidance, which restricted its operational use to ELINT gathering. The most significant modification was the installation of the APR-25 and

---

‡ In 1952 the Air Force introduced a new procurement system for handling urgently needed high-priority electronic equipment called the Quick Reaction Capability (QRC) program, which was designed to circumvent the long and tedious process of procuring electronic equipment under the AN number system.

TABLE 2.1. RB-66C ELECTRONIC WARFARE SUITE (CIRCA 1967)

ELINT	Qty	Type	Function
EWO Position One	1	APR-14	Direction Finding Receiver
	1	ALA-6	Direction Finding Receiver
	1	ALA-5	Pulse Analyzer
EWO Position Two	1	APR-14	Direction Finding Receiver
	1	ALA-6	Direction Finding Receiver
	1	ALA-5	Pulse Analyzer
EWO Position Three	1	APR-9	Radar Intercept Receiver
	1	ALA-6	Direction Finding Receiver
	1	ALA-5	Pulse Analyzer
	1	APR-25	S/X/C Band Radar Detection and Homing
	1	APR-26	SAM Launch Warning Receiver
	1	APS-44	Tail Warning Radar
EWO Position Four	1	APR-9	Radar Intercept Receiver
	1	ALA-6	Direction Finding Receiver
	1	APA-74	Pulse Analyzer & Camera
ECM (jamming)			
	1	ALT-18/ALT-6B	Transmitter
	1	ALT-16	Transmitter
	1	ALT-15L	Transmitter
	1	ALT-15H	Transmitter
	2	QRC-279 (V)*	Transmitter Controller
	1	QRC-279 (H)**	Transmitter Controller
	2	ALT-13/QRC-218	Transmitter/Modulator

*Vertical Polarization
**Horizontal Polarization

Source: Courtland C. Moore, "EB-66C Out-Country Electronic Reconnaissance, 1965–67, A Case Study," Report No. 3655, Maxwell Air Force Base, AL, 1968.

APR-26 radar homing and warning (RHAW) equipment (see table 2.1). The Big Sail RB-66Cs that arrived at Takhli began flying combat missions on May 4, 1965.[20]

The mission of the RB-66Cs sent to Southeast Asia was to gather ELINT data for the identification and analysis of radars in order to obtain a radar order of battle for North Vietnam. When they began flying over the North in May, the North Vietnamese air defenses were in an embryonic stage of development and tactical ELINT aircraft operated over most of the country with relative ease. Penetration and peripheral reconnaissance were performed in a continuing effort to monitor the growth of the North Vietnamese radar network. Prescribed flight paths were flown during these early missions, and hostile radar emissions were located and recorded for subsequent analysis. When a radar emission was intercepted, the EWO responsible for the frequency band in which it was operating took a series of relative bearings to the transmitter site. The time required to acquire the necessary data for location and analysis was normally between six and ten minutes. To establish the location of the enemy radars, the converging bearings obtained by the RB-66C were manually plotted on the ground after the aircraft landed.[21]

Throughout the Rolling Thunder campaign, RB-66Cs and B-66Bs were particularly effective against low-frequency early warning and acquisition radars. Jamming, however, always involved a compromise between effective jammer coverage and aircraft survival. While the effectiveness of jamming decreased as the distance from the radar increased, it provided greater protection from hostile fire. The attacking strike force was best protected when it was between the orbit of a noise-jamming aircraft and the target. To prevent this, the North Vietnamese began to position anti-aircraft weapons to prevent the RB-66Cs and B-66Bs from assuming an ideal orbit for jamming.[22]

Many factors made the task of jamming the North Vietnamese radars difficult. By the end of 1965 the air defense system over North Vietnam had more than twenty different radars, which made it impossible to degrade the whole system. A single aircraft deployed against a single radar ceased to be effective. Using a combination of jamming, chaff, and crossing tracks, several B-66s had to be utilized. Simultaneous jamming by two ECM aircraft reduced the effectiveness of the Fan Song surface-to-air missile radar but did not degrade it completely. As the war continued in the early part of 1966, U.S. intelligence estimated that the North Vietnamese had created 134 missiles sites, not all of which were equipped with missiles.[23]

The freedom of operation enjoyed by the RB-66Cs and B-66Bs (redesignated as EB-66Cs and EB-66Bs in 1966) came to an abrupt end in February 1966 when an EB-66C was shot down by an SA-2 missile near Vinh, 140 nautical miles south of Hanoi. This, according to Van Nederveen, "marked the beginning of a southward and westward extension of North Vietnamese surface-to-air missile (SAM) defenses." The EB-66s continued to be a thorn in the side of the North Vietnamese, so much so that they began dispatching MiGs against them in an attempt to prevent the EB-66s from interfering with their radar network. The problem of keeping the EB-66s out of effective jamming range became so important that in early January 1967 the NVA Air Defense Command sent four SA-2 battalions (twenty-four launchers) to an area northwest of Hanoi. On February 4, 1967, the 89th Battalion of the 274th Missile Regiment shot down an EB-66C over Bac Can province, capturing four crewmen in the process. The downed aircraft provided a treasure trove of documents including wide-ranging information on U.S. electronic warfare that were analyzed by the Air Defense Command, the Military Technical Institute, and the General Staff.[24]

A new variant, the EB-66E, arrived in Thailand in August 1967. It had a higher power output on its jammers that increased its ability to overwhelm hostile radars. Its jamming transmitters were also tunable, which enabled the EWOs to change frequencies during the flight and jam different types of radars simultaneously.[25]

CHAPTER 3
ELECTRIC SPADS, VOODOOS, SKYNIGHTS, AND ELECTRIC INTRUDERS

When Operation Rolling Thunder began, the only U.S. carrier-based aircraft equipped to provide stand-off jamming protection for the strike forces were the ten piston-engine EA-1F Skyraiders assigned to Detachment No. 1, Airborne Warning Squadron 13 (VAW-13). Subdetachments of two or four aircraft from the squadron were deployed on each of the carriers operating in the Gulf of Tonkin. The EA-1F was a modified version of the AD-5N Skyraider converted into a dual-purpose radar reconnaissance/radar countermeasures aircraft. Known colloquially as "Electric Spads," they were flown with a four-man crew consisting of a pilot and navigator in the cockpit and two ECM operators in the rear compartment. For the ECM mission, the plane was equipped with an APL-5B radar receiver, an APA-69A direction finding set, an ARR-13 panoramic radar receiver, chaff dispensers, and two stores-mounted ALT-2 radar noise jammers. "Although the ECM gear was antiquated by 1960 standards," as John Nichols and Barret Tillman assert, "it was effective when employed by skilled, experienced aircrewmen. Thorough knowledge and delicate, almost artistic, touch was necessary to make the equipment work well." The optimum jamming range was about twenty-five miles. Although the plane was armed with two 20-mm cannons, these were seldom used.[1]

Lacking the performance to survive in the areas of North Vietnam defended by MiG-17s, which began to appear in April 1965, the EA-1Fs were restricted to flying off the coast of North Vietnam, at altitudes between 8,000 and 10,000 feet, where they jammed the Whiff and Fire Can fire-control radars covering the ingress and egress points of the Navy's strike groups. Because the jammers were directional, the EA-1Fs had to fly directly toward or away from the enemy radar. The identification of a particular enemy radar was achieved when a line peak at the radar's frequency showed up on the radar receiver's CRT display. The relative bearing to the intercepted signal could then be determined from the image on the radar direction finder's display. Although there was no range indication on the screen, experienced operators could often provide accurate estimates based on the image and signal strength. To jam the radar, the operator manually tuned one of the two ALT-2 jammers to the superimposed image on the screen of the first display. Once the jammer was tuned to the radar's frequency, the jammer was turned on, filling the enemy's radar screen with "snow," making it impossible to locate the aircraft.[2]

Seeking a better solution to the limited capabilities of the EA-1F, Capt. Julius S. Lake, responsible for overseeing the Navy's electronic warfare systems for aircraft, came up with the idea of converting the now obsolete A3D-2 Skywarrior heavy bomber into a stand-off jammer. That the Air Force's electronically outfitted RB-66s were built on a similar airframe might have been a factor in his thinking. The idea for converting the Skywarriors came to Lake while he was still in the Far East. As Lake recalled in an interview with the historian Alfred Price: "I called on Adm. John Lacouture who was CO of *Saratoga* when I was executive officer; now he was the chief of staff to the Carrier Division Commander in the Tonkin Gulf. We agreed on the need for a plane to provide EW support. Putting the tanker package in the plane's bomb bay was no big deal, that could be done at squadron level. Putting the "E" into the plane took a bit longer."[3]

Conversion of the first five Skywarriors (redesignated as A-3 in 1962) into combined tanker/jammer aircraft was undertaken by Naval Air Rework Facility (NARF) at Alameda, California, in 1966 (thirty-four KA-3B aircraft were also rebuilt to this version later). The modifications included the addition of a hose-and-drogue refueling unit, the installation of electronic jamming equipment in the bomb bay, and countermeasures equipment in the forward section and modified tail cone. The new variant, designated as the EKA-3B, was initially described as the TACOS (Tanker-Countermeasures-Or-Strike aircraft), even though the bomb bay was full of jamming equipment and the bomb rack pinned shut. The TACOS name did not last long within the fleet, however, as this variant of the A-3B was soon dubbed the "Queer Whale."[4]

The EW suite in the EKA-3B consisted of two countermeasures systems: two ALT-27 jammers for disrupting air defense radars in the E/F bands and one ALQ-92 communication jammer. The former utilized a pair of steerable antennas covered by the new belly "canoe" faired into the fuselage. The ALQ-92 had two distinct antenna arrays. One took the form of a single large vertically polarized VHF blade antenna under the nose. This was used to jam North Vietnamese fighter control frequencies. The other was a set of horizontally polarized antennas housed in four large blisters on the side of the fuselage for use against air search radars such as the Soviet P-10 (Knife Rest). The plane also carried one ALR-28 X-band D/F receiver, one ALR-29 panoramic receiver, one ALR-30 panoramic receiver, one APR-32 SAM Launch warning receiver, and two each of the ALQ-41 and ALQ-51 deception jammers. The EKA-3B also carried two ALE-2 chaff-dispensing pods. The frequency range and power output of the equipment installed in the EKA-3B was considerably better than that on the EA-1F, and its capability to operate at higher altitudes increased the geographical area that could be covered.[5]

Airborne Warning Squadron 13, at NAS Alameda, was given the job of introducing the EKA-3B to the fleet. The aircraft carried a crew of three that included the pilot, a navigator/pilot and a Navy Flying Officer (NFO) who operated the EW equipment. The squadron received the first of these aircraft in May 1967, but the lack of trained A-3 personnel and inadequate parts support hindered the squadron's efforts to become proficient with the aircraft and its electronic gear. Nevertheless, a detachment of EKA-3Bs left California on board the *Ranger* (CVA 61) heading for the war zone on November 4, 1967. After they arrived in Southeast Asia, the EKA-3Bs operated on board various carriers in the Gulf of Tonkin, refueling Navy strike formations and providing stand-off jamming cover.[6]

On April 5, 1965, an RF-8 from the VFP-63 detachment assigned to the *Coral Sea* (CVA 43) flying a photo reconnaissance mission in the Hanoi area brought back evidence that revealed the distinctive "Star of David" road pattern of a Soviet S-75 Dvina (SA-2) missile site under construction fifteen miles southeast of Hanoi. The discovery confirmed what had been anticipated for some time and was considered important enough for Rear Adm. Edward C. Outlaw, the Task Force 77 commander, to fly to Saigon to discuss the photographs with Lt. Gen. Joseph H. Moore, commander of the Pacific Air Force's 2nd Air Division, which was responsible for all USAF units in Southeast Asia. Together they formulated a plan for a joint Navy–Air Force strike on the SAM site, which they sent up the chain of command. John T. McNaughton, assistant secretary of defense for international security affairs, ridiculed the need to strike the SAMs. "You don't think the North Vietnamese are going to use them!" he scoffed. "Putting them in is just a political ploy by the Russians to appease Hanoi." After what seemed like an inordinate delay, the proposal to attack the missile site was "disapproved" and the mission against the SAM site was cancelled because Secretary of Defense McNamara was worried that American ordnance might kill or injure some of the Russian technicians assisting the North Vietnamese.[7]

The SAM site under construction southeast of Hanoi was simultaneously discovered by an Air Force U-2 high-altitude reconnaissance airplane. The Air Force, well aware of the danger presented by the SA-2 (three U-2s had already been shot down by the missile, one over Cuba piloted by Maj. Robert Anderson Jr., Gary Power's U-2 over Russia, and one piloted by a Chinese Nationalist pilot killed when shot down over China), instructed SAC to install the System XII ECM equipment developed by Applied Technology for the U-2 that would warn the U-2 pilots if they were being tracked by the SA-2's Fan Song radar. The system used state-of-the art electronic miniaturization techniques and ignored military specification requirements. It was the most advanced radar warning receiver in terms of weight and capability. System XII units also recorded each radar-tracking sequence. Analysis of these recordings revealed changes in the Fan Song radar's characteristics that proved useful in designing ECM devices for U.S. aircraft operating over Vietnam. The System XII had limited jamming capability, however, so the pilots were directed not to fly within thirty nautical miles of known SA-2 sites until the improved System XV—based on a Navy ECM unit—was installed.⁸

The discovery of the SAM missile site necessitated an increased need to locate all the North Vietnamese air defense radars. The Navy led off on April 16 when several of its carrier-based EA-1Fs flew their first ECM mission over the North. The Air Force followed on April 29, using three RF-101 Voodoos, hastily refitted with QRC-160-1 ECM pods designed to jam the SA-2's Fan Song radar.⁹

The QRC-160 was the brainchild of Capt. Gerald R. Sensabaugh Jr. Captain Sensabaugh was assigned to the countermeasures office in the Pentagon in March 1961, when a letter came in from the commander-in-chief of the Pacific Air Force requesting that a more effective and smaller jamming pod be developed to protect his fighter-bombers from radar-directed anti-aircraft fire. This closely matched an unsolicited proposal that had been submitted by General Electric. Sensabaugh saw General Electric's proposal as a way of protecting fighter-bombers against surface-to-air missiles, but he thought it unwise for political reasons to push this capability. He had worked closely with the tactical air folks across the hall and knew that they claimed they did not need jamming pods to defeat the missiles because they could outfly them.¹⁰

To get around this obstacle, Sensabaugh sold the concept to the Air Staff as a countermeasure against the SON-4 and SON-9 fire-control radars. He told the staff that the pod was mainly to counter flak-control radars. "But oh, by the way, it also covers the same frequency band as that used by the Fansong [*sic*] missile control radar." General Electric received a development contract for the QRC-160, which began flight testing at Eglin Air Force Base in Florida on July 1962, when it was hung under an F-100. The success of the tests led to a limited production research and development contract for 150 units, which were delivered to the Tactical Air Command beginning in 1963. "The Tactical Air Command," according to the account written by Indonesian television network iNews, "refused to provide aircraft and crew for combat testing and tactical research. The Aviation Systems Division of the Air Force Systems Command, which was responsible for the engineering test project, did not integrate tactical research within the plan because there were no potential users."[11]

Unfortunately, "using electronic countermeasures," in the words of Col. Bud Voland, who was an Air Force Electronic Warfare Staff Officer at the time, "was not part of their [TAC's] mentality." To the fighter pilots, the added weight and drag of an external pod was objectionable because it degraded the performance of their aircraft. When the QRC-160 pods arrived, most of them were placed in storage, where they were all but forgotten until June 1965, when a small batch of QRC-160-1 jamming pods arrived from Okinawa for installation in the RF-101 Voodoos of the 15th Tactical Reconnaissance Squadron based at Tan Son Nhut airfield. The pods, which had been produced under the Quick Reaction Capability (QRC) program, never went through any fault testing. The people in charge of the pods thought that when they were needed, they would just bring them out of storage, mount them on an aircraft, and off it would go. But they failed to realize that the electronic systems of that era deteriorated when left unused for any length of time.[12]

Ground crews at Tan Son Nhut, working from unfamiliar manuals, fitted three of the RF-101 Voodoos with the QRC-160-1 that had been taken out of storage. Capt. Anton D. Brees, an electronic warfare officer on assignment to Tan Son Nhut, described what happened next: "The first mission was launched, and they got no jamming power. They couldn't duplicate the fault on the ground, and on the next mission, there still was no power. In the meantime, inoperable pods were collecting in the repair facility and nobody knew how to fix them. Somebody opened one up, and found several capacitors and resistors that had broken off the cards. The low frequency vibration, caused when planes taxied out over uneven concrete, had shaken them off. No wonder they couldn't get any power out of those pods!"[13]

A team of experts was dispatched from Wright-Patterson Air Force Base in an effort to figure out how to make the pods work. In an attempt to fix the problem, they filled the space holding the electronics with potting compound, hoping the solidified material would hold the electronic components in place. The repaired QRC-160-1 would work for seven or eight hours (about three missions) before it failed again.[14]

Although the QRC-160 pods appear to have been effective, they were extremely unreliable and difficult to maintain. After much effort, according to the information provided by Brees, the Pacific Air Force concluded the pods were not capable of sustained combat operations, were unreliable, and were not supportable in a combat theater. The pods were shipped back to the United States. This experience cast a shadow of suspicion over future ECM pods or additions.[15]

The reported discovery of the construction of SA-2 SAM sites in early April also resulted in an urgent requirement for active ECM. To help satisfy this need, a detachment of six EF-10B Skyknights from Marine Composite Reconnaissance Squadron 1 (VMCJ-1) were ordered to Da Nang Air Base in South Vietnam. The Douglas Skyknight was a transitional aircraft that filled the gap between the technology of propeller-driven aircraft and swept-wing jets. It was designed around a conventional straight wing and a tail group similar to that of the Skyraider with side-by-side seating for a pilot and a radar operator. The jet-powered Skyknight emerged from a Navy request issued to the aircraft industry in 1945 for proposals to produce a two-seat carrier-based night fighter with long range, good performance, and an airborne intercept radar. The specification called for a fighter with 500-mph performance at 40,000 feet and the ability to detect an enemy aircraft at 125 miles. The last requirement dictated the use of a large radar dish in the nose that added to the drag index. It also demanded a second crew member to operate the APQ-35 radar. The first aircraft, designated the F3D-1, flew on February 13, 1950, and entered operational service with VC-3 at Moffett Field, California, in December 1950. The airplane, which was equipped with Westinghouse J34-WE-36 jet engines rated at 3,000-lb. thrust, proved to be underpowered and only twenty-eight aircraft of the initial production run were produced. All remained stateside and were only used for training purposes. The F3D-2 superseded the F3D-1 and flew for the first time on February 14, 1951. It was powered by two upgraded Westinghouse J34-WE-36 engines rated at 3,400-lb. thrust.[16]

After the Korean War ended, the F3D-2s were pulled out of front-line service and assigned to Marine Composite Squadron 3 (VMCJ-3). Why they were assigned to VMC-3 has never been clarified, although it appears likely that they were intended to prepare the unit for conversion to jet aircraft that would replace their obsolete piston-engine mounts. Although no ECM jammers were employed by Marine ECM units during the Korean War, there were those in the Corps' ECM community who were farsighted enough to foresee the need to provide active EMC support for attacking aircraft in light of the proliferation of radar-controlled anti-aircraft weapon systems. A number of officers, including Maj. Thomas MacDonald and a few senior enlisted men, had served with the first EW units in Korea. These experienced airmen were familiar with the modifications they had made to the AD-4Ns during the war. In the late spring of 1955, someone serving in VMCJ-3 (possibly Major MacDonald) went to the Marine headquarters and suggested that the F3D-2s be converted into an ECM aircraft as a potential replacement for the piston-engine AD-5N Skyraiders then in use by all three Marine Composite Squadrons. The first F3D-2 selected for modification was bureau number 124620, which had served in Korea. A second F2D-2 (BuAer No. 125786) was also modified.[17]

The work in transforming the two aircraft moved swiftly during the summer of 1955 as ECM equipment identical to that used on the AD-4s was installed and tested on the F3D-2s. In May 1956, the two modified F3D-2s in VMCJ-3's inventory were designated F3D-2Q variants and their modifications used as the basis to convert thirty-five F3D-2s into ECM F3D-2Qs that were redesignated as EF-10Bs when the Department of Defense's Tri-Service Aircraft Description System took effect in 1962.[18]

When Lt. Col. Otis W. Corman, VMCJ-1's commanding officer, led the contingent of EF-10B Skyknights into Da Nang airfield on April 17, 1964, it became the first electronic warfare squadron of any service stationed in Vietnam. This squadron was not new to air operations in Southeast Asia, as it had been providing the Navy and Air Force with electronic countermeasure support in the region for some time. Although Colonel Corman's unit was administratively detached to Marine Air Group 16 (MAG-16), the unit's operations were directed by the 2nd Air Division.[19]

By the time VMCJ-1 arrived in Vietnam, the EF-10B's electronic warfare gear was already ten years old. And while its receiver set (ALR-8 panoramic surveillance receiver, APA-69A direction finder, and ALA-3 pulse analyzer) covered all emitter frequencies utilized by the North Vietnamese and was an invaluable electronic intelligence asset, its ALT-2 jammer lacked the capability to concentrate energy against a single emitter. Instead, the jamming signal—fed to high-gain antennas mounted in the nose—radiated in all directions so that the energy reaching the targeted emitter was too diluted to be effective. Because the EF-10B lacked an aerial refueling capability, its two underwing pylons were needed for external drop tanks in order for the aircraft to reach the northernmost parts of Vietnam. This limited the EF-10B's ability to augment its internal jamming system with additional jamming pods.[20]

When the Marine EF-10Bs arrived in theater, the 2nd Air Division was dealing with the Vietnam People's Air Force MiGs that now threatened U.S. aircraft during air strikes against targets in North Vietnam. To counter the MiGs, the EF-10Bs were tasked with the job of degrading the NVA's early warning and ground control interception (GCI) network in the area of U.S. air operations, which at that time encompassed all of North Vietnam south of the 20th parallel. The Air Force RB-66s jammed those radars directing anti-aircraft fire. "Thus, the Air Force," as Warrant Officer J. T. O'Brien, USMC (ret.), wrote, "was willing to take on the hot action and the Marines would be relegated to the less glamorous task, which was no surprise."[21]

The most pressing need for ECM support was to counter the radar-controlled anti-aircraft artillery (AAA) as the SAM threat had not yet materialized. On a typical mission, each EF-10B was configured with two internal ALT-2 noise-jamming transmitters, one chaff pod, one ALQ-31 pod containing one ALT-19 noise jammer, and one ALT-17 radar/communications jammer. EF-10Bs were launched at five- to ten-minute intervals and flew a prescribed track north from the DMZ at 30,000 feet, jamming early warning and GCI radars. Other missions were flown at 20,000 feet in a racetrack pattern along the ingress/egress routes of the attacking aircraft. The ALT-2 and ALT-17 jammers prevented or broke the lock of a Fire Can's radar until the attacking aircraft were in their low-altitude ordnance delivery modes. Chaff drops were also effective, but that meant trading off the external fuel tanks, which meant that multiple aircraft were required to provide an effective screening corridor.[22]

During the spring and summer of 1965, Marine EF-10Bs usually laid down the jamming barrages for Air Force strikes against the North. Because the EF-10Bs lacked aerial refueling equipment, they had to carry 300-gallon auxiliary fuel tanks when supporting attacks in the Red River delta in order to spend fifty minutes orbiting over the Gulf of Tonkin. As Air Force historian Bernard Nalty noted, "Marine fliers sometimes shut down one engine to conserve fuel while descending from their usual operating altitude of 30,0000 to 35,000 feet during the return flight to Da Nang." The rate of climb of a fully loaded EF-10B was so sluggish that the Marine Skyknights had to take off toward the sea in order to avoid small arms fire from Viet Cong guerrillas concealed along the airfield's inland perimeter.[23]

What began as a normal MiG combat air patrol mission over North Vietnam on July 24, 1965, dramatically changed the nature of the air war over Vietnam. SA-2s launched under the direction of Soviet advisors assigned to North Vietnam's first SAM regiment shot down an Air Force F-4C Phantom covering a strike force of F-105Ds attacking ground targets. The MiG CAP F-4C pilots flying at an altitude of 23,000 feet expected little resistance from the North Vietnamese forces, believing they were too high for anti-aircraft fire to be effective, and few MiGs had been seen so far in the war. Because the first emission from a Fan Song radar had been detected by an RB-66 the day before, it was essential that RB-66Cs were included in the operation. The mission had been uneventful until the warning "blue bells ringing, blue bells ringing," sent from an RB-66C accompanying the strike force, blared over the radio, indicating that a Fan Song SA-2 guidance radar was transmitting. A second call of "blue bells ringing" came five minutes after the first. One of several SA-2 missiles launched against the stacked F-4 formation exploded moments later. The blast from the 268 pounds of high explosive in the missile's warhead ripped apart the fuselage and control surfaces of the F-4C flown by Capt. Richard P. Keirn, instantly killing his back-seat radar operator, Capt. Roscoe Fobar. Keirn successfully ejected from the damaged aircraft and became a prisoner of war. Keirn's F-4C, which lacked any SAM warning devices, was the first of 110 U.S. Air Force aircraft lost to SAMs in Southeast Asia.[24]

President Johnson's order for a retaliatory air strike went through the chain of command, but by the time it got to the 2nd Air Division headquarters, the SAM batteries were long gone. Instead, dummy missiles had been placed at the site as a "flak trap." By the end of the year, fifty-six SAM batteries dotted the countryside around Hanoi and Haiphong. U.S. Air Force and U.S. Navy strike aircraft tasked with carrying out Operation Rolling Thunder were ill-equipped to counter North Vietnam's air defenses, which included cutting-edge Soviet early warning radars, radar-guided surface-to-air missiles, and anti-aircraft artillery.[25]

Skyknights would sometimes venture inland to jam Fire Can and Fan Song sets. These early Fan Song jamming missions were usually conducted during the night. The EWO on board these nighttime interdiction missions would listen for the SA-2's tracking and guidance signals, radio SAM warnings to the light bomber they were escorting, and then try to jam the Fan Song guidance beam. These missions ceased in late 1965 when the proliferation of SAM missile sites and the increased skill of SAM crews forced the EA-10Bs to orbit too far from most inland targets for their radar jamming to be effective. The EA-10Bs could disrupt radars along the coastline, but they were only marginally effective against targets more than sixteen nautical miles inland. From April until October the Marine EA-10Bs were the only aircraft in Vietnam capable of jamming the SA-2's radar until the Air Force Brown Cradle B-66Bs arrived in theater.[26]

In late October 1966, the first Marine EA-6As arrived in theater when six of these aircraft and an aircrew/maintenance cadre from VMCJ-2 joined VMCJ-1 at Da Nang. These "Electric Intruders" were a modified version of the Grumman A-6A all-weather strike aircraft that had been converted into an electronic warfare platform to replace the Marine Corp's aging EA-10Bs. In its original configuration the EA-6A was to carry a Loral ALQ-53 surveillance system housed in the characteristic "football" fairing atop the tail along with a mixture of ALQs and ALTs placed in eight underwing pylon-mounted pods. The first operational EA-6A, delivered to VMCJ-2 at Cherry Point in late November 1965, only had seven storage stations, with one centerline station, four inner wing stations, and two outer wing stations beyond the wing fold. These outer wing stations initially were used to carry the low-band receiver pods of the ALQ-53 EW receiving system. Later, these mounting points were used to support ALE-32 or ALE-41 chaff dispensing pods and the AGM-45 Shrike missile. The ALR-15 multiband threat warning system and ALQ-41 deception repeater jammer were also installed on the wing stations as needed. To make room for all the additional ECM equipment, an eight-inch plug was inserted into the forward fuselage.[27]

When reliability problems with the ALQ-53 threatened to delay delivery of the EA-6A, the Marine team at the Naval Air Systems Command managing the EA-6A program issued a contract to Syracuse University's electronic laboratory to design modifications to correct some of the problems unrelated to the jammer look-through problem. Look-through allowed the operator to observe the effect of the jamming on the subject signal. The effort to resolve the problems with the ALQ-53's look-through mode was solved by issuing a contract to Bunker Ramo Corporation to produce modification kits that would convert the ALQ-53s already delivered into the ALQ-86. The new ALQ-76 was also behind schedule, forcing Grumman to outfit aircraft being readied to deploy with vintage ALQ-31B pods, which had lower-powered ALT-6B noise jammers that were substituted to outfit aircraft being readied to deploy in the fall of 1966.[28]

CHAPTER 4
THE SAM THREAT

The Soviet Union began work on an air defense missile shortly after the end of World War II, exploiting the technology used by the Germans in the development of the Wasserfall surface-to-air missile. Heightened tensions surrounding the start of the Korean War in 1950 caused Stalin to speed up work on strategic defenses, including the design of a guided missile system to defend Moscow. The system, named Berkut (Golden Eagle), was designated as the S-25. The Berkut/S-25 system and associated radars were deployed primarily to defend Moscow and were deployed as two concentric rings of missile batteries, an outer ring fifty-three to fifty-six miles from Red Square to attrite any incoming bomber formation approaching Moscow, and an inner ring at a radius of twenty-eight to thirty-one miles to deal with any bombers leaking through the outer ring. As described by Steven Zaloga, ring roads were "designed to provide ready access to the launch areas, and these became the basis for these well-known features in the contemporary Moscow landscape. U.S. intelligence at the time estimated that the creation of the ring roads and launch sites around Moscow in 1953–55 consumed the equivalent of an entire year's production of concrete, which gives some idea of the scale and priority of this program." The construction program, which began in 1953, took five years to complete. When it was finished, there were twenty-two missile regiments, each fielding sixty missile launchers, around the inner road, and thirty-four regiments on the outer road, for a total of fifty-six launch sites. The Central Intelligence Agency plotted their locations after Carmine Vito's U-2 flight over the Soviet Union on July 5, 1956. Vito's mission took him directly over Moscow. Although the city was hidden by smog, the filters on the U-2's camera cut through the haze. Additional knowledge on Soviet activities in this area was collected from German engineers who were allowed to return from the Soviet Union after Stalin's death. This included information on the system's V-300 missile, which was designated the SA-1 by U.S. intelligence, as well as information on the B-200 fire-control radar and its A-100 search radar.[1]

Deficiencies with the SA-1 missile and its fixed-base limitation led to the development of a mobile anti-aircraft missile system based on the S-75 Dina missile (designated SA-2 by U.S. intelligence, NATO codename Guideline), designed by Boris Vasilievich Bunkin. The SA-2 was a road-transportable, two-stage, 35-foot-long missile that weighed a little over 5,000 pounds and had a maximum velocity of Mach 4. It had a solid-propellant booster that was 26 inches in diameter and 8.5 feet long. The liquid-fueled second-stage sustainer was 20 inches in diameter and 26 feet long. Its 430-lb. warhead contained 268 pounds of high explosives, providing a lethal yield radius of 150–200 feet.[2]

Two different radars were required to identify and acquire targets for the S-75 Dina missile and provide fire control. The P-12 (Spoon Rest) search radar was used to identify potential targets for the missiles. Once a target was ascertained, its location, course, and speed were passed on to the SA-2's fire-control team operating its Fan Song B radar, which was a track-while-scan radar that combined automatic detection and tracking, guiding the missile to the target using a system known as command guidance. To acquire a target, the Fan Song, cued by one of the Spoon Rest radars located throughout North Vietnam, used its search mode with its antennas locked in a wide-beam configuration that sacrificed accuracy in exchange for maximum detection. Once the target was acquired, the Fan Song was switched to one of several modes that would lock on to the target while tracking the missile via the guidance transponder located on its sustainer stage. The Fan Song's computer processed data on both the missile's trajectory and its flight path to the target, issuing commands to the missile by means of a guidance transmitter that used a different frequency from the track-while-scan radar, which had its own dish-shaped antenna. The guidance signal, which controlled the missile's trajectory, began no later than four seconds after launch. An electronic receiver in the base of the sustainer section picked up the signal, which was sent to the missile's control unit.

FIG. 4.1

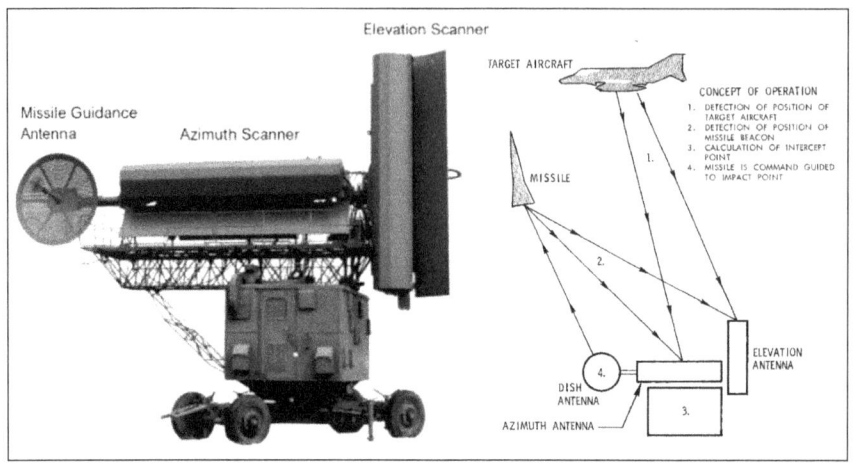

Fan Song radar showing radar signals used to guide SA-2 missiles to the target
Source: Bernard C. Nalty, *Tactics and Techniques of Electronic Warfare: Electronic Countermeasures in the Air War Against North Vietnam 1965–1973* (Washington, DC: Office of Air Force History, 1977), 19.

When the missile was about one thousand feet from the calculated point of interception, a command from the ground armed the warhead.[3]

The Fan Song radar used two modes of guidance to direct the SA-2 toward its target: the Treokh Tochek (three-point) mode and the Polavinoye Spravleniye (half-angle correction) mode. The former was a command line-of-sight method wherein the missile followed the Fan Song radar beam aimed at the target. While this method minimized the flight path and the time it took to reach the target, it required the SA-2 to make sharper and sharper corrections as it closed on the target, which could be difficult for the missile to achieve due to its aerodynamic limitations; the target could escape by making high-G turns away from the missile. The half-angle correction method was more sophisticated. It directed the missile to lead the target so that it needed less correction to hit it, but the Fan Song needed a clean, jam-free return to calculate the half-angle. The SA-2 could also be guided manually in a mode called "auto-track," but this also required a clean radar return. "The command link scheme," wrote Carlo Kopp, "and the need to carefully select control laws and radar modes resulted in the need for high levels of operator skill and a good understanding of engagement geometries." The crew required to operate the system consisted of a fire-control officer, responsible for tactical assessments and launch decisions; a guidance officer, responsible for guiding the missile during engagements; and three system operators, responsible for elevation and azimuth acquisition, tracking the target, and measuring range to the target.[4]

The Fan Song radar system was a mobile unit packaged in four vans that were towed into place using a general-purpose truck. The four vans were:

- The PV van, which contained the radar head, transmitter, and radar electronics.
- The UV van, which contained the consoles and stations for commander, guidance officer, and operators.
- The AV van, which contained the initial firing sequencing hardware, moving-target-indicator hardware, target and missile analogue tracking computers, uplink command generator, and uplink transmitter.
- The RV van, which was needed to run the diesel power generators and the electric gear.

A typical battery or "battalion" had four to six SA-2s on semifixed, single-rail launchers, positioned 200 to 350 feet apart in a star-like pattern surrounding the Fan Song trailers and the Spoon Rest radar. Six reload rounds on their articulated trailers were also included within the battery's site.

The Soviets' decision to send the S-25 system to North Vietnam occurred on November 17, 1964, when the Politburo—the supreme policy-making body of the Soviet Union—decided to send increased support to the North Vietnamese. In addition to the surface-to-air missile system and the advisors needed to support it, the aid also included aircraft, radar, artillery, small arms, ammunition, food, and medical supplies.[5]

Establishment of the first SAM regiment, the 236th, formed on January 7, 1965, was given the highest priority by the North Vietnamese military.

Officers scoured the armed forces, civilian universities, and technical schools looking for the best engineers, electricians, technicians, and mechanics to form the new regiment.[6]

Construction of the first missile site began at the end of March but proceeded at a very leisurely pace. The first seventy missiles, accompanied by a cadre of Russian air defense specialists to advise the North Vietnamese, did not arrive by ship until April 1965. Although the launch revetments were near completion by the beginning of May, no SAM hardware had been installed. The delay in completing the missile site was apparently caused by a dispute over who was to man the sites, and the route the missiles would take overland through China. "The evidence," according to the account provided by the Global Security website, suggests that throughout the spring the North Vietnamese "vacillated between yielding to Chinese pressure and thus deferring completion and activation of the SA-2 missiles until the fall, when the North Vietnamese cadres could complete their training in the USSR to operate them," or agreeing to the Soviet proposal to send eight battalions of SAMs and four thousand Soviet advisors and technicians through China by rail. The mounting pressure of the U.S. bombing campaign forced the North Vietnamese to pursue the latter course, and they were able to persuade the Chinese to permit a limited quota of Soviet SAM personnel to pass through their territory on the way to North Vietnam.[7]

While the North Vietnamese were constructing their initial SAM sites, U.S. tactical electronic aircraft continued to operate over most of the country with relative impunity. Penetration and peripheral reconnaissance were performed in a continuing effort to monitor the growth of the NVA's radar order of battle.

The introduction of the SA-2s was a watershed event that changed the nature of the air war over Vietnam by denying medium altitudes to the Alpha Strike Forces of Rolling Thunder. Unless an effective counter was found, the missiles could inflict unacceptable losses. Despite numerous pleas from the military, however, President Johnson failed to permit offensive action against the SAM sites until after the first aircraft, an F-4C piloted by Capt. Richard P. Keirn, was shot down by a SAM. When Johnson approved a retaliatory air strike against the offending sites, the Joint Chiefs of Staff quickly sent an order to Adm. U. S. Grant Sharp Jr., Commander-in-Chief, U.S. Pacific Fleet, who assigned the task to the Air Force. On July 27, 1965, Lt. Gen. Joseph H. Moore, commander, USAF 2nd Air Division, sent fifty-four F-105s escorted by twelve F-4Cs and eight F-104 Starfighters, under the code name Spring High, to attack SA-2 sites 6 and 7 along with the barracks at Cam Doi and Phu Nieu that were suspected of housing air defense personnel. The strike group was supported by three EB-66Cs, whose mission was to provide a warning of SAM radars, and six Marine EF-10Bs to provide jamming.[8]

"Those planning the attack," Price noted, "were entering uncharted territory, for nobody had attempted such an enterprise before. Their tactics were harnessed to the belief, almost universal at that time, that SAMs were close to 100 percent effective against aircraft flying any higher than very low." Thus the F-105s attacked at an extremely low level in the belief that the greatest threat would be from SAMs and radar-controlled anti-aircraft guns.[9]

The air strikes were an unmitigated disaster. Anticipating the attack, the 236th Air Defense Regiment, which had downed Keirn's F-4C on July 24, moved the two SAM batteries involved, replaced the missiles with dummies, and deployed 120 anti-aircraft guns around the area. U.S. aircraft flying into and out of the target areas faced intense ground fire for seven and a half minutes from 37-mm, 57-mm, and 85-mm anti-aircraft guns. Enemy gunners shot down four F-105Ds and damaged a third, which crashed into another as it limped toward home. "Six F-105s had gone down," wrote author Dan Hampton in *The Hunter Killers*, "three highly trained, experienced, and irreplaceable pilots had been killed." Worse, poststrike photo analysis showed that one site appeared to be designed to be a decoy flak trap and the other appeared to be empty. Had the attack been conducted immediately after July 24, it might have come without warning and succeeded. But the battalion's commander—a Soviet major advising the North Vietnamese—had repositioned his batteries away from the Son Tay region the night before the attack. "It was only then that sham site 6 and 7 were created with cannibalized missile bodies, and sometimes even logs propped up in revetments to resemble operational SAMs."[10]

Although all but one of the aircraft sent to destroy SAM sites 6 and 7 and the barracks at Cam Doi and Phu Nieu were shot down by conventional anti-aircraft fire, the continued proliferation of SAM sites was of considerable concern to the leadership directing the air war over Vietnam. Destroying the sites, which continued to be discovered, became a priority in the next series of Rolling Thunder strikes. On August 9, 1965, a force of twelve F-105Ds accompanied by aircraft flying MiG CAP, ECM, and ELINT, struck another new site, which was heavily defended by a variety of anti-aircraft guns ranging in size from 37 mm to 100 mm. While no aircraft were lost and only one F-105D was damaged, poststrike bomb damage revealed once again that the Americans had struck an empty missile site. This indicated that the North Vietnamese defenders were, in the words of Jacob Van Staaveren, "able to anticipate an attack and to disperse missiles and associated equipment quickly."[11]

Three days later, during the night of August 11–12, the Navy lost its first aircraft to SAMs. Two A-4Es flying off the *Midway* (CV 41) on an armed reconnaissance mission sixty miles south of Hanoi observed what appeared to be two flares glowing beneath the clouds fifteen miles north of their position. The pilots of the two A-4s watched the spots of light come out of the clouds before they suddenly realized they were looking at the glow of two SA-2 missiles heading directly toward them. Although they tried to maneuver away from the danger, their actions came too late. Seconds later the SAMs exploded, destroying the A-4E piloted by Lt. (jg) Donald H. Brown Jr. and damaging the A-4E piloted by Lt. Cdr. Francis D. Roberge. Shrapnel from the SAM's warhead punctured Roberge's drop tanks, causing his air-to-ground Zuni missiles to fire and igniting flares that were still in their wing racks. He jettisoned the remaining stores and managed to limp back and land on board the *Midway*, "his Skyhawk's belly scorched, wrinkled, and peppered with more than 50 holes."[12]

Even before the loss of *Midway*'s A-4Es, Admiral Sharp, after receiving approval from the Johnson administration via the Joint Chiefs, initiated an effort to find and destroy the North Vietnamese missile sites under the operational name of "Iron Hand," which, according to Vice Adm. Malcolm W. Cagle, was coined by Capt. Alton B. Grimes, then on duty in the Pentagon reconnaissance center. During the next two days, Navy planes from the carriers on Yankee Station in the Gulf of Tonkin flew seventy-six low-level missions in search of SAM sites. Five aircraft and two pilots were lost to ground fire and seven others were damaged, but they failed to locate even one SA-2 launcher.[13]

"The downing of a second U.S. aircraft by a SA-2 missile and the Navy's five losses during a search for the missile sites," according to Van Staaveren, "sent shock waves throughout the JCS and the services, forcing Air Force leadership to take action." Gen. John P. McConnell, Air Force chief of staff, directed Brig. Gen. Kenneth Dempster, deputy director of Operational Requirements and Development, to convene a high-powered Air Staff task force to answer one question: "What is the most effective means of neutralizing the threat posed by SAM missiles and the heavy anti-aircraft in the Southeast Asia conflict." Dempster's task force, which included representatives of the Air Staff, the major air commands, industry, and the scientific community, met between August 13 and August 18. One of the members described it as "ten or twelve guys who sat in smoke filled rooms and brought contractors in to figure out what to do." The committee recommended a list of requirements that was needed to combat the SA-2 that included: a warning system to alert aircrews when they were under enemy radar surveillance; better pinpointing of enemy radar locations; timely processing of intelligence data; prompt air strike decisions; adequate ECM for all fighter aircraft; precise navigation for aircraft flying at high speed and low altitude into a target area; and suitable tactics for strikes in areas defended by anti-aircraft weapons. They divided the remedial measures between those to be completed in the short term (six months) and the long term (six to eighteen months). Dempster reviewed each of the task force's recommendations with everyone in the chain of command, starting with the commander of the Pacific Air Force.[14]

One of the most significant recommendations made by the task force was for the development of a fighter designed specially to locate SA-2 sites and mark them for immediate attacks by accompanying strike aircraft. To expedite the availability of such an aircraft the task force decided to use existing off-the-shelf equipment installed in an existing aircraft. This concept came from a former SAC Electronic Warfare Officer (whose identity has never been determined) working at North American Aviation, and his proposal to modify a North American F-100F for this purpose was outlined on the back of an envelope during his flight to the meeting at the Pentagon.[15]

The Navy already had a program underway to address the SAM threat. After evidence of the first SA-2 site was discovered in April, Captain Lake began daily skull sessions with representatives from the Naval Research Laboratory, the Office of the Secretary of Defense, Naval Intelligence, and the Naval Air Systems Command on how to detect the Fan Song radar and defeat the SA-2. Lake realized that it might be possible to install ALQ-51 ECMs on A-4 Skyhawks. If such an installation were possible, it would greatly increase the aircraft's survivability in the increasingly dangerous environment in the North.[16]

The ALQ-51 was a deception repeater that employed angle-gate deception. It worked by retransmitting the pulses received from the Fan Song's track-while-scan radar so that it introduced errors into the radar's tracking system, which made it appear that the target aircraft was one thousand feet from its true position.[17]

To see if the ALQ-51 would fit into the A-4, which was a small airplane, Captain Lake traveled to the Navy's flight-testing facility at NAS Patuxent River (PAX), Maryland, where he crawled around the interior of an A-4 looking for a place to install the ALQ-51 electronics. The gun bay seemed the obvious solution, but Lake and the Naval Air Systems Command (NAVAIR) engineers at PAX were reluctant to deprive the plane of its guns. Lake and the NAVAIR engineers could not agree on how to do it. To solve the dilemma, Lake went to the Douglas plant in Long Beach and spent an entire afternoon on the flight line trying to figure out the best place to put the ALQ-51, using wooden boxes he had made up that mimicked the size and shape of the electronic gear. "In the end," Lake stated, "we found that there was room in the gun bay to fit the jammer, both cannons and half the ammunition load."[18]

Lake's proposal to add the ALQ-51 to the A-4 Skyhawk led to the establishment of Project "Shoe Horn," a NAVAIR-funded program to provide the engineering effort needed to formalize the modifications needed to equip the A-4s with the deception jammer. Sanders Associates received a contract to build a prototype to be installed by the plane's builder, the Douglas Aircraft Company. The prototype was then tested by Air Test and Evaluation Squadron 5 (VX-5), which evaluated the prototype against the "Flint Stone" radar, a Fan Song surrogate located at the Sanders test facility in Merrimack,

New Hampshire. The tests were successful and the new jammer, modified to increase its effectiveness, was designated the ALQ-51A.[19]

After the testing of the modified ALQ-51s was completed, Sanders Associates received an order for installation kits to retrofit the ALQ-51s, most of which had been held in storage, to the ALQ-51A configuration. This was accomplished by Sanders personnel, who along with Navy engineers began installing the upgraded ALQ-5As to the *Independence* (CA 62) air group's A-4Es in September 1965. The installation was accomplished by placing the two ALQ-51A electronic enclosures in the space beneath the Skyhawk's cockpit that had previously been filled by the plane's 20-mm ammunition. A smaller ammunition magazine that could only hold forty rounds of ammunition was installed above the gun bay. The A-4s also received ALE-18 chaff dispensers.[20]

On September 16, 1965, an A-4E pilot reported success in jamming a Fan Song radar that had been tracking him. Four days later, the A-4Es used their newly installed ALQ-51s to divert six SA-2s that had been fired at them during an early Iron Hand mission. The ALQ-51A could tackle only one emitter at a time, and the NVA radar operators (or their Soviet advisors) learned to beat it by differentiating false returns from genuine ones.

The successful introduction of the ALQ-51A led to its installation on additional Navy aircraft. A once classified chart produced by staff of the Commander-in-Chief of the Pacific Fleet (CinCPacFlt) indicates that ALQ-51A jammers were installed in the RF-8 photo-reconnaissance version of the F-8 Crusader in April 1966, in the F-8s themselves in the following December, and in F-4Bs starting in March 1967.[21]

Although Project Shoe Horn was considered a success by those people involved, the reliability of the ALQ-51A was initially poor due largely to deficiency in spare parts, test equipment, and training. "The guys didn't have any training in the U.S. because we didn't have any equipment back here," explained Lake. "Everything was out there. But they couldn't support it, they couldn't use it properly, they couldn't maintain it properly, they couldn't test it properly." Even so, the ALQ-51A proved its worth when it functioned properly. Price tells us that "For planes with operable deception systems, the loss rate to SAMs was about one plane per fifty missiles fired

TABLE 4.1. ECM EQUIPMENT INSTALLATION

INSTALLATION PERIODS OF ECM EQUIPMENT
IN U.S. NAVY AND U.S. AIR FORCE AIRCRAFT

U.S. NAVY AIRCRAFT	
A-4	ALQ-51 / APR-27 / ALE-29
F-4	ALQ-51 / APR-27 / ALE-29
F-8	ALQ-51 / APR-27 / ALE-29
RF-8	ALQ-51 / APR-27 / ALE-29
U.S. AIR FORCE AIRCRAFT	
F-105	ALQ-71
F-4	ALQ-71

J A S O N D J F M A M J J A S O N D J F M A M J J A S O N D J F M
1965 — 1966 — 1967 — 1968

UNCLASSIFIED / SECRET

. . . compared with one plane per ten missiles fired if no ALQ-51A was fitted or if the equipment malfunctioned."[22]

The incorporation of the ALQ-51A in the Navy's planes initially caused the loss rate in SAM-defended areas to drop significantly. "With study and practice, however, the Vietnamese missile crews [and their Soviet advisors]," as described by Merle Pribbenow, "developed procedures for distinguishing between false and actual targets. These methods included comparing differences in the signal quality and characteristics of each target, analysis of each target's *delta* rate (its rate of change in bearing and elevation), flipping the radar screen range-scale settings back and forth to detect anomalies, and briefly switching the radar antenna to the stand-by position." Using these sophisticated techniques, the NVA missileers were successful in overcoming the ALQ-51A's deception jammer during an engagement on August 13, 1967, and were able to bring down three Navy A-4s with their missiles.[23]

CHAPTER 5
ECM PODS, RHAWs, AND CHAFF DISPENSERS

In September, General Dempster, after conducting his investigation into potential means of countering the North Vietnamese air defenses, convened an Anti-SA-2 symposium at the Eglin Air Force Base in Florida. The symposium was attended by eighty people from the Air Force, the Navy, and the Department of Defense. The Navy officers described the new AGM-45 Shrike radar homing missile and the Shoe Horn program to fit the ALQ-51 into the A-4 and other tactical aircraft. Col. Joe Gillespie, head of the Electronic Warfare Test Division (EWTD) at Eglin, arranged for Ingwald "Inky" Haugen, a retired Air Force officer with twenty years of electronic warfare experience who now worked as a civilian project officer at EWTD, to present his ideas for defeating the SA-2 missile system. Haugen believed that four F-105s each carrying a couple of QRC-160-1 jamming pods could seriously degrade the SA-2 system if the pods were properly spaced in azimuth and elevation. Although Haugen's presentation was not received with enthusiasm, the Air Proving Ground Center and the Tactical Air Warfare Center at Eglin received orders to conduct an evaluation of the jamming formation tactic under the code name Problem Child.[1]

The QRC-160-1 pod, which was designed to be attached to a wing pylon, was 8 feet 4 inches long, weighed less than 100 pounds, and operated in the D/E radar bands. At the heart of its internal electronics were four voltage-tuned magnetrons that each produced a frequency-modulated, continuous-wave jamming signal. The magnetrons could be tuned to produce overlapping frequencies for barrage jamming or they could be set for spot jamming on four different frequencies. When successful, the jamming blinded the enemy's search and missile-control radars by filling their radar displays with noise, making them unable to detect or track targets. For power, the QRC-160-1 relied on a ram-air turbine that drove a 3.5 kVA alternator. The pod had two stub antennas and two slotted antennas on its underside to provide 360-degree jamming.[2]

When the pods arrived at Eglin, each was carefully checked and adjusted to ensure that it performed to specification. The tests at Eglin were conducted using the center's precision tracking radars and instrumentation. As Alfred Price, in his history of electronic warfare, noted, "The contrast between the Eglin series of tests and those carried out with the QRC-160-1 pods in Vietnam a few months earlier, could scarcely be greater."[3]

After the pods were carefully adjusted, two pods were fitted to each of the four F-105s that flew separated laterally and horizontally by 1,500 feet during a series of runs against Eglin's surrogate Fan Song radar. This positioning, which was the brainchild of "Inky" Haugen, Problem Child's technical advisor, produced the optimum jamming of the Fan Song radar operator's scope. Testing, which began in October 1965, showed that the QRC-160-1 pods would screen the formation from all aspects at most ranges until the formation came broadside to the radar site. The aircraft could then be seen by the radar for a short time. This was just long enough to launch a missile, but not long enough to guide it successfully to the target.[4]

To improve the QRC-160-1's capability, Eglin's engineers recommended that the unit's electronics be modified to vary the jamming strength in synchronization with the radar's scan pattern. General Electric agreed to modify the pods, which subsequently produced near stationary strobes on the radar's azimuth and elevation displays, which increased the difficulty of picking out individual aircraft. The modified QRC-160-1 was designated the QRC-160A-1. The physical dimensions of the pod, its structure and shape, were probably changed since the modified pod was stronger and heavier than the original.[5]

Testing of the QRC-160A-1 began in January 1966. "Like its predecessor, it was a self-contained, barrage noise jammer with four 75-watt magnetron components (soon replaced by 100-watt models)." These components were fitted inside an aerodynamic pod 7.5 feet long, 10 inches in diameter, weighing 200 pounds. Except for a 28-volt status light in the cockpit, all electrical power came from an integral ram-air turbine."⁶

By the end of January, preliminary evaluations of the QRC-160A-1's performance indicated that it was rugged enough for combat and could achieve full degradation of the SA-2 missile system provided that all four aircraft were present, in correct position, with at least one jamming pod working properly. But the tests also showed that there was a reduction in effectiveness if there were three jamming aircraft, and even a greater reduction if there were only two. If only one aircraft was jamming, the chances of surviving a missile attack were slim.[7]

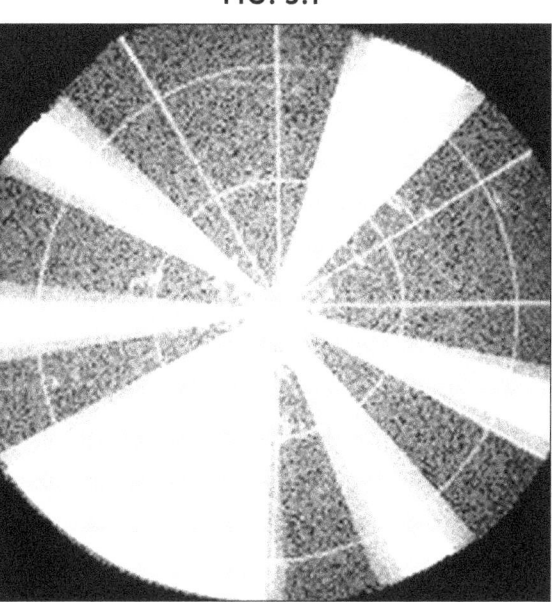

FIG. 5.1

Showing the effects of synchronized noise jamming on a jammed radar's display *United States Air Force*

While the four-plane QRC-160A-1 formation appeared to solve the SA-2 missile problem, efforts to obtain operational tests ran into a stone wall. For one thing, there was still a deep-seated opposition to the pods within the Southeast Asia theater based on past experience. There was also a prevailing mindset, according to one officer familiar with the issue, against electronic warfare systems or tactics within the F-105 community. The pilots were skeptical of the QRC pod and did not want to substitute pods for bombs. And the Seventh Air Force, which had now assumed command responsibility for most Air Force operations in Vietnam, was reluctant to spare four aircraft to practice the jamming formation in Thailand.[8]

No further action was taken until the 335th Tactical Fighter Wing flying out of Takhli airfield in Thailand experienced large losses during July and August. During a two-month period, the wing lost half of its fifty-four aircraft. On September 16, 1966, Lt. Gen. William A. Momyer, Seventh Air Force Commander, requested a combat evaluation of the QRC-160A-1 pods. Before the month was out, twenty-five of the pods had been sent to Thailand along with the technicians needed to maintain them and a test team to conduct Project Vampyrus, which was established to determine the value of the pods against Fire Can anti-aircraft fire control and Fan Song guided missile control radars.[9]

The pods were simple to operate. Before takeoff, the ground crew adjusted the controls on the outside of the pod to establish the center frequency and bandwidth that would jam particular kinds of radar. Once in the air the only action the pilot needed to take was to turn the transmitters on and off.[10]

The first of nineteen test missions using a four-ship flight of F-105Ds flying in a jamming formation with each plane carrying two jamming pods was flown on September 26. The effectiveness of the QRC-160A-1 pods carried by the F-105Ds was monitored by EB-66Cs that accompanied the missions conducted over Route Package 1.* The Electronic Warfare Officers on board first verified the number, location, type, and transmission characteristics of radars protecting the targets, then observed how these transmitters reacted to the pod-carrying F-105Ds. When necessary, the EB-66Cs jammed radars outside the frequency range covered by the self-protection pods.

On October 8, a veritable tour de force mission was flown by three four-plane flights of F-105Ds in a coordinated strike on the Nguyen Khe oil storage facility south of Hanoi. The first flight—code named Taksan—was tasked to provide flak suppression. It experienced problems before takeoff when Taksan 02 had mechanical problems and was replaced by an aircraft that had no jamming equipment. The flight divided into two elements

---

* Route Packages were the names given by the U.S. Air Force and U.S. Navy to describe areas of air operations over North Vietnam. See map on page 96.

ECM PODS, RHAWs, AND CHAFF DISPENSERS 51

FIG. 5.2

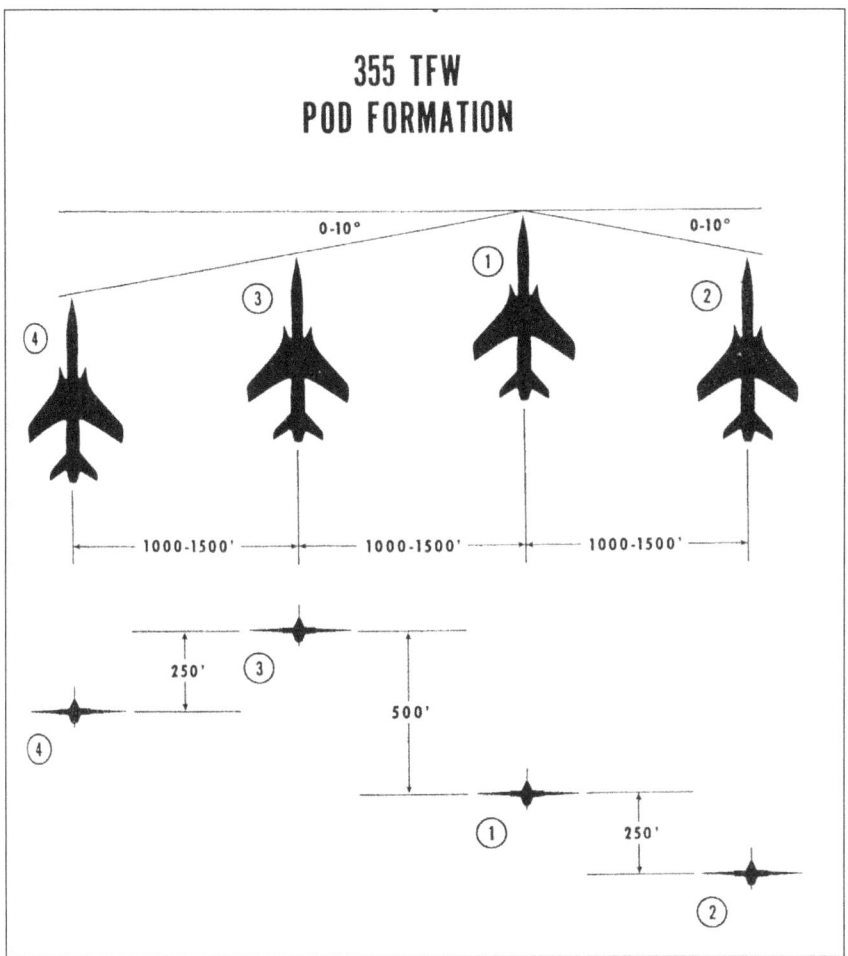

F-105D Pod Formation *United States Air Force*

just before entering the target area and the two pods carried by Taksan 01 immediately ceased function, leaving the two aircraft without any ECM protection. Within seconds, the pilot of Taksan 01 found himself in deadly danger when a MiG-21 unexpectedly dived toward him. As he eluded the MiG, three 85-mm shells burst close enough to punch holes in the skin of the F-105D. He jettisoned his bombs to gain speed and maneuverability, but his engine suddenly quit. No sooner had he restarted the balky turbine than an SA-2 came streaking toward him. He quickly nosed over in a diving turn that brought him within 2,500 feet of the ground. Anti-aircraft shells burst around him until he climbed out of range.[11]

Taksan 02, also without ECM protection, encountered intensive fire from radar-controlled anti-aircraft guns but escaped without damage. The other two planes in the flight turned on their QRC-160A-1 pods and experienced only light to moderate fire. None of the other F-105Ds on the mission, which all had working pods, reported being attacked by SAMs or radar-directed ground fire.

The tests conducted under Project Vampyrus showed that the QRC-160A-1, soon to be redesignated the ALQ-71, was suitable for combat operations, although their reliability remained poor. As Price noted, "The pod had originally been designed for one time use aboard a fighter delivering a nuclear weapon in a high-speed straight-and-level bomb run. Now the pods had to fly day after day, carried by aircraft employing the full range of combat maneuvers. The delicate electronic systems did not take kindly to being mounted on the outer wing pylon where there was vibration as the wing flexed. Added to this the pods experienced huge variations in temperature, from tropical heat on the ground and sub-zero conditions at altitude."[12]

The ALQ-71 had additional shortcomings that were not identified until the pods became standard equipment early in 1967. One such problem was the location of the control box. This was installed to the F-104D pilot's right rear, where it was difficult for the pilot to see the light indicating that the ALQ-71 was no longer working; the pod might quit functioning without the pilot's realizing it. This problem was solved by moving the control box. Another recurring problem sometimes happened when the jamming pod reverted to standby as the aircraft was pulling out of a dive following weapon release, leaving the aircraft exposed just when the equipment was needed the most. The cause of the problem was determined to be the power relays, which were liable to snap open during a high-G pullout. It was solved by redesigning the mounting bracket that rotated the power relays 90 degrees. Later, when radar homing and warning (RHAW) gear was installed, it was discovered that the ALQ-71's jamming signal prevented the pilot from using the RHAW system. A modification to the wiring fixed this problem and allowed the pilot to interrupt the jamming for a few seconds, long enough for the RHAW gear to react.[13]

By the end of November 1967, all F-100Ds flying into SAM-defended areas carried the ALQ-71 jamming pods. To produce the required jamming footprint on the ground, however, the pilots had to maintain a maximum bank angle of 15 degrees. If the bank angle exceeded this, the effectiveness of the jamming was reduced. The formation, recalled Maj. Gen. John A. Corder,

> was awkward to fly. . . . If you were in the high position and the leader went down or turned into you, you couldn't roll upside down and pull down like you normally would to keep positive G on the airplane. If you did that, your jamming would point way into the sky. You had to push on negative G to hold your place in formation, and that was uncomfortable. If the formation turned, it was either very slow and gentle, or else it was very aggressive and hard. Everything else was done with wings level, either pull back or push down. . . . [You didn't] get closer than 500 feet from the next guy, and [didn't] get further away than 2,500 feet.

To maintain an effective jamming formation a flight had to avoid banked maneuvers and maintain formation integrity.[14]

A study conducted by the Pacific Air Force to investigate the effectiveness of the ALQ-71 from September through December 1966 found that during November and December losses of F-105s carrying jamming pods flying over Route Package 6, in the most heavily defended area of North Vietnam, were less than one-third those suffered in the same area before the introduction of pods. "Crews," according to Price, "reported seeing missiles start to guide toward their formation. But then, instead of singling out an individual aircraft, they usually flew erratic paths and sped past the formation before exploding harmlessly when well clear."[15]

The addition of the jamming pods was extolled by the commanding officer of the 388th Tactical Fighter Wing at Korat Royal Thai Air Force Base, which had also received the pods. "The introduction of the QRC-160A-1 pod to the F-105 weapon system," he wrote to the director of operations at Seventh Air Force Headquarters, "represents one of the most effective operational innovations I have ever encountered. Seldom has a technological advance of this nature so degraded an enemy's defensive posture. It has literally transformed the hostile air defense environment we once faced, to one in which we can now operate with a latitude of permissibility."[16]

The increasing use of the QRC-160-1, which now accompanied almost all U.S. strikes, was so effective in blinding the North Vietnamese radars that on one occasion one of the SAM batteries was so desperate to overcome the heavy jamming that it shot down one of its own MiGs. The jamming was so heavy during the attacks on targets in Hanoi conducted between April 25 and April 29, 1967, that it was difficult for the North Vietnamese to fire any missiles. "Many battalions," according to the official Air Defense Command history, "experienced great confusion when trying to identify targets through the interference." By October, the North Vietnamese missileers were resorting to firing massed barrages of SA-2s in an attempt—with some success—to down U.S. aircraft. They also began sending MiGs after the EB-66s in order to drive them out of effective range.[17]

While efforts were being made to correct the deficiencies of the QRC-160-1, which had not passed through the normal procurement process, engineers with the Systems Engineering Group at Wright-Patterson Air Force Base were working on several programs to improve the jamming pods. One of the programs involved new versions of the QRC-160: one to be used against X-band radars that would later enter production as the QRC-72; a version that would be used against acquisition radars, designated as the QRC-160-4; and a more powerful jamming pod that covered the S and C radar bands.[18]

A small quantity of QRC-160-8 pods, an advanced model using a backward-wave oscillator, were sent to Southeast Asia for testing in the latter part of 1966. They worked well, except for the built-in receiver that was designed to recognize specific threat signals and automatically select the appropriate jamming response. Unfortunately, the receiver frequently had difficulty differentiating between a Fan Song radar and a surveillance radar. This problem was never solved, and the receiver had to be disconnected in the production version, designated the ALQ-87, which entered service at the end of 1967. In addition, the pod often went into the transmit mode when there was no direct threat. After a short time in service, it was discovered that the ALQ-87s ram-air turbine that ran the pod's electrical generator could not stand up to the high-speed maneuvering encountered in combat, which forced the pod to be tied into the plane's electrical system.[19]

Besides laying down a continuous jamming barrage, the ALQ-87 had a sweep modulator that could introduce a random burst of reinforcing noise in a so-called "pulse power option" and could simultaneously perform any two of the following at one time: jam a Fire Can fire-control radar; jam a Fan Song guided missile radar, or interrupt the SA-2's guidance in a process called "beacon jamming." The latter, also called "down-link" jamming, interfered with the signal that enabled the SAM controller to follow the missile on radar in order to correct its trajectory.[20]

Bernard Nalty, who served as senior historian in the Office of Air Force History, states that "Both the ALQ-71 and the ALQ-87 enjoyed impressive success with this [down-link jamming] technique." From December 1967, when down-link jamming began, until the bombing north of the 19th parallel ended on April 1, 1968, SAM batteries launched some 495 SA-2s at Iron Hand F-105s, but downed only three planes, two of which had been jamming the tracking radar instead of the down-link.[21]

The ALQ-51 and ALQ-87 jammers were not the only important innovation in electronic warfare introduced in the early part of Rolling Thunder. In August 1965 the Navy began installing the APR-23 radar-homing receiver in some of its A-4Es.[22]

The APR-23 was developed by the Melpar electronics company of Fairfax, Virginia, in competition with RCA and Westinghouse, in response to a Navy request during the height of the Cuban Missile Crisis to come up with a radar-homing receiver that could be used to locate Fan Song radars. By the time the fly-off competition was over, the crisis had passed. Nevertheless, some APR-23s were produced and were routinely installed in 1965 until an improved version, the APR-27, became available in early 1966.[23]

In the interim, the Navy ordered fifty APR-24s from Melpar in the second half of 1965 and the early part of 1966 to provide four or five

F-4Bs in each carrier squadron with a radar-warning device, the intention being that aircraft equipped with the APR-24 would serve as pathfinders for hunter-killer teams assigned to locate and destroy SAM sites. The APR-24 functioned much like the APR-23, except that unlike the APR-23, which only operated in the C-band, the APR-24 detected signals in the C-, S-, and X-bands used by the Fan Song, Fire Can, and MiG radars. When the APR-24 detected a radar signal in any one of these bands, it sounded an audio signal that alerted the pilot and his backseater and displayed an electronic strobe in the direction of all the radars that were illuminating the F-4. The strobe originated in the center of a 3-inch diameter CRT display. Both the strobe and the audio signal were coded differently to indicate which radar was trained on the aircraft. The strobe grew longer and the audio got louder the closer the F-4 got to the radar.[24]

The first APR-24s were installed on four F-4Bs of Fighter Squadron 14 (VF-14) on board the USS *Franklin D. Roosevelt* (CV 42) while in transit to Southeast Asia. Eight of the squadron's other F-4Bs were configured with the APR-27, another warning system that alerted the crew with an audible alarm when a SAM battery had activated its tracking and guidance radar. In the course of the next thirteen months, at least twenty-four F-4Bs in eight different squadrons were modified with the APR-24 system.[25]

Unfortunately, the APR-24, which was rushed

FIG. 5.3

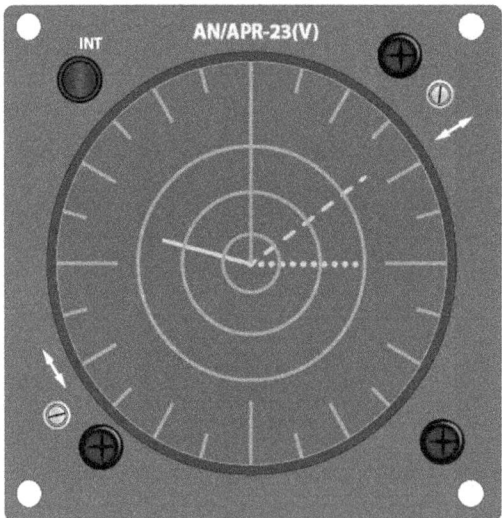

The CRT display supplied with the APR-25. The dotted line radiating from the center indicates the direction of the nearest radar threat (based on signal intensity). Other radars shown on this display are presented as a straight and dashed lines. *Phantom Phacts*

into service without a carefully worked out development program, was temperamental and difficult to keep working. Inadequately trained personnel and a critical shortage of parts also contributed to the problems associated with the APR-24. "The equipment," according to the VF-14 command report, "due to its low reliability, lack of spare parts and trained AO/AT's to maintain it, and coupled with extremely 'dirty' EW environment of North Vietnam, provided little or no improvement in the operational capability of the F-4 weapons system." As an interim solution, Applied Technology APR-25 warning systems were installed as a stopgap measure in aircraft of VF-142 and VF-143 on the USS *Constellation* (CVA 64) for its 1967 Vietnam cruise until the APR-30 could be developed. The APR-25 provided 360-degree warning of S-, C-, and X-band radar signals and consisted of several small electronics boxes, four spiral antennae, a small panel of lights, and a three-CRT threat display.[26]

The APR-27 was a missile warning system that searched for SA-2 missile and guidance signals to alert the pilots (or aircrew in the F-4Bs) of the missile threat or actual launch of an SA-2 so that they could visually obtain it and outmaneuver the missile. John Sherwood, in *Afterburner: Naval Aviators and the Vietnam War*, describes how the APR-27 functioned: "Two levels of threat are detected: missile alert and missile launch. When the APR-27 recognized the presence of missile guidance signals in the normal (or rest) position, it put out a missile alert tone. When a missile actually launched and began receiving steering commands, a red light flashed rapidly, and the audio tone of the system changed from a low-frequency warble to a much higher frequency."[27]

The APR-27 only told the pilot or crew that a missile had been launched. In order to take appropriate action to avoid the missile, they had to check the APR-25's strobe for its direction, then look in that direction for the missile. Both systems working together gave the pilot plenty of time to react to a SAM launch.

A-4 Skyhawks were the first of the Navy's aircraft equipped with the APR-27, beginning in April 1966. F-8 and RF-8 Crusaders followed in November, but it was not until a year later in November 1967 that the F-4B Phantoms got theirs. In the interim, F-4B crews had to rely on a Rube

Goldberg–like setup based on the CMR-312, known as "Little Ears." The Singer model CMR-312A was a miniature portable microwave receiver about the size of an automobile radar detector (24 x 24 x 14 inches) that weighed only 4.3 ounces. Who approved its installation and how it was introduced into the F-4B community remains unknown. The CMR-312 was attached to the aircraft's canopy using suction cups fastened to the unit and a navigation antenna attached to an aluminum bracket on the Radar Intercept Officer (RIO)'s glare shield. A Y-cable plugged into the unit was tied into the RIO's helmet so he could hear the audible warning given off by the CMR-312 when it detected radar emissions. There were different sounds for the different frequencies the radars operated at, so with training and practice the aviator could determine what kind of radars were in the area, but the system did not indicate the direction or distance to the radar and did not indicate when a missile was launched. Naval aviator John Nash, who later became VF-161's commanding officer, stated that Little Ears lasted about two weeks in his squadron, although it was in the F4-B piloted by Lt. William M. McGuigan when he and his RIO, Lt. (jg) Robert M. Fowler, downed a MiG-17 on July 13, 1966.[28]

When the first APR-27s were installed in A-4Es under Project Shoe Horn in April 1966, they also received the first ALE-29 chaff dispensers.[29] The ALE-29 was an electromechanical device that ejected chaff cartridges or flares mounted on the aft end of the engine bulge of each engine. The chaff cartridges contained shredded metal strips that formed a cloud of reflective surfaces when ejected that blinded enemy radars. Each unit held thirty cartridges containing flares, chaff, or a combination contained in plastic buckets that were loaded into the dispenser. The countermeasure material could be ejected in single or multiple salvos consisting of a series of single bursts ejected at specific intervals. As indicated in table 4-1, ALE-29 chaff dispensers were also installed on the Navy's F-4Bs, F-8s, and RF-8s by the end of the year.[30]

TABLE 5.1 RHAW GEAR

Aircraft	Equipment	Comment	Source
A-4	APR-23	From Cuban Missile Crisis	Price II, 295
	APR-23	Indicated direction of radar	Davies. A-4, 28
	APR-25	Replaced by APS-107	Davies. A-4, 28
	APR-27	Launch warning system	Davies. A-4, 28
A-4	APR-23	Use in Oct 65	Price III, 38
F4B	APR-24	Similar to APR-23, Jun 66 to Feb 66, F-14 Four planes configured, unreliable	France, AN/APR-24
	APR-27	Eight planes configured with	France, AN/APR-24
F-4B	APR-25	VF-142 & VF-143 1967 cruise	ATI Phantom Phacts
	APR-27	Missile warning set (similar to APR-26)	ATI Phantom Phacts
	APR-27	Missile warning, audible only no direction	APR-27 Phantom Phacts
	APR-25 vs APR-27	See Sherwood, 107-108	
F-100	Vector	Similar to APR-23, similar to Syst. XII	Price III, 44
	Vector	Dempster wanted to put one in every plane Became APR-25	Price III, 53-54, 66
F-100	IR-133	Panoramic scanning and homing	Price III, 45-46, 48, 66-67

CHAPTER 6
SUPPRESSING SAM SITES

The loss of the Air Force F-4C on July 24, 1965, and the Navy A-4E on the night of August 11–12 to the North Vietnamese SA-2s added new emphasis for the need to locate and destroy the growing number of SAM sites in North Vietnam. In an effort to find and destroy the enemy missile batteries, the carriers of Task Force 77 in the Gulf of Tonkin launched seventy-six low-level missions to search for the sites on Thursday, August 12, and Friday, August 13. None of the aircraft had any specialized equipment or armament to attack the missile sites, and poor visibility forced many of the strike aircraft to operate below 500 feet. No missile sites were located and the Navy lost five planes and two pilots to conventional anti-aircraft fire, which damaged seven other planes. The thirteenth, in the words of Vice Adm. Malcolm W. Cahill, "was truly a black Friday."[1]

The failure of the Iron Hand* missions conducted on the 12th and 13th demonstrated the difficulty in finding the highly mobile SAM sites, which could be packed up and moved by truck in less than three hours. Moves to new launching sites were executed at night. The total time spent in moving from one site to another over a distance of thirty to forty kilometers was ten to fifteen hours. Moves were made during the course of a single night to ensure the viability of the radar coverage and equipment. The North Vietnamese moved a missile battery whenever it was overflown by a U.S. reconnaissance plane, making it extremely difficult to locate SAM sites.[2]

---

* The term "Iron Hand," which was first used as the operational name for the first air strikes launched against the SAM sites, became the generic name for both the missions and aircraft dedicated for this purpose.

The Air Force was already working on the problem of how to locate and obtain more data on the SA-2 missile system. Although the Central Intelligence Agency had accumulated considerable information on the system—including the use of a microwave radio fuze to detonate the warhead after it was armed by the Fan Song radar's guidance link—a considerable amount of detail on the missile's associated electronics was still not fully known. Steps to locate SA-2 sites for follow-on attack by fighter aircraft had begun under Project Left Hook, a combined operation authorized by the JCS involving the Strategic Air Command, the Pacific Air Force Command, and the National Security Agency. The latter was responsible for providing signals intelligence within the Department of Defense. In the summer of 1965, two Ryan Aeronautical Model 147D drones the Air Force had bought for so-called Long Arm electronic-intelligence operations over Cuba in 1962 were pulled out of storage and shipped with three Model 147Es to Bien Hoa Air Base in South Vietnam.[3]

The Long Arm project was originally conceived during the Cuban Missile Crisis as a method of collecting electronic data on the Fan Song radar installations using Ryan Model 147B drones equipped with direction finding radar receivers capable of picking up Fan Song radar transmissions. The expendable drone relayed its ELINT data to specially configured RB-47H Stratojets flying outside the lethal range of the SA-2s. The delivery schedule for the high-altitude Model 147B was several months out, however, so the Air Force also ordered seven Model 147Cs, a production version of the 147A that had an extended wingspan of 15 feet and incorporated a no-contrail system, as an interim solution. Two of these, designated Model 147Ds, were modified with an ELINT system for the top-secret Long Arm missions. Because the drone's fiberglass construction produced an aircraft with a low radar profile, SAC designed a radar repeater using a traveling wave tube to amplify and rebroadcast the Fan Song radar signal when it was irradiated by a Fan Song radar. The Model 147Ds and the two modified RB-47Hs, designated as Long Arm aircraft, were ready for operation in December 1962. By then the crisis had passed and the two Model 147Ds were put in storage at Eglin Air Force Base, Florida.[4]

SAC also acquired three Model 147Es that were similar to the Model 147Ds except that they were based on the Model 147B† airframe, which had a 27-foot wingspan that enabled it to fly as high as 62,000 feet. Both D and E model drones were 23 feet long and were powered by a 1,700-lb. thrust J69-T-29 turbojet engine that provided a top speed of about 600 mph. All the Model 147Es were also placed in storage after the JCS directed SAC to cease work on the Long Arm project on January 8, 1964.[5]

The Long Arm drones pulled out of storage and delivered to Bien Hoa were flown to the target area and launched from underwing pylons attached to modified C-130 cargo planes designated DC-130As assigned to SAC's 4080th Strategic Wing (SW). The DC-130 carried at least one launch control officer, an airborne control officer, and an airborne technician who used a UHF transmitter to control the drones as needed. The drones had autonomous navigation equipment and generally flew a programmed flight for most of their mission, though they were often controlled remotely during the Model 147E's launching phases. The DC-130A was the ideal platform for launching the drone; it was reliable, had great range and endurance, and had the internal power and roomy interior to host the command guidance equipment. Once in the air, the drones were monitored by one of the two specially modified Long Arm RB-47Hs from the 55th Strategic Reconnaissance Wing (SRW). The signals emitted from the Model 147D and E drones were picked up by the EWOs in the RB-47H who recorded and analyzed them. The first Model 147D accompanied by RB-47H (tail number 296) was launched on August 20, but it was knocked down by ground fire before any data could be obtained.[6]

Photo-reconnaissance missions over North Vietnam with the Model 147B photographic-equipped Lightning Bugs continued in parallel with the Long Arm flights. These had begun from Bien Hoa in October 1964. A recovery detachment from the 4080th SW deployed to Da Nang Air Base retrieved the drones that survived their flights over North Vietnam and parachuted to earth.[7]

---

† Dubbed the "Lightning Bug" for security reasons to distinguish it from the Model 147A "Firefly."

In mid-August 1965, Admiral Sharp, in an attempt to get better dissemination of electronic intelligence on the SAM sites, directed the Air Force and Marine Corps to collaborate more closely in employing one or more Marine EA-3B Skywarriors with Air Force tactical reconnaissance and strike missions. Sharp also asked the Air Force to undertake periodic joint strike and SAC reconnaissance drone operations to determine if they would activate Fan Song radars, leading to more precise location of the SAM sites. On August 21 a drone was sent to a target area accompanied by a Marine EA-3B along with Air Force photo and strike aircraft. The Marine aircraft twice succeeded in "exciting" Fan Song radars, but the short duration of the signals ceased before the aircrew could flash their location to the photo and strike aircraft.[8]

The first successful mission took place on August 21, when "two drones," according to Van Staaveren, escorted part of the way by twelve F-105s, were sent on an Iron Hand mission. One of the drones was the second Long Arm Model 147D discussed by Cahill. As the Long Arm drone began its coast-in period, the Long Arm RB-47H began recording the signals from a Fan Song radar. After an eleven-minute break, Fan Song signals were received continuously for one hour and twenty minutes. This data permitted three fixes on SAM site installations within a five-mile circle. An SA-2 destroyed the drone. A flight of F-105s was quickly directed to the suspected location but was unable to locate the site, underscoring the continuing difficulty experienced by Air Force and Navy crews attempting to destroy the SAM sites.[9]

After the second Long Arm Model 147D was expended, ELINT operations using the Model 147E drones were continued under a new project named United Effort. "Six RB-47 sorties named Old Bar," Cahill wrote, "were flown by the end of September, likely training and telemetry check missions to ensure the Model 147Es could pass data to the RB-47Hs." The first operational mission of Project United Effort's Model l47E was launched on October 16, 1965. Unfortunately, the drone's sensors and traveling wave tube failed. "It went on to fly a leisurely loop around North Vietnamese air space," according to author David Axe, and parachuted to a safe recovery.[10]

In August 1965 the Navy, recognizing the difficulty in visually locating SAM sites, began a major effort to utilize the homing capabilities of the APR-23 combined with the nominally accurate ELINT obtainable from a Brigand-configured EC-121 and the ability of a Big Look–configured EC-121 to detect Fan Song radar emissions. Unlike prior Iron Hand operations, the new missions, which began on August 11, relied on relatively small number of aircraft to probe and search the defended area, or were held offshore to act as "pouncers" if a SAM attack developed. The A-4Es flying at varying altitudes from 50 to 5,000 feet used their APR-23s and captive Shrike radar homing missiles (see chapter 8) to try to localize SAM installations. The next day, seven aircraft hunting for SAM sites were lost to conventional anti-aircraft fire in forty-five minutes without locating a single site. "A better solution," wrote Peter Davies, "was found by installing the Sanders AN/ALQ-51 track-breaker equipment in several A-4Es."[11]

The first successful Iron Hand strike conducted by the Navy's carriers took place on October 17, 1966, when four A-4E Skyhawks from the *Independence*'s air group attacked the SAM sited assigned number 32 near Kep airfield, fifty-two miles northeast of Hanoi. The site had been detected two days earlier by a reconnaissance drone. The A-4Es were guided to the site by an A-6A intruder flown by Lt. Cdr. Cecil "Pete" Garbera of Attack Squadron 75 (VA-75). The A-6A was a carrier-based, all-weather attack plane that had its combat debut when VA-75 began operations from the *Lexington* (CV 16) three months earlier on Yankee Station in the Gulf of Tonkin. It carried a crew of two—a pilot and a bomber/navigator (B/N)—arranged in side-by-side seating and was equipped with the unique navigation/bombing system named DIANE, an acronym for Digital Integrated Attack and Navigation Equipment. DIANE was a revolutionary avionics package that enabled the crew to fly missions at low altitudes in rough terrain at night in all weather conditions. The system included ground-mapping radar, track radar, an analog computer, and an inertial navigation system that allowed the crew to attack prescribed locations or targets of opportunity without the crew having to look outside the cockpit. Ordnance could be released manually by the B/N, visually with the aid of radar, or via the fully automated system. In the latter method, the B/N designated the target by radar and the system computed and released the ordnance. These features, coupled with the extra eyes provided in the cockpit via the addition of the B/N, made the A-6A the ideal aircraft to lead and guide the Alpha Strikes of the Iron Hand missions.[12]

As the A-4Es were leaving the area after the successful strike on the missile site, Cdr. Harrison B. Southworth, commanding officer of VA-74, reported seeing three separate fires among the radar vans, ten vehicles burning, and two missiles broken and burning.[13]

The second and third United Effort missions were conducted on October 20 and November 5. Each ended with the drone's destruction, but they were partially successful capturing data on some signals, but not the critical fuzing signals. The second drone was lost without capturing the Fan Song radar frequencies used to track the missile's transponder or the arming signal. The latter was difficult to acquire because of its extremely short duration. Robert Schwanhausser, drone manager for Ryan Aeronautical, did some troubleshooting on the drone's electronics and determined they were overheating. Operations were halted after the second drone was lost and the remaining Model 147E was returned to the United States for maintenance and upgrade.[14]

In between these two attempts to obtain additional data on the SA-2 radar signals, the Navy conducted another hunter-killer experiment in collaboration with the Air Force. On October 30, 1965, Lt. Cdr. Trent R. Powers piloting an A-4E assigned to VA-164, embarked on the USS *Oriskany* (CVA 34), was assigned the task of leading a two-division flight of Air Force F-105s to locate and destroy two SAM sites in an area near Hanoi that was heavily defended by anti-aircraft guns and missiles. To accomplish this task, Powers' A-4E was fitted with one of the APR-23 radar-homing receivers. Powers was instructed to fly his A-4E Skyhawk to the 355th Tactical Fighter Wing at Takhli Royal Thai Air Force Base to make plans for the coordinated attack. The next day Powers led the eight-plane group over more than six hundred miles of unfamiliar, cloud-shrouded, mountain terrain to the target area on the Kep highway bridge thirty-five miles northeast of Hanoi, where they were joined by a Navy strike group from two carriers that were assigned to take out the bridge.[15]

While Navy jets hit the Kep bridge, Powers picked up a Fan Song radar from one SAM site and then observed two SA-2s fired from a second site about two miles away. After directing the F-105s to that site, he headed back to the first site, making a low-level attack on the target with the A-4E's MK-82 Snake Eye bombs. During the attack his airplane was hit by ground fire and went out of control, forcing him to eject from the burning airplane. He parachuted to safety and was captured by the North Vietnamese and died in captivity. Both missile sites were severely damaged. For his actions that day, Powers was posthumously awarded the Navy Cross. According to Alfred Price, this was the first and only mission in which an APR-23–equipped A-4E was used as the lead on an Iron Hand mission.[16]

With the exception of the October 30 mission described above, the numerous search and destroy missions based on the APR-23–equipped A-4Es that began in September were not successful in locating any other occupied SA-2 sites. Problems with the "APR-23 ranged from saturation from too many S-band Fan Song and Fire Can radars in an area to Fan Song operation so brief and intermittent as to preclude localization."[17]

Nevertheless, the Navy and Air Force continued to hunt for and attack SAM sites. Between November 5 and 8, Air Force and Navy aircraft found and attacked four SAM sites. On November 6, for example, two Iron Hand A-4Es bombed a SAM site close to the Me Xa highway bridge near Nam Dinh, destroying two Fan Song radar vans and two SA-2s from the 63rd Missile Regiment, the same unit that had downed the F-4C on July 24.[18]

They were aided by new tactics and the addition of the five Brown Cradle B-66Bs sent to Takhli Royal Thai Air Force Base in October after a Joint Staff review reported serious deficiencies in the ECM and ELINT resources needed to provide timely warning of SAM firings. The B-66B (redesignated EB-66B in January 1966) possessed twenty-three jammers configured to counter all known North Vietnamese air defense radars. The new tactics called for the use of B-66B jammers within fifteen nautical miles of a SAM site using a minimum of two B-66Bs assigned to each ingress route while another was placed in orbit outside the SAM ring and a fourth penetrated the target's air space.[19]

While U.S. forces in Southeast Asia were trying to locate and destroy the growing number of SAM sites in North Vietnam, scientists and engineers in the States worked on modifications to the lone Long Arm Model 147E drone that had been returned to the United States. They worked to improve its capabilities to acquire the elusive data on Fan Song's beacon-tracking system and fuzing signal via the installation of the CIA-developed System XVII. The wide-frequency system covered the C- and S-bands and also had a direction finding capability intended for use in the CIA's U-2 flights to collect data on the Soviet ABM program. Development of the system was authorized in July 1964, and a contract for the production of two prototypes and some ground support equipment was awarded to the HRB Systems Division of Singer Corporation on September 1, 1964.[20]

The idea to use System XVII was suggested by an Air Force lieutenant colonel—whose identity remains secret—serving as the chief the Plans and Field Activities division of the Office of Special Activities (OSA) within the CIA. In a "Memorandum for the Record," dated August 13, 1965, he recommended that the System XVII ELINT package be modified to collect more information on the SA-2 missile's operation. The OSA, established in 1962, was formed for the purpose of conceiving, developing, producing, and operating integrated covert photographic and electronic intelligence collection systems utilizing sophisticated scientific and technological instrumentation. The organization was composed of a small contingent of scientists and engineers working closely with support specialists and industry for the purpose of achieving maximum speed and efficiency.[21]

The modifications requested in the Memorandum for the Record included deletion of the C-B elements plus two telemetry transmitters so that the drone would weigh less and fly at a slightly higher altitude. The modified system provided the data needed to determine

- the frequency of the S-band fuzing signal and when it turned on;
- the frequency of the beacon;
- the time of the L-band arming signal.

The S-band scanning receiver from System XVII was required because of the extremely short time (200 milliseconds) that the arming signal was on before impact with the target.

"If the present Long Arm drone were to be flown," the memo's author stated, "we would probably get only the time of arming." With the proposed modifications they would probably get some X-band, CW, and pulse data as well. "It seems logical," he continued, "that we should seek the good counsel of Intelligence Division and C&FE/OSA along with OEL to arrive at a recommendation for the exact configuration of the Long Arm package and a recommendation for its employment."[22]

No record has been found of the modifications made to the Long Arm drone returned to the United States, but it seems likely that these changes were made to the Model 147E before it was returned to Vietnam for use in the fourth United Effort mission. On February 13, 1966, the Model 147E was guided to a surface-to-air missile site south of Hanoi. As it approached the site, the battery's Fan Song radar locked on the drone, and the missile crew launched an SA-2 missile. What happened next was elegantly described by Carl Schuster in his article "Lightning Bug War over North Vietnam." "In the seconds it took the SA-2 to reach its target, the drone detected, recorded and transmitted all of the missile's vital electronic signals to an aircraft flying over the South China Sea. In an instant, U.S. intelligence had captured the deadly missile's tracking, acquisition and guidance signals, and the sequence in which those signals appeared during an engagement." In the milliseconds before it was blown out of the sky, the drone's sensitive listening gear recovered the frequency and operating characteristics of the warhead's proximity fuze, which was instantly relayed to a Long Arm RB-47H. Eugene Fubini, the assistant secretary of defense for research and engineering, described the February mission as "the most significant contribution to electronic reconnaissance in the last 20 years."[23]

CHAPTER 7

BLACK BOXES AND WILD WEASELS

Because Brig. Gen. Kenneth Dempster's office oversaw the Quick Reaction Capability program, with authority to procure new equipment on an accelerated basis, his office was undoubtedly aware of the QRC-153-2 radar homing equipment that had been installed in the F-100Fs that had participated in the Gold Fire I exercise conducted between the end of October and the beginning of November 1964. The series of maneuvers conducted during this time period were designed to evaluate the Air Force's ability to enhance mobility of the Army. The Air Force was to test and practice concepts and doctrine related to command and control, reconnaissance, close air support, and assault airlift.

The exercise, conducted over more than two million acres of wooded rolling terrain and farmland in the northern Ozark region south of Fort Leonard Wood, pitted the units of Task Force Ozark, the friendly side, against Task Force Sioux, the opposition representing the aggressor. Before the exercise began, several F-100Fs assigned to Task Force Ozark were equipped with QRC-253-2 radar homing receivers. When the war games started, the F-100Fs were ordered to penetrate deep into enemy territory to seek out and destroy the opposition's Hawk air defense missile sites. "One Super Sabre," according to an article that appeared in the March–April 1965 issue of the *Air University Review*, "destroyed six missile launchers and several missiles" (these obviously meant simulated attacks).[1]

Based on the recommendations made by the Air Staff task force charged with solving the SAM problem, the staff in Dempster's office began to formulate a plan to provide tactical aircraft with a means of hunting down SAM sites by homing in on their radar emissions. Although Bendix had proposed installing such a system in F-100 Super Sabres in the spring, "The Air Force," according to Larry Davis, "rejected the device stating that *there is no requirement for such a system.*" The new look into the hunter-killer idea was encouraged by the demonstration given to Dempster's task force by the Applied Technology company's Vector radar receiver.[2]

Applied Technology, Inc. (ATI) was started in April 1959 by William E. Ayer, formerly a senior researcher at Stanford University's Systems Engineering Laboratory. ATI quickly established itself as a producer of specialized jammers and radar-warning receivers. In the early part of 1965, Edward Chapman, a recent hire who had served as an Electronic Warfare Officer in Air Force B-52 Stratofortress bombers and was familiar with the B-52's APS-54 tail warning radar, came up with the idea for a new type of radar-warning receiver that had a CRT display to show the relative bearing of the threat. He intended to sell the device to SAC.[3]

Building on Chapman's idea, one of ATI's engineers redesigned the System XII (developed for the U-2) to cover three radar bands in the 2–12 GHz range and added a 3-inch CRT to show the direction and strength of the intercepted signal as a strobe originating from its center. The type of strobe—solid, dashed, or dotted line—indicated which radar band was being received. An audio output allowed the operator to listen to the radar's scan pattern, which aided in identifying it. The company named the new device the Vector receiver.[4]

The Vector was considerably smaller and more effective than the B-52's APS-54 tail warning radar that Chapman was familiar with. Recognizing that the Air Force would likely buy it once they saw how well it worked, Chapman, using one of his Air Force contacts in the Pentagon assigned to the B-52 Project Office, arranged for a company presentation. By the time he received authorization for the visit on August 5, 1965, Dempster's Air Staff task force, heavily involved in investigating various radar homing and warning systems, were eager to see what ATI had come up with. When Chapman and his presentation team showed up at the B-52 Project Office with suitcases full of electronic equipment, they found a note on the door directing them to a conference room on the third floor, where they found General Dempster and twenty-five officers eagerly awaiting the Vector presentation. Chapman later described what happened next to Alfred Price during an interview conducted years later as part of Price's research for his *History of US Electronic Warfare*: "We set up a signal generator and antenna system on a pedestal in the middle of the conference table. I gave an introduction saying what we were going to do. Then Bob [Johnson, the engineer who designed the Vector] proceeded to tell them how the Vector receiver worked. Then we walked around the room with the prototype, to demonstrate the equipment's direction finding capabilities."[5]

The presentation went well and after a few questions they picked up their equipment and left. Based on the Air Force's interest, ATI's management decided to build five Vector systems using company money. The company, according to John Grisby, then VP of engineering, "figured that if the Task Force didn't come up with anything better than what we had that we might be the likely candidates to supply Radar Warning Systems. If we started early on our own risk, we could be 'Johnny at the rat hole' when a decision was taken by the Air Staff."[6]

In addition to their interest into the Vector IV, Dempster's team had begun to focus on the merits of ATI's IR-133 panoramic radar receiver. On August 27, Maj. Irwin Joel "Pierre" Levy, an electronic warfare expert who worked for Dempster, called Grisby at his office as Grisby was preparing to leave for home. Levy wanted to buy a couple of IR-133 receivers that one of ATI's salesmen had been pushing. The units were priced at $40,000 each, but he had only enough money for one. After consulting with ATI president Bill Ayers, Grisby told Levy they had a deal. ATI would provide two units for $40,000. The IR-133, as they discussed, had the following specifications: it covered the Fan Song's S-band, had a sensitivity of -70 dbm to -80 dbm, handled 0.3-microsecond-wide pulses, and had automatic and manual scanning. It had a 1-inch by 3-inch panoramic CRT display, the main box could be a B-1D case, and the indicator needed to be small, as it had to fit on a fighter's instrument panel. The IR-133 received radar pulses through a set of antennas located symmetrically around the nose of the aircraft. The signals were analyzed for frequency (which told the crew whether the signal was from a SAM, anti-aircraft gun emplacement, or some other type of radar) and the repetition rate, which indicated whether the radar was in a search, tracking, launch, or guidance mode. By comparing the signal strengths on each side of the aircraft's nose, the EWO instructed the pilot to turn right or left to home on the signal. At first it was thought that range could be determined through triangulation as the aircraft turned slightly away from the radar signal, using the CRT display to indicate the offset, but translating the minute changes in the CRT into range proved to be unworkable and the technique was abandoned. Henceforth range calculations would have to be made based on operator judgment, which in reality could only be determined if a Weasel crew actually saw a missile launched from a camouflaged site. The range problem would plague the F-100F Wild Weasels throughout their deployment.[7]

Grisby considered ATI's proposal to be quite reasonable until Levy told him that the first one had to be delivered in thirty days and the second one fifteen days later. "Pierre," said Grisby, "these things don't exist; they are paper-ware, advertising flyers used to try to stir up some interest in the reconnaissance and surveillance community." If Grisby agreed, the Air Force contract would be ready the following Monday. Grisby said okay and, instead of going home, went up to the engineering floor to get the ball rolling on the detailed design work that was be necessary to meet the thirty-day delivery date.

On Monday, August 30, 1965, Grisby was in Washington to sign the IR-133 contract and attend to other business. In the morning Ed Chapman took him to Andrews Air Force Base, where the Air Force Systems Command contract office was. Grisby describes what happened next:

> There we were met by a little old lady who said, "I've never seen anything happen so fast in all of my life." She had typed up the contract and the technical specifications for the IR-133, with multiple carbon copies, on Monday so that they were ready for signing on Tuesday morning. Remember that this is in the days of carbon copies, and Ditto- and Mimeograph-machines, not xerographic copiers. The specifications took a grand total of one and one-third pages and essentially said build us a couple of S-band radios to the numbers Pierre and I had discussed on the telephone the previous

Friday evening. So[,] contract AF18(600)-2879 was issued for two VECTOR-IV and two IR-133 Systems at an "order of magnitude" Firm Fixed Price of $80,000 with delivery dates of 9/28/65 for the first system and 10/13/65 for the second system plus some undefined field/installation support. The in-house project number assigned was 10105. Thusly, we had a "home" for the first two VECTOR systems that we were building for inventory.[8]

A few days later, Chapman was summoned to a meeting in the Pentagon with Major Levy and his boss, Col. William B. "Willie" Williamson. It was then that he first learned of the Air Force's plan to equip two-seat F-100Fs with radar homing equipment to seek out and destroy SAM sites using both the Vector warning receiver and the IR-133 homing receiver. The top-secret project was given the name Wild Weasel. The first portion of the name, "Wild," reflected the nature of the mission, and it also described the personalities and attitudes of the crew members who volunteered to fly such missions. The second portion, "Weasel," was selected because the aircraft "were supposed to 'weasel' their way into enemy territory at low altitude, to sniff out electronically the position of SAM sites, and effectively mark those sites so that accompanying bomb-laden fighter-bombers could visually acquire and destroy them." Most sources claim that the name Project Ferret had been considered at first, but Harold Johnson, who was a crew member of the F-105 Wild Weasel III group sent to Southeast Asia in 1966, claims the program name was changed from "Mongoose" when that was discovered to have been a clandestine World War II project. Both claims were partially true, as Dan Hampton explained in his book *The Hunter Killers*. Air Force officers in the Pentagon, according to Hampton's account, initially named the project "Ferret" after a World War II radar killing program. He goes on to explain that they thought a new, fresh name was needed. "'Mongoose' was chosen after the ferocious little animal that killed deadly snakes. It seemed appropriate, but had also been previously used by the CIA, so . . . it became the 'Wild Weasel' program."[9]

The two-seat F-100F, it turns out, was the ideal candidate for the Wild Weasel role. In addition to the second seat, which was needed for the Engineering Duty Officer assigned to operate the radar warning and homing gear, the F-100F had similar flying characteristics to the F-100D that was optimized for ground attack. It was also fast and had space to accommodate the system's electronics. The F-100F was a relatively inexpensive aircraft too, and was readily available. To meet Dempster's requirement, four low-flying-time F-100Fs were selected from the 27th Tactical Fighter Wing (TFW) at the Cannon Air Force Base in New Mexico and flown to the North American Aviation facility in Long Beach, California, where they were to be modified under an Air Force contract issued to the company's Space and Information Division. Upon arrival, the F-100F bearing the tail number 58-1231 was taken into a closed hangar, gutted of its wiring, and had all of the nonessential instruments in the backseat panel removed to fit the new equipment.[10]

The R&D model of the Vector prototype was delivered to Long Beach, where North American's engineers and technicians, aided by two Air Force EWOs assigned to the project, assessed the modifications needed to install the new system. These included the installation of new wiring and cables, the "black boxes" housing the system's electronics, and the specialized antennas needed to capture the radar signals. The Vector IV,* as it was dubbed, had a 3-inch CRT display that presented a strobe indicating the direction of the threat and a panel of small lights, called the Threat Panel, that indicated the different types of radar threats present. Four antennas (two mounted under the nose intake facing forward on each side of the aircraft centerline and two facing rearwards on the trailing edge of the vertical fin) were connected to an electronic receiver mounted in the nose. Although the IR-133 was still in development, ATI knew the size of the receiver and the panoramic display. This information, along with the types of cables used and the location of antennas, was provided to North American personnel so they could prepare the F-100F for installation of the IR-133 receiver.[11]

The first Vector IV was delivered on September 12, 1965, and installed in the first F-100F selected for modification. This aircraft, with a North

---

* Named for the four antennas that provided a vector to the radar threat.

American Aviation crew, was flown for the first time on September 16. Ed Chapman later disclosed that the Vector warning system did not work at all during the initial test. "It wasn't picking up signals and it wasn't giving the proper indications." The problem, which was diagnosed by Bob Robinson, was the type of cable North American had used for the installation, which was not the kind specified by ATI. The Vector system depended on balancing the output from the antennas and cables on the front and back of the aircraft. To verify that the cable installed by North American was the problem, ATI hooked some of their own cables to the box in the cockpit and ran them out the open canopy. "When we switched [it] on," recalls Chapman, "we showed the guys there wasn't any problem with our system. 'It's your cables. . . . Rip it apart, put the right cables in and everything will be fine.'"[12]

Back at ATI, development of the IR-133 proceeded rapidly and the company was able to deliver the first unit on September 28, twenty-nine days after starting the design. The third Vector IV and the second IR-133 were delivered to North American Aviation on October 12. In the meantime, William C. Doyle, ATI's director of systems engineering, had come up with the design a for new warning receiver "which, if it worked the way he hoped, would give the air crews a few seconds warning prior to a SA-2 'Guideline' missile coming off the rails of the launcher-transporter vehicle." Doyle got the idea when he came across a piece of data on the SA-2's L-band guidance signal while visiting a potential military client. On the flight home he sketched out the block diagram to show how it would work. Doyle discussed the idea, which later became the WR-300, the next day. It looked promising enough for them to give General Dempster a briefing over the telephone as to what they had in mind and what it might do for aircrew survivability. Dempster said to "get started on your own nickel and if it's worth really pursuing we will cover you." And that is what ATI did. They began design work on September 23 with the expectation of receiving two detection warning systems in two weeks at an anticipated value of $18,000 (equivalent to more than $180,000 today). The WR-300 detected the launch of an SA-2 by monitoring the SAM's guidance and control frequency. The launch of a missile was indicated by a characteristic shift in the power of the guidance signal when the missile was fired that caused the

WR-300 to activate a red warning light on the aircraft's instrument panel. "This light," stated Grisby, "got the infamous name of the 'Oh, Shit' light, meaning that a Guideline Missile was on the way." ATI delivered the first unit on November 8, 1965.[13]

After the modifications to the four F-100Fs borrowed from the 27 TFW were finished—a process that took ten days—they were flown to the Tactical Warfare Center at Eglin, where they were united with the flight crews that had been "volunteered" to fly them in combat. The pilots were drawn from other F-100 units; the EWOs were from SAC B-52 or EB-66 squadrons. "The crews were to be mated and trained together. But a couple of small problems arose," as aviation author Larry Davis wrote in his often-quoted work *Wild Weasel*. "First, some of the pilots had almost never heard of an EWO . . . and second. [sic] All of the EWOs came from SAC bombers—big, multi-jet beasts that rode the skies like airliners. None of the EWOs had much, if any, single engine fighter time. And most 'were reluctant to bounce all over North Vietnam in a single engine fighter with a wild-eyed, hot dog pilot at the controls.'" Capt. John E. "Jack" Donovan's reaction to the job was typical of the EWOs once they found out what they had volunteered for. "You want me to ride in the back seat of a two-seat fighter with a teenage killer in the front seat? You gotta be shittin me," which when converted to "YGBSM" became the traditional motto of the Wild Weasels that often appeared on their insignia.[14]

When training began in October, there was no prepared program of instruction, no classrooms, and no intelligence briefs that were normally used in an Air Force program. As Dan Hampton put it, "The pilots and the EWOs were the experts; they were supposed to sort it out somehow." Maj. Gen. Benjamin B. Putnam, commander of the Tactical Warfare Center, told them to just make it work. The training flights, which were conducted as part of the Wild Weasel 1 Operational Test and Evaluation program, began on October 11 and continued through November 18. After a few familiarization flights, the four Wild Weasel prototypes began flying test missions against the Soviet Air Defense Simulator #1 (SADS-1) that had been fabricated and delivered to Eglin by the Army's Harry Diamond Laboratory. It was a working surrogate of the Fan Song Model B radar, built to conform to the best available intelligence on the system. Unlike the mobile, van-mounted Soviet model, the surrogate's electronics were housed in a building on Okaloosa Island. To protect the antenna from the salt air and the weather, and to hide it from the Soviets, it was enclosed in a large, white rubber dome. The system was continually upgraded as new information on Soviet radars was obtained from the CIA's program of precision measurement carried out by Air Force aircraft and crews that had begun collecting highly specialized radar data during the summer of 1963. When training operations began for the F-100 Wild Weasel crews at Eglin, the CIA had conducted numerous ELINT missions and acquired signal intelligence on Fan Song, Spoon Rest, Knife Rest, Flat Face, Back Net, and Bar Lock radars.[15]

During these training exercises, which took place over Eglin's test ranges, the EWOs detected the SADS-1 radar using the IR-133 panoramic scanning receiver, in the manual tuning mode to identify and analyze the signal. The IR-133 provided an initial azimuth that the aircraft followed until the signal was strong enough for the shorter-range Vector IV set. The F-100F crews found that the IR-133 worked best when flying at a medium altitude, following the beam directly toward the transmitter. On one such occasion the panoramic scanning receiver picked up the tracking signal 107 nautical miles away. As they closed with the transmitter, the EWO in the back seat had to rely on the 3-inch Vector CRT display mounted in the center of the rear cockpit to locate the SAM site. A strobe in the CRT indicated the direction of the signal and was divided into three concentric rings that could be used to approximate the distance to the radar based on the length of the strobe. Once they came within SAM range, the crew had to search visually for the site before it could be attacked.[16]

Although the practice sorties allowed the Wild Weasel crews to figure out how to use and gain experience with their electronic equipment, they were all conducted against the SADS simulator, which looked nothing like the real Fan Song. The simulator's white building, topped with its distinctive radome, could be easily seen from miles away, and it never moved. This provided a false sense of accomplishment for the Wild Weasel crews that would be quickly shattered when they tried to locate the well-camouflaged, heavily defended SAM sites in Vietnam.[17]

After completing their specialized training, the four F-100Fs under the command of Maj. Gary A. Willard Jr., took off from Eglin's runway at 1000 hours on November 21, 1965, headed for Korat Royal Air Force Base, Thailand. Delayed by bad weather and a layover in Hawaii, the flight did not arrive in Korat until late in the day on November 25. Upon arrival, Willard's detachment was assigned to the 2nd Air Division's 6234th Tactical Fighter Squadron and began a sixty-day operational trial that ran from November 28, 1965, to January 1966. The trial period to evaluate equipment and tactics used by the Wild Weasels was based on the following objectives:

1. To determine the warning capability of the radar homing and warning equipment installed in the Wild Weasels.

2. To investigate the effect of jamming by friendly aircraft on the Vector IV and the IR-133 equipment.

3. To determine the homing accuracy of the radar homing and warning equipment and the capability of the crew to place the aircraft within visual range of the target.

4. To determine tactics for employing the Wild Weasel aircraft against SAM defense systems.

5. To determine maintenance requirements and reliability of the radar homing and warning equipment.

6. To determine the organization and manning requirements for Wild Weasel operations.

7. To determine the training requirements for flight crews and maintenance personnel.

8. To test any additional equipment which became available.[18]

During the first three days of the operation, from November 28 to November 30, the F-100F Wild Weasels flew orientation missions with the F-105s of the 388th Tactical Fighter Wing. The purpose of these missions with EB-66s near the North Vietnam border was to provide the crews with area orientation in a high-threat environment, verify the capabilities of their radar homing and warning gear (RHAW), and to observe the effects of jamming on the Vector and IR-133 equipment. The F-100Fs "would troll along the border, monitoring their scopes and listening to the various radar tones emanating from North Vietnam."[19]

Bad weather delayed the first Wild Weasel mission until December 19, when a flight of two F-100Fs led by Maj. Gary A. Willard Jr. took off in an attempt to locate SAM sites in North Vietnam but were unable to pick up any Fan Song signals. The F-100F Wild Weasels in Willard's flight were armed with two LAU-3 canisters of twenty-four 2.75-inch rockets and two hundred rounds of ammunition for each of its two M-39 20-mm cannons. A pair of 355-gallon drop tanks provided enough fuel to complete their mission, which was to locate the Fan Song radar and mark its location with the white phosphorus (Willy Pete) rockets, or with a combination of high explosive and phosphorus rounds, to identify the site for the accompanying F-105s. This, as Peter Davies and David Menard observed, meant that the SAM-hunting F-100Fs "were going to have to fly ahead of the strike formations over some of the most heavily defended targets in military history." Once they marked the target, the F-105Ds that accompanied Willard's flight, which were armed with four LAU-3 canisters loaded with six to eight CBU-24 cluster bombs and an internal M-61 20-mm cannon that spit out six thousand rounds per minute, would attempt to knock out the site. Neither of Willard's Wild Weasels, however, were able to detect any Fan Song signals, and the flight returned to Korat without engaging the enemy. On the following day, the Wild Weasels suffered their first combat loss when an F-100F piloted by Capt. John J. Pitchford was downed by anti-aircraft fire while leading an unsuccessful attack on a SAM battery near Kep airfield. Both Pitchford and his EWO, Capt. Robert D. Trier, bailed out of the stricken aircraft. Pitchford was captured and remained a prisoner of the North Vietnamese, and Trier was shot and killed while trying to resist capture.[20]

The Wild Weasels were hampered by an excessive list of restrictions contained in the rules of engagement established by the Department of Defense in Washington, DC, in accordance with President Johnson's wishes that controlled all aspect of the air campaign. In addition to avoiding attacks on SAM sites within a thirty-nautical-mile circle from the center of Hanoi and a ten-nautical-mile circle around Haiphong, attacks within thirty nautical miles from the Chinese border were prohibited. The most restrictive rule confronting those trying to hit the SAM sites, however, was the one prohibiting the suppression of SAMs and gun-laying radar systems in populated areas and attacks on North Vietnamese air bases from which attacking aircraft might be operating.[21]

Despite these restrictions, on December 22 a Wild Weasel piloted by Capt. Allen T. Lamb, with Capt. Jack Donovan as his EWO, conducted the first successful SAM strike led by a Wild Weasel after they detected a Fan Song radar 100 miles from the target. Captain Lamb described how he closed the target and marked it for the accompanying F-105Ds:

> I kept the SAM site at around "ten o'clock" so he wouldn't get the idea I was going after him. When I could, I dropped into shallow valleys to mask our approach. Now and again, I'd pop up for Jack to get a "cut." After breaking out of the Red River Valley I followed the strobe on the Vector IV and turned, keeping the river alongside us. At this point the IR-133 strobes started "curling off" at "12 o'clock," both to the right and left of the CRT, and I knew we were right on top of the site. I started climbing for altitude and Jack kept calling out our SAM positions literally left and right. My rockets hit short, but as I pulled off the target there was a bright flash.[22]

The F-100F's rockets had hit one of the SA-2's fuel tankers, which clearly marked the target. Having successfully identified the SAM site, the two airmen watched as the F-1005D Thunderchiefs engulfed the target area with seventy-six 2.75-inch rockets, sending smoke and dust three to four hundred feet in the air. The flight expended its remaining ordnance on the AAA defenses guarding the site.[23]

From December, when the first Wild Weasel/Iron Hand mission was flown, through February 5, 1966, six SA-2 installations were overflown by the hunter-killer teams of Wild Weasels and F-105Ds. The aircrews learned that in heavily defended sectors of North Vietnam the best tactic to use during SAM search and destroy missions was for one F-100F to lead three F-105Ds into a suspected SAM target area at 8,000 feet with five miles of visibility. This was above small caliber anti-aircraft and small arms fire, but left enough maneuvering to dive should an SA-2 launch appear imminent. When a Fan Song signal was picked up, the Wild Weasel would either home in directly, at altitudes between 4,500 and 8,000 feet, or drop down for terrain masking. Instead of flying directly over the emitting radar, the Wild Weasels tried for an offset of 100 to 500 feet so that it would be easier to acquire the target visually. Once the installation was found, the F-100F would try to mark it with rockets for the F-105D strike aircraft. When the terrain contained numerous ridges and valleys and interfered with radar reception, the aircrews developed a low-altitude tactic using the terrain to shield them from SAMs or anti-aircraft fire until they were directly over the site. When using this tactic, the flight flew at normal search altitudes until a signal was located and a bearing determined. Then the flight descended below the line-of-sight altitude, flying up valleys and over ridges, popping up to obtain another bearing, and then descending again. The greatest advantage of this tactic was the element of surprise.[24]

A number of shortcomings in the Wild Weasel's equipment were revealed during these initial operations. While the Vector IV was capable of detecting the emissions of the Fan Song radars, it was unable to determine whether or not a particular radar was tracking the Wild Weasel. It also had a high false alarm rate. The crews also found that the IR-133 could not be used for homing during the frequent maneuvers to avoid anti-aircraft fire or when the receiver was saturated with numerous signals. Although the SA-2 presented the most apparent threat to the Wild Weasel flights, the difference in the speed of the F-105 and the F-100F was a key detriment that affected some missions.[25]

Because the F-100 was originally designed as a supersonic, high-altitude fighter-interceptor, it suffered from both poor maneuverability and poor performance at low altitudes, making it vulnerable to anti-aircraft fire. Another factor was its radar cross-section, which, according to one former F-100F pilot, was larger than the accompanying F-105s. The Iron Hand five-ship formation was the only one in the sky, and the F-100F stood out distinctively from the others, alerting the North Vietnam defenders of the Wild Weasel's presence and making it easier for the crews of the radar-controlled anti-aircraft guns defending the SAM sites to focus on the F-100Fs. Although this fact has never been revealed before, it only added to the hazards faced by the F-100F pilots and may have contributed to their high loss rate. In any case, the loss of a second F-100F Wild Weasel to anti-aircraft fire convinced the 2nd Air Division that the F-100F Super Sabres were too slow when fitted to carry external stores to survive in the most heavily defended areas in the North and they ceased flying above the Red River delta at the end of March. By then, three of the first nine F-100Fs had been lost to ground fire and one overstressed its airframe beyond repair. The remainder were so damaged that to ensure that enough Wild Weasels were available to support operations in Route Packages 5 and 6, the next batch of Wild Weasels based on the F-105F airframe were rushed to the theater. Once these arrived, many of the initial Wild Weasel pilots were sent back to the training program established at Nellis Air Force Base as instructors. The arrival of the F-105Fs, which had the same performance as the D models, marked the end of those Iron Hand formations designed to compensate for the difference in the speed of the F-105 and the F-100F.[26]

CHAPTER 8
AGM-45 SHRIKE

POTENTIAL GAME CHANGER

As SAM sites proliferated in the fall of 1965, both the Air Force and the Navy realized that not only was it difficult to locate them, but better weapons were needed to attack and destroy the Fan Song radar sites while keeping the Wild Weasels away from the heavy concentration of anti-aircraft weapons guarding the SA-2 missile sites. Fortunately, the Navy had already developed such a weapon: the AGM-45 Shrike, an air-to-surface, stand-off missile designed to destroy or suppress Fan Song radars by homing on their electronic emissions. Operational evaluation of the new missile began in August 1965.[1]

In 1956, Lt. Cdr. William J. Moran was stationed at the Naval Ordnance Test Station (NOTS) in China Lake, California. He was on assignment as senior officer for air-to-air weapons when NOTS began working on Corvus, a large, air-to-surface, nuclear-armed, antiradiation missile. In addition to being a passive weapon, guided on the emissions from a threat radar, it also could be used as an antiship missile guided in a semiactive mode by an illuminator on the launch aircraft. A data link embedded in the missile's electronics enabled it to receive midcourse guidance commands during the semiactive mode until it was close enough to detect the target's radar and switch to passive mode. It was complicated, expensive, and powered by a liquid-fueled engine that was not favored for a weapon carried on board an aircraft carrier.[2]

Lieutenant Commander Moran did not think Corvus would fix the tactical problem of how to destroy the enemy radars he had encountered in Korea. He wanted something that would allow a pilot to hit his target without himself becoming a target "painted" by enemy radar. As NOTS historian Elizabeth Babcock wrote, "Bill repeatedly expressed his strong opinion that Corvus would not solve the tactical problem because of its complexity and cost." Moran discussed his concept for a tactical antiradiation missile with Barnard "Barney" Smith, the newly appointed head of the Weapons Development Department (Code 40). Smith felt the concept warranted looking into and put together a team, led by the head of Code 40's Aeromechanics Division, Leonard T. "Lee" Jagiello, to work on the idea.

Jagiello's group put together a plan to produce a proof-of-concept ARM (antiradar missile) that was as simple as possible using beefed-up Sidewinder* sensors to control four moveable cruciform wings. Simulations conducted on the station's electronic analog computer showed that a simple "bang-bang" (on-off) control system would work. All it had to do, said Jagiello, was to "lock on their gun or missile-directing radars and get close enough to damage the antenna. . . . We conceived the idea of the pilot pulling up to a certain angle, predetermined by where the range of the target radar is, and launching."[3]

Two proof-of-concept missiles were assembled using a Sparrow airframe, Corvus seeker, Sidewinder servos, and four Mighty Mouse 2.75-inch rocket motors for propulsion. These were fired from an F3D Skynight against an SCR-584 S-band radar. The first round malfunctioned, but the second round struck close enough to the target to confirm the feasibility of the proposed new weapon.[4]

Having proved that the concept would work, and with a feasibility study for the proposed ARM in hand, Moran and Jagiello traveled to Washington, DC, to present their idea to the chief of the R&D Division within the Bureau of Ordnance, Capt. Edward A. "Count" Ruckner. With Ruckner's support, Congress authorized $250,000 "to provide in a relatively short time a cheap and simple missile with limited antiradiation capability" while canceling the Corvus missile program "in an effort to maintain balance within the economically feasible overall Navy development program." The name Cobra, which the historian Cliff Lawson described as a "snake on a

---

* AIM-9 heat-seeking air-to-air missile developed by NOTS.

par with the Sidewinder in terms of lethality," was attached to the weapon. Designing the missile proved to be a greater challenge than anticipated, as Duane J. "Jack" Russell, who was in charge of the guidance-section design, remembered: "Using that Corvus guidance system was a monstrosity, a lot of problems in the servo, we had to get going on the design of the warhead and fuzing system, we had to come up with a better rocket motor."[5]

By the fall of 1961, the redesigned missile was now named "Shrike," "after the bird, which according to folklore would peck out its enemy eyes," according to NOTS historian Elizabeth Babcock. The overall weapon system, including support equipment, was now known as Weapon System W-115 with the missile designated XASM-N-10. The missile's designation was changed to AGM-45 on December 11, 1962, when the Department of Defense issued Directive 4000.20 "Designating, Redesignating, and Naming Military Rockets and Guided Missiles." In 1963 the designers produced a C-band seeker that could be used in place of the Corvus S-band seeker. The missile used an 8-inch Sparrow airframe powered by a Rocketdyne MK 39 solid-fuel rocket motor. Aerodynamic control was provided by four cruciform wings. The missile was 10 feet long, weighed 390 pounds, and had a nominal range of at least 10 miles. The 145-lb. warhead was detonated by a dual-mode (proximity and impact) fuze and had a kill probability of 0.95 at a miss distance of forty feet. The warhead, as described by one of the project's pilots, "was composed of a high explosive charge around which were stacked thousands of three-sixteenth-inch steel cubes in such a way that when detonated by the target detecting device, there would be a distribution of steel cubes such that at least one cube would penetrate every square foot within the target effective range." It projected a 70-degree cone of projectiles in front of the target. "Extensive analysis of the potential targets as well as the various endgame-encounter geometries resulted in the incorporation of five different charge-to-mass ratios into a single warhead. This design yielded five distinct fragment velocities that would work most effectively for the expected range of warhead-target encounters." The fragment size was optimized for doing maximum damage to the radar antenna.[6]

When the idea for the Shrike was first conceived, it was believed that the pilot would be briefed before the flight on the location of a SAM site previously located by aerial reconnaissance. The attack pilot approached on a straight line to the target (presumably pinpointed by the APR-23), pitched up to about 30 degrees, and launched the missile before the Shrike had acquired the Fan Song's radar signal. The Shrike then followed a long loft pattern and descended into the "basket" where it picked up the threat radar's signal and began homing on it. The problem with this tactic, as they discovered during test runs off the Cuban coast during the Cuban Missile Crisis, was that the radar environment was so crowded that the pilots could not single out any particular target. To allow the pilot to locate the target, NOTS came up with a fire-control system to assist the pilot in determining the proper launch point based on a CRT radar scope that indicated to the pilot—based on the frequency of the radar detected—what kind of a target it was and where it was. Unfortunately, the NOTS history that described this tactic did not provide either the AN number or any other information that could be used to identify this equipment, and I have been unable to find any other reference to the equipment or how it functioned.[7]

Sources differ as to when the Navy began using the AGM-45 Shrike in Vietnam. The *Dictionary of American Naval Aviation Squadrons* indicates that the AGM-45 made its combat debut during a mission conducted by VA-23's A-4E Skyhawks on April 25, 1965. Van Staaveren, in his history of the Vietnam War from 1965 to 1966, states that the Navy began testing the AGM-45 in joint strikes against suspected SAM sites in August. Both sources may be correct, although I have been unable to locate any record or description of the April 25 mission.[8]

There was more to the Shrike weapon system than the missile itself. As with any new air-launched missile, specialized equipment was needed to integrate the weapon and its sensors so that the pilot could locate the target and fire the missile within acceptable launch parameters. On the A-4E this included the APR-23 radar homing and warning system, the AJB-3 low-altitude bombing system, and the on-board electronics package that connected the missile to the aircraft.[9]

The pilot used APR-23 to tell him, based on the frequency of the target, what kind of a target it was and where it was, using the strobe length displaying signal intensity to determine the approximate range to the SAM radar. The A-4 pilot picked up the active emitter signal while in level flight and aligned the axis of the aircraft toward the emitter. He then lowered the nose of the aircraft to point directly at the radar (presumably using the A-4E's APG-53 to determine the range to the target). He then calculated the so-called "dip" angle from range tables provided by NOTS. This was the launch angle needed to provide the missile with enough energy to reach the target. To launch the missile the pilot pulled up to the launching angle and hit the pickle switch on the control column, releasing the weapon. As soon as the missile had passed its peak altitude and started to come down, its seeker detected the site's radar emission and homed on it.[10]

The "dip loft" maneuver approach, according to Cdr. George M. "Bud" Biery II, was overly complex and disliked by the many of the pilots, who preferred to use the "down the throat" approach. To launch the missile in the correct attitude he relied on the A-4E's AJB-3 low-altitude bombing computer, which used data on the Shrike's aerodynamic characteristics to determine the optimum launch attitude. The pilot centered the needles on the AJB-3 display and just fired. Exactly how this worked is not clear. On a "dumb" bombing run the AJB-3 relied upon the ground attack mode of the APG-53A to provide the correct launching range. The AJB-3 marked a center point on the cockpit screen that showed the ideal flight path. A second blip showed the actual flight path. The pilot's task was to fly the A-4E so that the ideal blips never diverged and then pull the trigger when the computer generated a beep in his helmet. Ernest Mares, in his recollections of firing Shrikes at China Lake, mentions using the crossed yellow needles that appeared in the AJB-3 vertical gyro without going into further details. The scarcity of technical documents on the subject—most of which were likely classified and now destroyed—and the lack of oral descriptions or the availability of living pilots who used the Shrike leaves the question unanswered of exactly how and when the pilot knew when to launch the AGM-45. It may have been, as one veteran pilot told me, that the Shrike was launched as soon as possible in order to avoid being hit by an SA-2. The SA-2 had an 18-mile range and could be launched at the attacking aircraft well before the Wild Weasel could launch the Shrike, which had

had less than half the range (about seven† miles) and was slower than the SA-2 (Mach 2 compared to Mach 3.5). In his words, it was trying to use a pistol against an adversary armed with a rifle.[11]

During the AGM-45's initial test period, which ended on September 25, the Shrike failed to live up to expectations. Pilots "found it difficult to judge the Shrike's performance because of the long release distance, the need for aircraft to take quick evasive action after firing it, and the possibility that the missile would err and follow a wrong signal." The small Shrike display was also a problem because of the difficulty in distinguishing individual targets in a high-threat environment. The Shrike's radar seeker was not gimbled and its limited three-degree view required a precise attack vector, which was difficult to achieve given the small display. Of the twenty-five AGM-45s fired by the Navy's pilots during the test phase, there were seven probable and two possible hits, two probable misses, and fourteen unknowns. Testing ceased because of the limited number of missiles available. What few missiles remained were set aside as a reserve.[12]

To improve the AGM-45's performance, NOTS developed a Shrike Improved Display System (SIDS) using the nine-inch APG-53 radarscope on the A-4E to display light spots representing the emitting radar. This was achieved by using an interface box that allowed the APF-53 to display on a time-sharing basis. NOTS sent 288 units to the fleet in 1967. An even better system designated as the APS-117 was developed between and 1965 and 1967. Known as the Target Identification and Acquisition System (TIAS), it employed a dedicated aircraft-mounted receiver and signal processor. TIAS provided rear-lobe detections of threat radars at ranges of twenty to thirty nautical miles and handled multiple threats in high-density radar environments without ambiguities or false alarms. Installation in A-4F Skyhawks began in 1969. The nose radar was replaced by a Shrike seeker head coupled to radar screen in the cockpit. Capt. Denny Sapp, veteran A-4F pilot, explained that he used it to "bore sight the radar van by looking at our cockpit radar screen."[13]

---

† The range using the high-altitude lofted trajectory was ten miles, according to the data listed in Andreas Parsch's AGM-45 data listed in his online "Directory of U.S. Military Rockets and Missiles."

FIG. 8.1

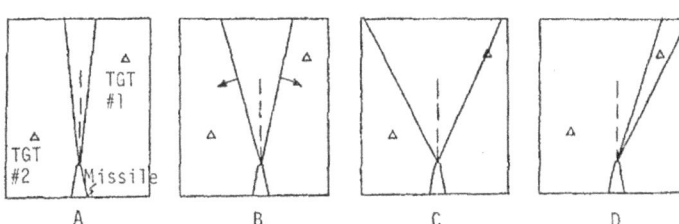

Angle gating operation. At control activation, the angle gates are centered on missile boresight as in figure 8.1A. If no target is detected, the gates begin to open as in figure 8.1B until a target is detected as in figure 8.1C. The angle gates then close around the target as in figure 8.1D and the missile alters course to place the target on boresight. Throughout the maneuver the angle gates remain "closed" around the target, thus preventing spurious targets from being detected. *McMaster, "AGM-45-7A Shrike: Final Test Report"*

An improved Shrike, the AGM-45A-3, became combat ready early in 1966. To increase the radar seeker's target identification, NOTS added angle gating to the AGM-45A's guidance package. "Angle Gating," as Michael McMaster explained, "allows the SHRIKE missile to lock-on the target closest to boresight and prevent extraneous targets from being acquired" (see figure 8.1).

The Navy began testing the AGM-45A-3 at the beginning in February 1966 and continued to use it against SA-2 SAM and other anti-aircraft radar sites into March, but damage assessment was difficult due to poor weather conditions and the mobility of the SAM sites.[14]

The Air Force was eager to obtain Shrikes for the Wild Weasel force, but AGM-45s were in short supply and the Navy was reluctant to part with any of the new missiles. Major Levy, General Dempster's electronic warfare expert, claims that he persuaded Julian Lake to part with some Shrikes in exchange for some Air Force APR-25s (the production version of the Vector IV). According to Levy, the Navy was having difficulties with its new homing and warning receiver. This would have been Melpar's APR-24. The APR-24 was troublesome (see chapter 5), but it did not enter service until June 1966, and the APR-25 did not see Navy service until it was installed on the F-4Bs assigned to VF-142 and VF-143 for their cruise on the *Constellation* that began in April 1967. Whatever the reason, the Navy did agree to provide a number of Shrikes to the Air Force.[15]

Shrike was first used by the U.S. Air Force on April 18, 1966, when one of the missiles was launched from an F-100F, one of the three that arrived in February to replace those lost in the first group sent to Southeast Asia. The F-100F was on an SA-2 search-and-destroy mission just North of the DMZ when it detected the emitter from a Fire Can radar six miles northwest of Dong Hoi. The pilot launched the missile hoping to track it to the radar in order to deliver additional ordnance in the area identified when the missile exploded. The pilot lost it in the haze and was unable to reacquire it visually. The Fire Can went off the air, however, and did not come back up for the remainder of the mission. The mission, therefore, and the performance of the missile was considered a Wild Weasel success.[16]

NOTS continued to modify and improve the AGM-45 throughout the Vietnam War. The first of these was the development of a G-band seeker and a white phosphorus warhead. The latter could be substituted for the standard blast fragmentation warhead to mark the SAM site for follow-on

FIG. 8.2

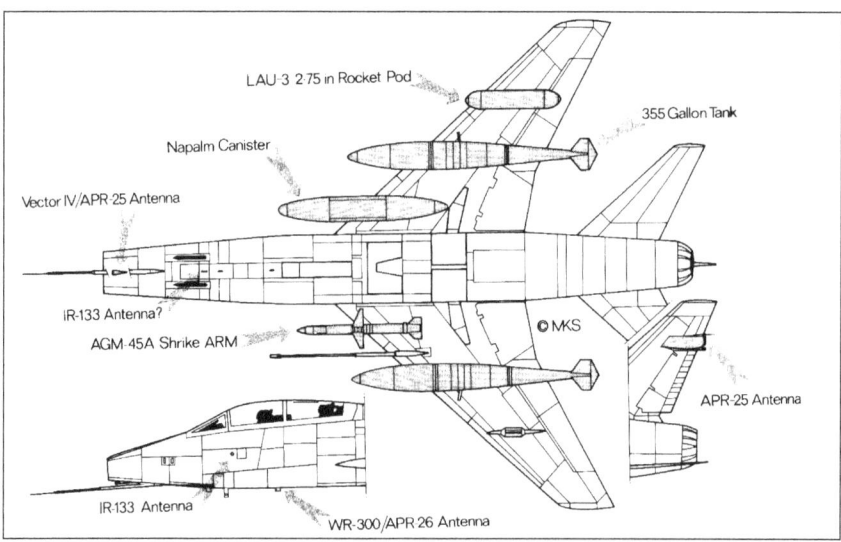

F-100F Super Sabre Wild Weasel I © *Martin Streetly, used with permission*

Iron Hand aircraft. The next major development, introduced in 1966, was the addition of angle gating, used on all subsequent variants of the AGM-45. An I-band receiver was introduced in 1967. Extended-range versions, designated by the suffix "B," entered service in 1969. These missiles were fitted with an Aerojet Mk 52 rocket in place of the Rocketdyne Mk 39, extending the missile's range from 10 miles to 28.8 miles.[17]

TABLE 8.1 AGM-45 VARIANTS AND PRODUCTION DATES

Designation	Guidance Section	Seeker Type	Produced Between
AGM-45A-1	MK 23 MOD 0	E/F-Band	1963 and 1966
AGM-45A-2 AGM-45B-2	MK 22 MOD 0, 1, or 2	G-Band	1963 and 1966
AGM-45A-3 AGM-45B-3	MK 24 MOD 0, 1, or 4	Broad E/F-Band with angle gating	1963 and 1969
AGM-45A-3A AGM-45B-3A	MK 24 MOD 2 or 5	Narrow E/F-Band with angle gating	1963 and 1969
AGM-45A-3B AGM-45B-3B	MK 24 MOD 3	E/F-Band with angle gating	1963 and 1969
AGM-45A-4 AGM-45B-4	MK 25 MOD 0 or 1	G-Band with angle gating	1964 and 1968
AGM-45-5	Development discontinues		
AGM-45A-6 AGM-45B-6	MK 36 MOD 1	I-Band with angle gating	1965 and 1970
AGM-45A-6 AGM-45B-6	MK 37 MOD 0	E/F-Band with angle gating	Cancelled May 1967
AGM-45-8	Discontinued in favor of Harm		
AGM-45A-9 AGM-45B-9	MK 49 MOD 0	I-Band with angle gating	Began November 1973
AGM-45A-9A AGM-45B-9A	MK 49 MOD 1	I-Band with angle gating, G-bias	?
AGM-45A-10 AGM-45B-10	MK 50 MOD 0	Broad E- to I-Band with angle gating	Initial procurement 1976*

*Cost per missile in 1976 was $34,835.

CHAPTER 9
WILD WEASELS II AND III

On October 14, 1965, while the F-100F Wild Weasels were undergoing evaluation at the Air Proving Ground and Tactical Warfare Center at Eglin Air Force Base, Col. James T. Johnson in the Service Engineering Division of the Aerospace Command received a call from Col. William B. Williamson in the Pentagon initiating Project Wild Weasel A1. Williamson told Johnson to install one of the Vector IVs on an F-105D and to conduct a flight test to see if the F-105D equipped with the radar homing and warning gear could also be used in the Wild Weasel role. If this could be done, all members of the Iron Hand team would be flying the same aircraft and the F-105D hunters would no long have to weave back and forth to avoid overriding the slower F-100Fs. ATI delivered one of the first five Vector IVs built to Sacramento Air Material Area (SMAMA) at McClellan Air Force Base, Sacramento, California, and helped install it on an F-105D bearing the tail number 62-4291. It was ready to fly by October 2. While this aircraft was being modified, another F-100D next to it was being modified to take the so-called "MAX-DIX" DPN61 radar warning system. This was a joint venture of Maxson and Bendix that Bendix later developed into the AN/APS-107. According to one account, the MAX-DIX installation included "several black boxes that were installed on top of the cockpit instrument shield and positioned from one side of the shield to the other, that required observation and probably switch action too."[1]

Capt. (later Lt. Col.) Joe W. Telford, one of the project test officers assigned to the initial Wild Weasel 1 training and Operational Test and Evaluation program at Eglin, remembers being told that General Dempster "had requested a very, very quick test to be conducted on ATI, Bendix-Maxson, Melpar on Loral radar warning equipment. The test results would be compared and large quantities of equipment would be quickly bought from the preferred company, above and beyond the already procured Wild Weasel I systems."[2]

The two F-105Ds modified by the SMAMA were sent to Eglin for flight testing and evaluation. Toward the end of the testing, Captain Telford ran into Maj. D. E. "Dough" Whatley and Maj. H. P. "Hap" Maree, the two pilots assigned to test the airplanes after they had each flown a mission in the F-105D. Whatley snapped at Telford, who had red hair. "Red," he said, "this was the last mission with the APS-107 [the MAX-DIX installation] where I am going to give you test people my undivided attention of looking from one side of that cockpit to the other in order to operate it." Whatley went on to explain that while he was homing on the A-7 Fan Song simulated radar site at a low altitude, he was concentrating so much on the electronics that he did not notice that he had rolled inverted. Although Whatley was able to right the aircraft, he told Telford he would never fly that system again while giving it his full attention.

While this heated conversation was taking place, General Dempster, who was visiting at the time, walked into the hangar where the discourse was taking place, heard the subject under discussion, and immediately began asking the two pilots questions about the equipment and their experiences. Major Whatley's negative input was undoubtedly responsible for General Dempster's decision to buy and install large numbers of ATI's Vector IV and WR-300s for use in all strike aircraft flying in Southeast Asia. Not long thereafter, on November 19, 1965, ATI received a contract for five hundred APR-25 and WR-300 systems, and the latter was soon designated the APR-26.[3]

The testing, as Air Force historian Bernard Nalty put it, "reinforced a lesson already learned": that searching visually for a SAM site and operating the RHAW gear while flying an aircraft was more than a pilot could manage. This venture was quickly shelved and a new project initiated to convert the two-place F-105F into what was termed the Wild Weasel II. The F-105F was originally intended as a trainer, with two crew members seated in tandem under separate canopies. Both cockpits were provided with dual controls so that either crew member could fly the airplane. To prepare the F-105F for the Wild Weasel role, an F-105F was taken out of storage and modified by installing the new radar homing and warning equipment in wing-tip pods instead of locating them inside the fuselage. Flight testing of this configuration revealed that the heavy pods created dangerous vibrations that necessitated strengthening the wing and restricting the aircraft to no more than 300 knots. The restricted performance characteristic plus the added cost of strengthening the wing doomed the Wild Weasel II as a replacement for the F-100F.[4]

On January 7, 1966, General Dempster ordered the two-seat F-105F into production as the Wild Weasel III. ATI personnel with the assistance of engineers from Republic Aviation, builders of the F-105s, hurriedly modified the F-105F bearing tail no. 62-4416 with the same electronic gear that had been installed in the Wild Weasel I. It took just eight days to convert the prototype, which was unofficially designated the EF-105F, for electronic fighter. The aircraft made its first flight on January 15 equipped with the APR-25 radar homing and warning system, the APR-26 launch warning receiver, and the IR-133 (soon designated the APR-36) panoramic receiver. Five additional F-105Fs were outfitted with the same gear before test flights with the prototype revealed problems with the installation of the RHAW equipment. All six EF-105Fs were returned to McLellan for repairs. The prototype returned to Eglin on March 12 and successfully resumed testing.[5]

While the F-105s were undergoing repairs at McClellan, ATI continued to work on the new system suggested by Colonel Johnson that would help the Wild Weasel III pilots visually locate the target radar even if it were heavily camouflaged. The Azimuth-Elevation (AZ-EL) system was an adjunct to the IR-133 that put a "pipper" dot on the gunsight reticle on the pilot's heads up display that could be used to identify the target radar. This eliminated the need for the pilot to look at another display somewhere on the instrument panel. "If everything worked out the way we wanted it to," Grisby later explained, "the pipper would stay on the target even with some maneuvering including inverting flight."[6]

The AZ-EL was installed in one of the F-105Fs and flown to Eglin in mid-April 1966. During the test flight, the pilot, Capt. Robert L. Tidwell, discovered an anomaly that occurred when flying from the water to over the land. The pipper would initially "ping" low into the water rather than on the radar site. Then, after the airplane transitioned to the land environment and the range to the target was reduced, the pipper would "walk up" to where it was actually pinging on the radar site. There was not enough time to locate the cause of the anomaly, and the F-105Fs, desperately needed in Vietnam, had to go "as is."

While the AZ-EL tests were being performed, the first set of Wild Weasel III crews underwent training. Where and how this took place has not been recorded. A second group of F-105 pilots and EWOs was also established that spring. We know about the training this group received thanks to recollections of Col. Edward T. Rock, whose memories of his experience as a Wild Weasel III pilot were published in *First In, Last Out: Stories by the Wild Weasels.*[7]

Rock was an F-105 instructor pilot at Nellis Air Force Base, Nevada, when he received orders for the Wild Weasel program, which was still classified. He was told only to report to a motel in Long Beach, California. Upon his arrival he was directed to an old hangar at Long Beach International Airport that had been converted into an impromptu training center. The training consisted of some classroom lectures from former SAC Electronic Warfare Officers and flying time in the simulator, which had RHAW equipment and a signal analyzer. To operate the radar, he was assigned Capt. Curt Hartzell, an ex-SAC B-52 EWO, who became his "back seater."

Training shifted to Nellis when the modified F-105s became available. Since there were no simulated SA-2 radars at Nellis, training flights were conducted against radars of opportunity. Most often they were one of two SAC radar scoring sites in the vicinity that Rock considered to be poor physical substitutes for the real thing: "they were not transportable or mobile and never moved, were not camouflaged, didn't shoot back, and were very unrealistic in appearance compared to enemy SAM sites."[8]

The first group of five F-105F Wild Weasels deployed to Korat Royal Thai Air Force Base on May 22, 1966, where they were formed into the fifth flight within the 13th Tactical Fighter Squadron (TFS) of the 338th Tactical Fighter Wing (TFW). Upon arrival at Korat, the Wild Weasel crews "began an intense schedule of orientation lectures on procedures, rules of

FIG. 9.1

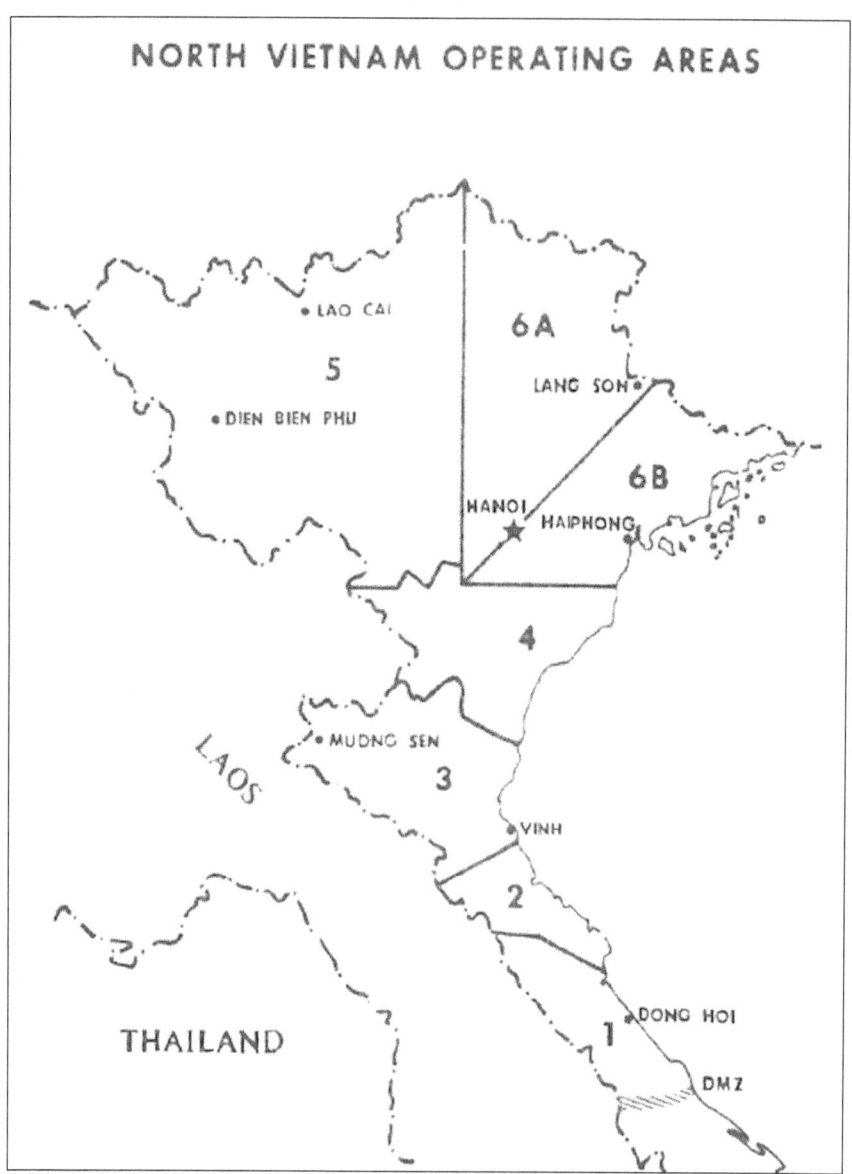

Route Packages were operating zones within North Vietnam established to reduce mission interference between land-based Air Force and carrier-based Navy aircraft.
U.S. Air Force History and Museums

engagement, enemy capabilities, orders of battle and escape and evasion tactics."[9] They began flying orientation flights on June 3. Combat operations began two days later with missions flown along Route Package 6, one of the six[*] operating zones established by Air Force–Navy coordinating team in December 1965 to reduce mission interference between land-based Air Force and Navy carrier aircraft operating over North Vietnam.[10]

In June a new passive warning system produced by the Loral Corporation, the ALR-31 "SEE-SAM," arrived at Korat along with the technicians needed to install it. SEE-SAM was designed to tell the crew whether or not a particular radar was targeting them. The idea for the device was suggested by Capt. Robert A. Klimek Jr., who proposed that the EWO lock onto the scanning pattern of the radar and adjust a sweep circuit so that the receiver scan pattern was synchronized to the radar. If it was in the center of the scan then their aircraft was probably being targeted. ATI decided that the concept was too "iffy" because the Fan Song operator did not have to center the beam to track the target, and they declined to proceed with it. Captain Klimek took the idea to Loral, which then developed it for the Air Force. Combat testing of the ALR-31, installed in Korat's F-105s, began on July 1 after the equipment had been ground tested. The set enabled the EWOs to monitor the SA-2 signal in much greater detail than the APR-25/APR-26 combination, and although it experienced minor problems, SEE-SAM ALR-31 seemed to work all right.[11]

Toward the end of June, six F-105Fs of a second group of Wild Weasels IIIs arrived at Korat and were sent to Takhli Royal Thai Air Force Base, where they landed on July 4, 1966. After arriving at Takhli, Captain Rock found that he and his aircraft had been assigned to the 355th Tactical Fighter Wing. Despite the urgent need for Wild Weasels, the commander of the 355th, Col. William H. Hold, decided that Rock's squadron would get a better understanding of the local procedures and the strike pilots wants and needs if their first combat flights were flown as strike missions. So on July 11, 1966, Rock found himself in a single-seat F-105D

---

[*] A seventh Route Package was added by Admiral Sharp in April 1966 by dividing Route Package 6 into A and B sections.

configured with two 3,000-lb. bombs flying the number two position in a flight of four F-105Ds heading to the Hanoi area. "The mission," as Rock later described it, "was unremarkable except it was the first time I had seen the radar homing and warning gear light up like a Christmas tree from all of the radars trying to track and destroy us with real bullets and real missiles. . . . The lights and noise were enough to scare the hell out of you regardless if the enemy was shooting or not." Rock dropped his two bombs and returned to Takhli unscathed.[12]

Two weeks later, on July 23, 1966, the first of the 355th TFW's Wild Weasel IIIs was lost from a barrage of SAM missiles during a strike in the Hanoi area. The plane flown by Maj. Gene Pemberton was hit and immediately went down. Two more Wild Weasels from the 355th were shot down over North Vietnam on August 7, and another was so damaged by anti-aircraft fire that it never flew again. After forty-five days at Takhli, only two of the eight crews accompanying the F-105Fs and just one aircraft was available for duty. What was left of the unit was sent to Korat to continue flying Wild Weasel missions with the Wild Weasel III-1 group.[13]

By mid-August only four of the eleven converted F-105Fs sent to Thailand remained flyable. Although the 388th TFW had lost only one of its five Wild Weasels, four of the six aircraft assigned to the 355th TFW had been shot down and two more damaged beyond repair. In October, six replacements arrived and were divided between the wings so that each had five aircraft. For the remainder of Rolling Thunder, the number of Wild Weasels serving with each wing varied from as few as four to as many as twelve.[14]

The decision to split the small number of Wild Weasels sent to Southeast Asia between two tactical fighter wings had negative consequences according to the SEAD study written by James Young Jr., who believed that many of the Wild Weasel losses were due to a lack of doctrine regarding the suppression of SAMs. Although the assignment to the two wings simplified their ability to support both strike wings, he felt the negatives outweighed the advantages. "By separating the Wild Weasels squadrons," he wrote, "7th Air Force made maintaining these specialist fighters more difficult by requiring maintenance crews at two bases rather than one." The geographical separation, he contends, also prevented the rapid exchange of tactics and slowed equipment upgrades and improvements. Young claims that shortcomings decreased combat efficiency during Operation Rolling Thunder but provides no hard evidence to support this contention.[15]

What Young fails to address is the urgent need for Wild Weasels to counter the SA-2 threat facing all the Air Force strike aircraft operating over North Vietnam. The Shrike-equipped Wild Weasel was the only answer to the SAM problem according to Col. Monroe S. Sams, commander of the 388th Tactical Fighter Wing. "WILD WEASEL aircraft accompanying a strike force into SAM-defended area," he wrote in his end of tour report on August 6, 1966, "offer the best solution to the SA-2 problem. . . . The Shrike increases the flexibility of WILD WEASEL operations due to the fact that it can be launched against an installation without visual acquisition."[16]

While Colonel Sams touted the importance of the Shrike, its performance proved less effective than hoped. From April 18, 1966, when the first Shrike was launched from a Wild Weasel, through July 15, 1966, F-100Fs and F-105Fs fired 107 AGM-45 antiradiation missiles. Of these, one was a confirmed hit and thirty-eight were "probables," yielding a "kill" rate of just 36 percent. Various factors contributed to this unimpressive record. The small size of the weapon's warhead was one. To be effective it required a direct hit. But the principal reason for the low number of successful attacks was the difficulty of putting the missile on target. Because the electronic gear installed in the Wild Weasels could not determine the precise range to the target and because of the need to loft the missile because of its range limitations, Wild Weasel crews had to rely on reference tables to determine the loft angle for a given speed, dive angle, and altitude.[17]

In order to aim the Shrike, the crew had to fly directly toward the target while referring to the range tables before pitching up to the loft angle. These actions or the launch itself, which was easily identified on radar, alerted the radar operators to the imminent danger from an antiradiation missile. The North Vietnamese radar operators (or their Russian advisors) quickly learned that they could thwart the Shrike by shutting down their radar. Once this was discovered, they needed to devise a tactical compromise that enabled them to engage the attacking American aircraft while offering only a fleeting target to the Shrike.[18]

Wild Weasel crews soon found out that if the targeted radar shut down before the Shrike had launched, then it seemed likely that North Vietnamese radar operators or their Russian advisors had learned to recognize the four-plane hunter-killer teams that relied on the Wild Weasels to identify the SAM sites. If the radar ceased transmitting after the Shrike had been launched, then the missile itself had been detected. Instead of transmitting for ten or twelve minutes before the introduction of the Shrike, Fire Can and Fan Song radars now remained on the air for three minutes or less. As these enemy tactics evolved, Wild Weasel crews found they could no longer cruise about and locate radars with their own detection gear. Instead, they had to rely on intelligence reports on specific transmitting sites in order to engage the target. Sometimes those radar signals were so brief that the crew did not have time to use the range table and had to estimate how sharply to pull up. If the enemy had already launched an SA-2, the crew had to made a best guess of the range and fire the Shrike in the hope that the firing would cause the Fan Song operators to shut down the radar, which would cause the SA-2 to go ballistic and miss them. Despite these limitations, the threat of the Shrike was in itself successful in deterring the North Vietnamese missilemens' ability to track strike formations accurately with their Fan Song and Fire Can radars.[19]

Between the summer of 1966 and the spring of 1967, the Wild Weasel's primary objective changed, as radar-suppression missions gradually took precedence over the hunter-killer-type missions, where they visually and electronically searched out SAM sites, identified them, and aided in their destruction. During these Iron Hand missions the Wild Weasels provided an electronically guarded corridor through which the strike force could pass using the Wild Weasel's radar-warning capability and the strike threat for protection. Although the Iron Hand flights, together with the wide-scale use of the QRC-160-A jamming pod by strike aircraft, did suppress North Vietnam radars and allow the destruction of more anti-aircraft and SAM sites (1,923 AA/AAA and 227 SAM sites were declared destroyed or damaged by the end of the year), the increasing sophistication of the NVA air defense system, coupled with intermittent barrage missile firings, continued to severely harass the Seventh Air Force strike force.[20]

Lt. Col. R. C. Reynolds, Seventh Air Force director of Combat Tactics, summarized Shrike's impact on the SA-2 problem: "Since the SHRIKE's introduction, the percentage of SAMs fired with radar guidance had been greatly reduced. The far greater number are fired in salvo and are unguided because the ground sites cannot afford to stay on the air. This has greatly reduced their effectiveness and had been a big boost to the moral of the strike pilots."[21]

The Wild Weasel, Shrike-equipped, Iron Hand missions greatly affected the operation of the North Vietnamese SAM sites. As one retired Central Intelligence Vietnamese Language and Operations Officer explained, "The impact of these attack on the missile crews was devastating. . . . Entire missile units wavered, afraid to attack." Major General Nguyen Xuan Maut, observing the operation of a missile battalion near Haiphong, was so frustrated that he shouted at the launch of the missiles: "Even my old eyes can see the target on your screen."[22]

The Shrike, however, could not, because of its small warhead, guarantee the destruction of a SAM site. Air Force evaluators conducting a study on the effectiveness of the Shrike during the period from March 1, 1967, through March 31, 1968, found it difficult to assess the damage caused to enemy radars. They maintained that a cessation of signals after a Shrike launch could not be regarded as proof that the radar had been hit, and they found evidence indicating that many of the missiles had been launched outside the optimum maneuver envelope. The report concluded that "only 5 percent of the Shrikes launched caused the destruction of a Fan Song radar or inflicted damage sufficient to put it out of action for more than one day. Nevertheless, if the Shrike was launched within its maneuver envelope, and the Fan Song remained on the air throughout the flight, the chances of a 'kill' were assessed at 40 percent or higher." The data also indicated, however, that Shrike attacks reduced SA-2 firing rates by as much as 90 percent. A follow-on weapon, the air-launched AGM-78 Standard Anti-Radiation Missile, promised to combine the destructive potential of the cluster bomb with the accuracy of the Shrike.[23]

Toward the end of 1966, the Seventh Air Force became so concerned with the MiG threat that it ordered all F-105s operating over North Vietnam to carry a least one AIM-9 sidewinder air-to-air missile on one of the aircraft's outer pylons. The dictate to carry an AIM-9 automatically reduced the number of Shrikes that could be carried from two to one, greatly reducing the offensive capabilities of the F-105F Wild Weasels. By the time the Weasel crews were able to convince Seventh Air Force Headquarters to rescind the order to carry a least one Sidewinder, the Seventh Air Force issued a directive ordering that all tactical aircraft conducting missions over North Vietnam be equipped with an ECM pod. By then the ALQ-71 noise jammer (full production model of QRC-160-A) had become standard equipment on the F-105s.[24]

The order to equip the F-105Fs with a noise-jamming pod was not received well by the Wild Weasel crews since it necessitated the removal of one of the two Shrikes normally carried by the aircraft. This reduced the offensive capability of the Wild Weasels that heretofore had carried two Shrikes, two cluster bombs, and a 600-gallon centerline fuel tank as the standard mission loadout. Worse, the ALQ-71s jammed the Wild Weasel receivers in addition to the enemy radars. The Wild Weasel crews complained bitterly about "this extremely insane order to cut their effectiveness in half," but their complaints fell on deaf ears. To get around this problem, some of the Wild Weasel units produced a field fix that enabled one Weasel to carry two ECM pods on one pylon so that a least one other aircraft could have the full complement of two Shrikes. They also tried using the Navy's ADU-315 dual Shrike rack, but it vibrated so badly when one missile was fired that it was quickly discarded.[25]

The universal adoption of the jamming pods by strike aircraft forced the North Vietnamese missileers to begin launching multiple SA-2s at each flight. It became common practice for them to launch one missile at the flight hoping the flight would become preoccupied with avoiding the first missile while other batteries launched additional missiles in order to catch the Americans by surprise.[26]

CHAPTER 10

AGM-78 STANDARD ANTIRADIATION MISSILE

During the summer of 1966, the Department of Defense began investigating alternative methods of increasing the capabilities of the Shrike. On December 1, 1966, the Naval Air Systems Command was authorized to proceed with the development of a more powerful, longer range ARM for use by both services. The command issued a contract to General Dynamics to develop an air-launched ARM based on the RIM-66 Standard air-to-surface missile that the company was producing for the Navy. The new missile, designated the AGM-78A Standard ARM, combined a Standard missile airframe and motor with a Texas Instruments seeker head from the Shrike AGM-45A3. The seeker, which was designed to monitor the SA-2's frequencies, was interfaced with the ER-142 receiver on board the aircraft. The 1,370-lb. AGM-78A carried a 220-lb. warhead detonated by both impact and proximity fuzes. Its Aerojet Mk 27 Mod 4 solid-fuel rocket motor allowed the missile to reach a maximum velocity of Mach 2.5 and a range of up to fifty-six miles. Because no completely new components had to be designed, development progressed quickly so that flight testing began during the summer of 1967. Although the AGM-78A had many advantages over the Shrike, the Navy, due to the missile's high cost, considered it complementary to and not as a replacement for the Shrike, which the Navy felt would suffice in many instances.[1]

While tests of the AGM-78A were being conducted at the White Sands Missile Range, both the Air Force and the Navy began to modify a small number of aircraft to handle the new weapon. The Grumman Aircraft Engineering Corporation stripped out the ground attack equipment from ten of the A-6A Intruders they had built for the Navy and fitted them with an ER-142 receiver, an APS-1078B radar homing and warning system furnished by Bendix, and the various antennas needed to support this equipment. These aircraft—designated A-6Bs—were also equipped with LAU-77 launching racks and the associated wiring necessary to connect the AGM-78s on the inboard pylons. Three more A-6As were modified to fire the AGM-78 using a system developed by the Applied Physics Laboratory. Another six were modified in 1970 with the IBM TPS-118 Target Identification System.[2]

By the beginning of 1968, each Navy attack squadron heading for Vietnam had received two or four A-6Bs, giving the air wing a much-improved Iron Hand capability. Attack Squadron 75 was the first to field the Standard ARM. The squadron, which was attached to Carrier Air Wing 11 on the *Kitty Hawk* (CVA 63), departed San Diego on November 18, 1967, and began flying combat sorties on December 23. The first operational use of the AGM-78A occurred on March 6, 1968, when two of the squadron's A-6Bs fired four of the missiles for the first time. Six more Standard ARMs were fired at SAM sites near Hanoi, Haiphong, and Nam Dinh the next day. Unfortunately, the results of these missions are unknown.[3]

While Grumman was modifying the A-6Bs, six F-105F Wild Weasels were pulled from combat units assigned to the 357th Tactical Fighter Wing and modified with the electronics needed to fire the new missile and also equipped with Loral's QRC-317A SEE-SAMs. The QRC-317 scanned the frequency spectrum used by the various Fan Song radar tracking signals and monitored the lower frequency band used for missile guidance. Four additional F-105 Wild Weasels, also modified to fire the AGM-78, were transferred to Southeast Asia from the Wild Weasel College at Nellis in December 1967. The 357th TFW scheduled its first Standard ARM mission for March 4, 1968, but it was cancelled due to bad weather, which allowed the Navy to be the first to use the missile in combat two days later. On March 10 a flight of four Wild Weasels from the 357th identified as Barracuda flight got their chance while covering an F-4 strike against a barracks near Hanoi.[4]

One of the pilots assigned to fly the F-105s was Maj. Warren J. Kerzon. Major Kerzon and his EWO, Lt. Col. Scott "Scottie" W. McIntire, were selected for this mission because of their previous experience performing flight tests of the AGM-78 at Eglin. It was their fourth combat mission and their third to the highly defended Hanoi area. As they approached Hanoi at 13,000 feet, McIntire began to pick up every type of radar signal from the multitude of early warning, anti-aircraft, and SAM radars that were part of the North Vietnamese air defense system protecting Hanoi. As Kerzon, who described the operation in detail in his memoir, noted, "the Bad Guys knew that we were inbound. . . . Above the hum of our J75 engine, the only sounds inside our trusty Thud were the 'squeaks-and-squawks' of the incoming hostile signals and our steady, heavy breathing." Because of bad weather over the target and because a large number of his aircraft were experiencing jammer pod malfunctions, the strike force commander aborted the primary target and headed for their backup secondary targets in Route Package 1. The MiG CAP stayed with the Wild Weasels as they continued to troll for SAM sites. They approached the south side of Hanoi, close enough to get the Fan Song radars to come up, but not so close as to spook them to turn off the radar. The nearest SAM site was just about twenty-five miles away—about the maximum range of the SA-2—when the AGM-78As, two of which were carried on the pylons of each of the F-105Fs, were launched.[5]

Barracuda 2 took the first shot. Kerzon saw it release from its pylon, but the missile failed to ignite and he watched as it fell away until it disappeared into the jungle canopy below. Next, Barracuda 3, after obtaining a strong lock on a Fan Song, released its first missile, which accelerated in front of Barracuda 3's aircraft and started a steep climb to gain range, then broke apart as it snap rolled over and over.

The Barracuda flight was getting too close to the SAM site, so they performed a full go-around. As he rolled out of the turn, Kerzon's APR-25 display showed multiple signals for the ten o'clock through two o'clock position. McIntire in the back seat said that he had a solid lock on a Fan Song at one o'clock. Kerzon pressed the missile release button on the control stick and heard a thump as the missile fired and began its up-and-over trajectory before it too began to snap roll and break up. Three launches. Three bad missiles. But the next five launches ignited and followed their normal trajectory, climbing to 80,000 feet before diving on the target at high speed. From analysis of the missile timing, the audio tapes that were now installed in the Wild Weasels to record the radar signals, and human intelligence sources, the Air Force determined that the AGM-78As launched from the Barracuda flight had hit three targets with two other "probables." This was three times as good as the Shrike. Months later, Kerzon was told that his second missile "had scored a direct hit on the target SAM site van, totally destroying it and killing its crew which included four Soviet air defense advisors."[6]

Air Force records for the Project CHECO report on "Second Generation Weaponry in SEA," compiled by Melvin F. Porter, indicate that only eight AGM-78A missiles were reported to have been fired by the U.S. Air Force during 1968. All of them appear to have been launched on May 10. The reasons for the small number of Standard ARMs used during this period have never been revealed. It may have been related to the poor reliability of the AGM-78A, or perhaps the limited number of missiles available, or they might have been waiting for the much improved "B" model that was about to become available. It might also have been related to the bombing halt ordered by President Johnson. On March 31, Johnson, in a televised speech to the nation, announced that in an effort to persuade the North Vietnamese to end the war, he was restricting the bombing of North Vietnam, halting targets north of the 20th parallel, including Hanoi. Because of the progress being made in the peace talks taking place in Paris, he explained, "I have now ordered that all air, naval, and artillery bombardment of North Vietnam cease as of 8 a.m., Washington time, Friday morning." This marked the end of Rolling Thunder and with it the necessity to go after the large number of SAM sites that had been established in the Hanoi-Haiphong area, and it alleviated the need to expend large numbers of expensive AGM-78s.[7]

When the bombing was prohibited above the 20th parallel, the air war shifted to interdiction missions along the Ho Chi Minh Trail that ran through Laos and the staging and supply areas in North Vietnam close to the DMZ. To protect these routes and supply areas, the North Vietnamese had moved SA-2 missiles into these areas too. This required the continuing use of the Wild Weasels for suppression on a more limited basis, using Shrikes as the primary method of attack. Typical of these missions was one flown by Barracuda flight on the night of May 19, 1968. Their job was to provide SAM suppression to protect the F-4 strike aircraft assigned to search for any supply traffic on the river or roads heading south in the area around Don Hoi, about one hundred miles north of the DMZ.[8]

During the briefing before the mission, Barracuda flight learned from wing intelligence that their flight profile would put them within range of the three active SAM sites, four active radar-controlled 85-mm anti-aircraft gun positions, and the usual batch of 57-mm and 37-mm guns. The mission was to be composed of their two Weasel Thuds and a four-ship flight of F-4 Phantoms. For this mission their F-105F would be configured with three external fuel tanks and two Shrikes.

As they approached the target area, the strobe on the APR-25 display in each cockpit appeared on the 3 1/2 ring, indicating they were almost on top of the SAM site. "Kerzon," writes aviation author Warren Thompson, "had never before seen a 3 1/2-ring Fan Song signal, so he knew he was in trouble and had to react fast. He had to do two things at once: point his aircraft directly at the site to aim the AGM-45 and build his airspeed to be able to maneuver away from the anticipated SAM launch. Rolling left toward the tracking signal emanating from the SAM site, he pushed the throttle into afterburner for maximum power." Halfway through his turn, as the strobe showed the SAM radar site was right in front of them, McIntire yelled, "Get 'em! Get 'em," and Kerzon hit the pickle button, launching the Shrike on the port inner pylon. It blasted off the wing and dipped sharply earthward homing on the strong radar signal. "As Kerzon's first missile disappeared, he rippled off his second Shrike." Then he immediately maneuvered wildly to avoid two SA-2s heading toward his aircraft, making a rolling left pull at the last instant. As he was avoiding these two SAMs, he saw a third missile coming his way. This was a basic SAM tactic, firing multiple missiles at the same aircraft, causing it to expend energy avoiding the first and second missiles, then, low on energy, be hard pressed to maneuver away from the third. He avoided this one too, just as the radar-tracking signal went off the air, indicating the likelihood of a successful kill, later confirmed by intelligence.[9]

In the fall of 1967, the Tactical Warfare Center at Eglin Air Force Base began a program to upgrade the F-105F. One of the first problems to be tackled was how to move the ECM function of the pylon-mounted ALQ-71 jamming pod internally in order to free up the pylon so the F-105F could carry more ordnance. Westinghouse and Republic Aviation solved the problem by splitting the components of Westinghouse's ALQ-101 ECM pod and putting them inside two fairings on either side of the fuselage just above the bomb bay doors. The fairings each had one receiving antenna and two transmitting antennas. The new system was designated the ALQ-105. In addition to providing the integral jamming capability, the Wild Weasel's APR-25/26 system was replaced by APR-36/37, an improved radar homing and warning system, and a productionized version of the SEE-SAMs was installed. The new equipment was installed on the Wild Weasel F-105Fs in order to cope with the different frequencies now used by the SA-2 radars as a safety measure to "keep the crews 'relatively' safe until" a countermeasure to improved Fan Song radars could be developed.[10]

In addition to these changes, the ER-142 panoramic receiver was replaced by the new Itek APR-35, which provided automatic direction finding and homing used to cue the AGM-78B (Mod 1) Standard ARM. The AGM-78B used a Maxson wide-band receiver that enabled the Standard ARM to be used against many different types of radars without the need for pretuning. The new seeker was also gimballed, which permitted a wider range of maneuver by the launch aircraft and eliminated the characteristic diving approach needed for the Shrike. It also had an SDU-6/B red phosphorus target and an electronic bomb damage assessment (BDA) subsystem. The BDA consisted of a radar pulse repeater housed in the dorsal fin of the AGM-78B and a computerized display in the launch aircraft. The repeater picked up the target's radar signal from the missile's seeker head and retransmitted it as a coded echo at the target's own PRF*. If no sign

---

* PRF stands for pulse repetition frequency of the radar system and is the number of pulses that that are transmitted per second.

was received from the seeker head, the repeater transmitted a self-generated pulse train at a unique PRF. The BDA subsystem contained sophisticated computerized circuitry to discriminate the real signal from noise and to predict the impact time as well as other features that allowed it to further verify the selected target. The computerized display in the cockpit removed much of the guesswork required to determine if the missile had worked or not (see table 10.1).[11]

The AGM-78B also had a digital control module that contained a memory circuit that allowed the missile to home on the radar even if it had stopped transmitting. The "B" model's control module, however, was not compatible with the AGM-78A's analog module. Although the launcher adapter and the suspension gear were common to both missiles, the modified F-105Fs (soon to be designed F-105Gs) were unable to fire the AGM-78A, and the unmodified F-105Fs were unable to fire the AGM-78Bs. To solve this dilemma the Air Force planned to modify the remaining AGM-78As after the modification program was completed.[12]

Many sources claim that the F-105G designation was the result of RHAW equipment changes made to the F-105F in the electronic control module needed for the AGM-78. This is only partially correct. According to a letter from Gen. John D. Ryan, Air Force chief of staff, the letter "G" was instated to "ensure that WILD WEASEL assets [could] be identified in appropriate programming documents and resources allocated accordingly."[13]

TABLE 10.1 AGM-78B BOMB DAMAGE ASSESSMENT LOGIC DISPLAY

AGM-78 Performance	BDA Logic
No Acquisition	No homing signal received throughout missile flight. Missile never tracked. Radar.
Loss of Acquisition	Missile tracked target radar for some time, but was not tracking just before impact. No valid hit or miss assessment.
HIT	Both homing signal and radar transmission received during flight; both stopped after impact.
MISS	Both signals received during missile flight; homing signal stopped after impact, but target radar continued to transmit.

FIG. 10.1

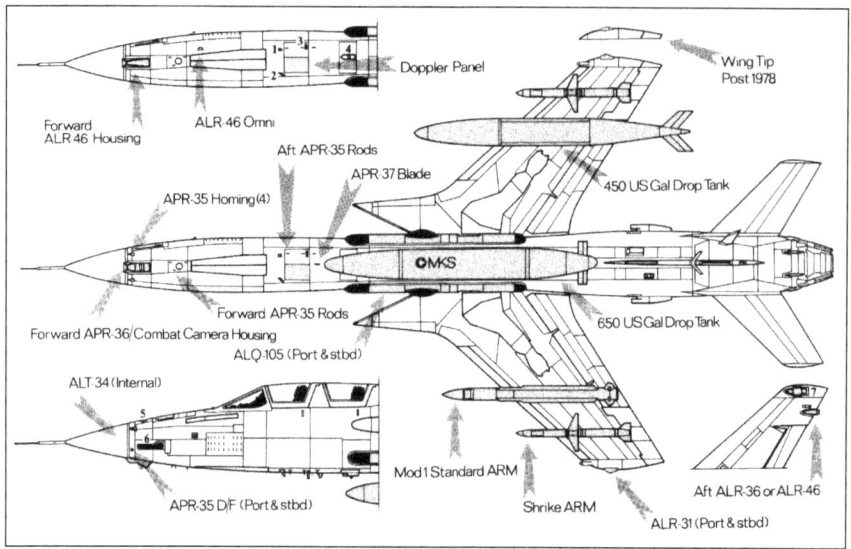

F-105G Wild Weasel III. 1 ARN-62 Tacan blade antenna. 2 VHF communications blade antenna. 3 Unidentified blade antenna. 4 Electronics bay cooling-air scoop. 5 APX-37 IFF antenna. 6 Muzzle port for M61 Vulcan 20mm rotary cannon. 7 RHAW pre-amp © *Martin Streetly, used with permission*

CHAPTER 11
DEALING WITH THE MIG THREAT

When American bombing efforts over North Vietnam began in August 1964, the North Vietnamese had a rudimentary air defense system consisting of approximately 1,421 anti-aircraft guns, twenty-two early warning radars, and four fire-control radars. The thirty-six MiG-17s donated by the Soviet Union and their North Vietnamese pilots were still in training in the People's Republic of China, and the North Vietnamese had yet to receive any SA-2 surface-to-air missiles. After Operation Pierce Arrow, the Soviet Union and China began to supply the North Vietnamese with a large number of anti-aircraft guns, ammunition, and fire-control and search radars. To support this equipment, according to one former director of operations for the U.S. Pacific Air Force Command's Joint Intelligence Center, "hundreds of men, women, and even high school students entered air defense training."[1]

When Operation Rolling Thunder began on March 2, 1965, intelligence officers within the command estimated that the North Vietnamese possessed as many as 1,057 medium (85-mm) anti-aircraft guns, 253 light (57-mm) anti-aircraft guns, and up to 286 (37-mm/23-mm) automatic weapons. Because the effective altitude limits for most of these guns was below 5,000 feet, they did not severely restrict U.S. strike aircraft that flew above the lethal range of these weapons. Targets were relatively easy to acquire, and aircraft ran into defensive firepower only on that part of their bomb run below 5,000 feet. "But," as Lt. Col. Robert Burch wrote in his CHECO report, "this condition was not to endure; the days of a relatively threat-free environment over [North Vietnam] were numbered."[2]

Following the Gulf of Tonkin crisis, the 2nd Air Division (which became the Seventh Air Force on March 28, 1966) requested SIGINT support for missions north of the DMZ. The Air Force had already been operating a pair of C-130B-II SIGINT aircraft in theater flying missions under the code name Silver Dawn. The C-130B-IIs, which first arrived in July, had been operating under the mission's name Queen Bee, which was changed to Silver Dawn when two more C-130B-IIs arrived in September. The specially instrumented C-130B-IIs were part of the Air Force's Airborne Communication Reconnaissance Program (ACRP), a highly secret program under Air Force management with the National Security Agency (NSA) providing collection guidance and technical support. The aircraft in the ARCP program were flown by SAC crews in SAC aircraft crewed with members of the U.S. Air Force Security Service (USAFSS), who had exclusive access to the SIGINT equipment. The USAFSS was a secretive and tight-knit branch of the Air Force tasked with monitoring, collecting, and interpreting military voice and electronic signals from countries of interest. Though part of the Air Force, they worked closely with the NSA, which usually assigned missions and objectives of the aircraft conducting SIGINT missions.[3]

During the Silver Dawn missions, which were flown twice daily, the C-130B-IIs shuttled back and forth along the North Vietnamese coast in the Gulf of Tonkin gathering intelligence, mostly radio transmissions related to the burgeoning North Vietnamese air defense system. They were always escorted by a MiG CAP flight of four F-4 Phantoms, each armed with four AIM-9 Sidewinder and four AIM-7 Sparrow air-to-air missiles. Two of the flight flew a moving orbit around the aircraft while the other two stayed with a tanker until the first flight ran low on fuel.[4]

In November, one month after the Silver Dawn missions started, the USAFSS 6924th Security Squadron, at Da Nang, established Project Hammock, a reporting system designed to disseminate the intelligence gathered from the communications transmitted by the North Vietnamese air defense system. The top-secret security personnel at Da Nang manually plotted pertinent radar tracks derived from information relayed from ARCP platforms operating over the Gulf of Tonkin and information obtained from a USAFSS site on Monkey Mountain on the end of a peninsula near the city

DEALING WITH THE MIG THREAT   113

of Da Nang. The site for which had been set up in late 1962 to intercept VHF communications generated by Vietnam People's Air Force (VPAF) and the North Vietnamese air defense network. They converted the tracks into the normal U.S. lateral-tell format, which provided the illusion that the tracks came from U.S. radar sources. This information was sent to the Air Force's Tactical Air Control Center (TACC) at Tan Son Nhut Air Base over a secure teletype and to the Ironhorse data processing/display/forwarding system, also at Monkey Mountain. A cross-tell link within the Ironhorse system was set up with the Seventh Fleet carriers in the Gulf of Tonkin so the warnings could be passed to the Navy's air controller, known as Red Crown, and to the Marine Tactical Data System. The data received at Monkey Mountain was plotted and relayed by secure teletype to the Seventh Air Force Headquarters, where it was manually plotted for display to the command officers.[5]

FIG. 11.1

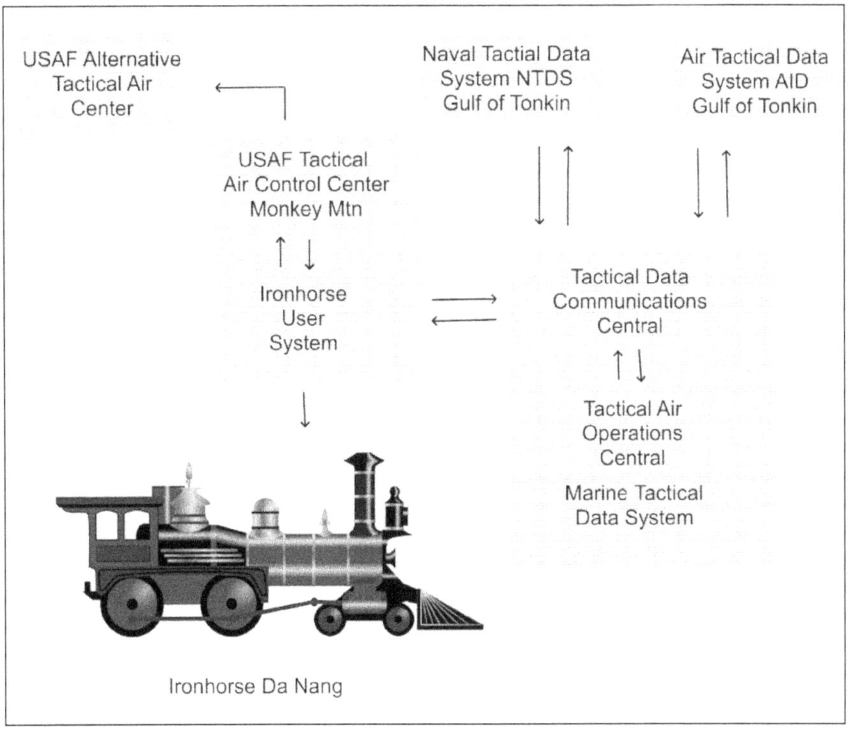

The Ironhorse system diagram *National Security Agency*

This led to duplication in which Tan Son Nhut and the Control and Reporting Post on Monkey Mountain received separate plots from Da Nang. In some cases, the information was different. To solve this problem, the TACC at Tan Son Nhut was dropped, while a new one was established at Monkey Mountain—the Tactical Air Control Center–North Sector (TACC-NS)—which assumed complete control of air operations over North Vietnam. The TACC-NS was staffed with technicians cleared for the SIGINT coming from the adjacent USAFSS site.[6]

The system was inherently slow due to the time needed to plot the data manually and the number of steps required to transmit the data down the line. The warnings could take anywhere from twelve to thirty minutes to reach the pilots. Furthermore, the warnings were passed to the pilots over their communications guard channel, which was already close to overloaded. Warnings would be transmitted, but in the confusion and clutter of radio communications, they could be missed or ignored, such as what happened on April 4, 1965, when North Vietnamese MiG-17s assigned to the 921st Fighter Regiment shot down two F-105s that were attempting to bomb the Thanh Hoa bridge.[7]

The 921st "Sao Do" (Red Star) Fighter Regiment—the first unit in the VPAF equipped with jet fighters—was established on February 3, 1964, at Mong Tu Air Base in China. Its aircraft complement, furnished by the Soviet Unit, consisted of thirty-six MiG-17s, an aerodynamically improved versions of the MiG-15 that entered service in the Soviet Air Force in October 1952. Although older, slower, and outgunned by U.S. fighters, the MiG-17 was lighter and more maneuverable. It performed well at low-speed engagements and could outturn every U.S. fighter, a characteristic of the aircraft that MiG pilots used to great advantage.[8]

On August 6, 1965, a day after sixty-four U.S. aircraft struck naval bases and petroleum storage areas in North Vietnam, the 921st Fighter Regiment moved from Mong Tu to Phuc Yen Air Base, the first jet-capable airfield recently completed immediately north of the Noi Bai International Airport nine miles north of Hanoi. The unit's commander, Lieutenant Colonel Dao Dinh Luyen, immediately initiated an aggressive training program to prepare the 921st pilots for combat with the American flyers who were bombing their country. As Chris Hobson and David Lovelady noted in their article on North Vietnamese MiGs, "the training was methodical and thorough and concentrated on the use of a network of ground-controlled interception radars that was being built in North Vietnam."[9]

On April 3, 1965, the 921st made its first operational sorties against a U.S. Navy strike force attacking the Thanh Hoa bridge. Four of the regiment's MiGs were vectored to a flight of four F-8E Crusaders from the aircraft carrier *Hancock* (CVA 19) that were providing flak suppression for a strike group of eight A-4Cs. The flight leader had completed his Zuni rocket attack, pulled off the target, and established a tight orbit when he observed tracers passing his left wing and felt hits. He thought the tracers were ground fire as no report of MiGs had been received. He turned hard right and exited the area climbing to 18,000 feet. As he neared the coast, he again saw tracers passing along the left side of his aircraft. He located a MiG-17 at seven o'clock, 2,000 feet behind him, engaged the afterburner, accelerated to Mach 1.0 and turned hard left, but was unable to locate the MiG. Though badly damaged, he managed to land at Da Nang.[10]

The next day, VPAF ground controllers using the extensive network of ground control intercept (GCI) radars that had been established throughout North Vietnam vectored four 921st MiG-17s behind four F-105Ds assigned to the 354th TFS preparing to attack the Thanh Hoa bridge for the second time. Despite the presence of eight F-100Ds that provided combat air patrol, the MiGs, hidden by haze and clouds, surprised the number one and two aircraft in Zinc flight. "They came out of the haze," explained Capt. Vern Kulla, who flew in the number four position, "adjusted their gun sight, fired and disappeared into the haze." The MiG's 23-mm and 37-mm cannon fire struck the rear of Maj. Frank Bennett's F-105 (Zinc 01) and blasted Capt. James Magnusson's F-105 (Zinc 02) behind the cockpit. Both aircraft were lost and their pilots killed.[11]

The MiG's success was due in large part to the GCI system that had vectored them exactly behind the two F-105s without having to maneuver themselves behind the Americans. This system consisted of a controller on the ground who relayed target and strike information to a flight of defending North Vietnamese interceptors. A senior controller at Bac Mai airfield, headquarters for North Vietnam's air defense system, assigned targets to a subordinate controller at Phuc Yen (as more airfields were added, the information was sent to subordinate controllers at these fields too). The GCI controllers used VHF voice communications to direct the MiGs to the threat area. There could be as many as four controllers at an airfield, all of whom had specific functions. There was an airfield controller who handled flight activity around an airfield, which, on occasion, included GCI. There was also a tower controller who directed takeoff and landing operations for aircraft. A third controller, the direction finding controller, provided navigational information to pilots, especially those returning from combat activity.

The fourth controller's job was to direct the fighters to the area of the hostile aircraft. The GCI controllers were the heart of the North Vietnamese fighter defense system. The controller told the MiGs the locations of the attacking aircraft and positioned them behind the U.S. aircraft so as to set them up with the advantage of surprise and position. The GCI controllers often warned MiGs when they might be attacked, making it difficult for U.S. pilots to ambush MiGs.

Hanoi's controllers had the advantage of the information from its extensive radar coverage of the region. Knowing also the locations of its own aircraft, they could see the entire combat situation come together on their own plotting boards and radar screens. Since American radar coverage could penetrate only part way into North Vietnam, Hanoi had a distinct advantage when Rolling Thunder began.

When the MiGs first sortied on April 3, 1965, the North Vietnamese possessed thirty-one early warning radars. By September the number of early warning radars had increased to fifty-six. The radars were a mix of Soviet-supplied P-15 (Flat Face) long-range search radars, PRV-11 (Side Net) height finder radars, and P-12 Spoon Rest medium-range search/acquisition radars. The Spoon Rest, which provided targeting for the SA-2 SAMs, was also used independently as an early warning radar. It was a highly mobile system installed on two all-terrain vehicles: an antenna/shelter truck and a truck for the generator. A secondary vehicle-mounted radar for the IFF system was also included in the Spoon Rest radar site.[12]

Spoon Rest had a maximum range of 125 miles (target at 50,000 feet) and relied upon an antenna composed of a six-bay, two-stack arrangement of twelve horizontal Yagi arrays. The entire assembly was forty-one feet long and eleven feet high. The antenna constantly rotated 360 degrees. The radar had an IFF interrogator, which indicated whether or not the target was an enemy, and a target speed indicator. Using the Spoon Rest's CRT display, an operator manually plotted the target's approach onto a large glass table in one-minute intervals. This information was sent to the Air Defense Headquarters that had been established at Hanoi's Bac Mai airfield in January. The Air Citation Center there received and processed this information and issued advisories to the Air Weapons Control Staff, also at the Air Defense

FIG. 11.2

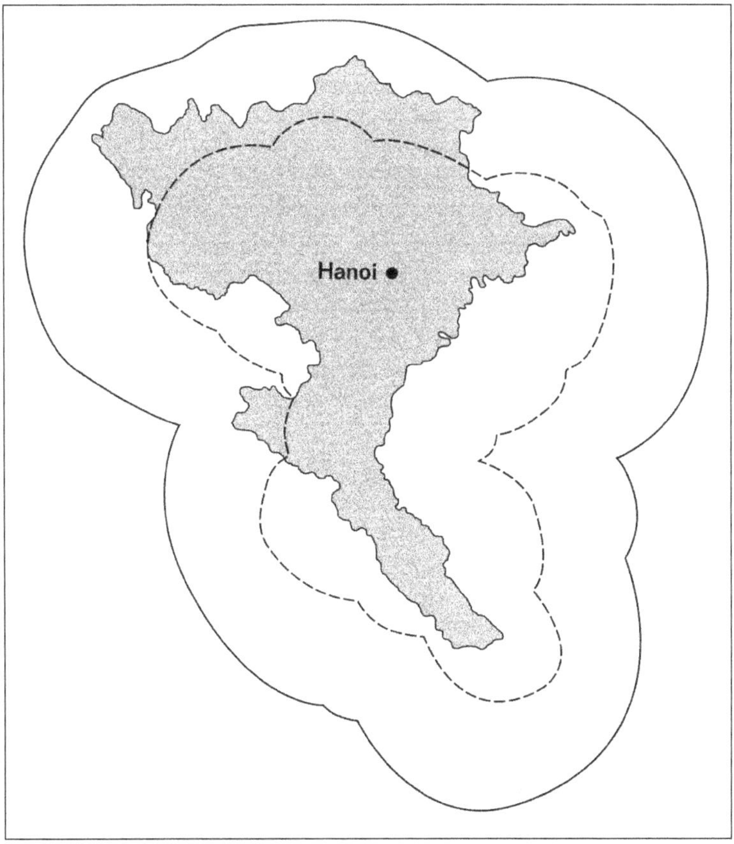

North Vietnamese early warning radar coverage (circa 1966). The solid line indicates the limit of detection for an F-105–size aircraft flying at 15,000 feet. The broken line indicates the limit of detection of the same type of aircraft at 5,000 feet. *United States Air Force*

Headquarters, and other parts of the air defense system. The Air Weapons Control Staff acted as a clearinghouse for the surveillance information. Staffed with representatives from the various elements of Hanoi's air defense system, the staff assessed the situation reports received from the surveillance system, plotted the threat tracks, and assigned targets to defensive forces, SAM units, AAA batteries, or the various fighter regiments.[13]

The loss of the F-105s on April 3 highlighted the difficulty of trying to protect strike aircraft from enemy aircraft that operated with an air defense system that had GCI and could give radar warnings to its own aircraft. The difficulties posed by North Vietnam's GCI was summarized in an Air Force report issued three days after the F-105s were shot down: "Commitment of MIG CAP against enemy aircraft engaged in a determined attack against OUR strike force is extremely difficult. In the Hanoi complex, for example, he has the advantage of GCI support plus the element of surprise. Moreover, he is able to initiate his attack with an advantageous speed differential over our MIG CAP aircraft that are in orbit."[14]

The need to provide MiG warning for Rolling Thunder formations had been recognized before the two F-105s of Zinc flight had been shot down, and plans were already underway to address the problem. In March, representatives of the headquarters of the Air Defense Command and the 552d Airborne Early Warning Wing, which would provide the necessary planes, met in Hawaii to work out the details of a JCS-directed deployment with the Pacific Air Force. As a result of this meeting, a decision was made to send a detachment of EC-121D Warning Star aircraft, designated as the Big Eye Task Force, to Southeast Asia to extend the coverage of the existing radar network, which at the time consisted of stations at Monkey Mountain and at Nakhon Phanom in Thailand. Because of this advance planning, the first three EC-121Ds of the Big Eye Task Force left McClellan Air Force Base, California, on April 4 heading to Tan Son Nhut Air Base, Republic of South Vietnam, where they would begin an initial thirty-day trial period.[15]

The EC-121D was a radar picket version of Lockheed's model 1049 Super Constellation that was similar to the Navy's EC-121M. It was designed and optimized for overwater radar detection to provide seaward extension of the North American Air Defense Command's contiguous radar coverage. The aircraft's radars were housed in two large radomes: one atop the fuselage for the APS-45 height-finding radar and one on the bottom side of the fuselage housing the APS-95 search radar. The latter was a high-power, long-range search radar with a theoretical range of 250 nautical miles but only reliable out to between 160 and 180 nautical miles. Another critical piece of equipment was the APX-49 IFF system, which allowed the crew to identify U.S. aircraft by a distinct coded response to the APX-49's electronic query. The crew usually consisted of a pilot, copilot, two navigators, two flight engineers, a radio operator, and the EWOs in the electronics compartment that included the senior director, a duty controller, a crew chief and his assistant, four radar operators, an intercept control technician, and two radar maintenance specialists.[16]

Missions over the Gulf of Tonkin from Da Nang began September 5, 1965 (two days after the first of the Navy's EC-121Ms assigned to VQ-1 that had been flying missions over the Gulf of Tonkin since July 1964 were deployed to Da Nang on a full-time basis). The mission of the Big Eye Task Force was to provide radar coverage of the Red River Delta, a pie-shaped sector with its tip northwest of Hanoi, broadening out to the irregular coastline stretching from Haiphong approximately sixty miles to the southwest. The crews had to overcome the shortcomings in the aircraft's APS-95 search radar, which was designed for use over water and could not cope with ground clutter.[17]

A normal mission profile (denoted as the Bravo Track) was flown at medium altitude. Although the APX-49 IFF coverage was excellent, the massive ground returns encountered blanked any aircraft radar returns. To overcome the ground clutter problem, a second aircraft (denoted at the Alpha Track) employed a low-altitude technique that the EC-121D crews had learned from their experience during the Cuban Missile Crisis when it was discovered they could eliminate most of ground clutter by flying close to the water. Flying over the Gulf of Tonkin as low as fifty feet caused the search radar beam to reflect off the surface of the water, raising the line of sight and avoiding the ground clutter. Flying at such a low altitude, however, was dangerous, as a gust of wind or a slight misjudgment in the controls could drop the aircraft toward the water, which forced both pilots to remain in their seats whenever they were at low altitude. One report told of pilots skimming the water while passing through tropical showers trying to avoid the masts of fishing vessels

120 CHAPTER 11

hidden by the downpour. The EC-121D during such missions, according to one former crew member, "was flown on autopilot because it was more reliable and steadier and it could hold altitude better than its human pilots. It also had quicker reaction time, especially in rough air. Even though we flew on autopilot, one pilot had to rest both hands on the control wheel with his forefinger covering the autopilot cut-off switch in case it malfunctioned."[18]

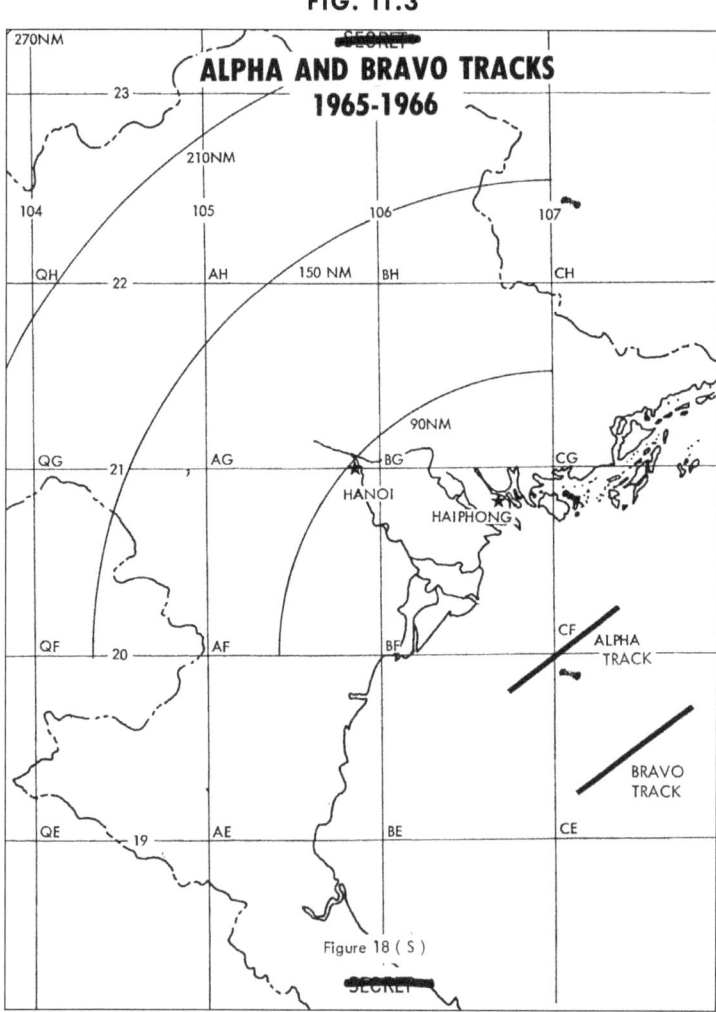

EC-121D mission tracks *United States Air Force*

The low-altitude radar approach limited the effective ability of the EC-121D's ability to point the exact location of hostile aircraft. Because most contacts took place beyond the range of the APS-45 height finder, Big Eye could not determine the altitude of the North Vietnamese MiGs. The best they could do was to determine the general location and heading of the MiGs, which were then broadcast over the common emergency channel (known as the Guard frequency) in the blind. The MiG warnings were only of limited value, according to one fighter pilot who claimed that "All Big Eye can tell you is that the MiG's are up and in what general area; after you hear the first call, subsequent MiG calls accomplish nothing more than clutter the radio."[19]

Despite these limitations, the EC-121Ds were sometimes useful in facilitating the destruction of North Vietnamese MiGs, as was the case on July 10, 1965, when the crew of an EC-121D flying in the Gulf of Tonkin detected two MiG-17s. William P. Reboli Jr., senior EWO on the Big Eye, issued a MiG warning that enabled two Air Force F-4Cs piloted by Capt. Kenneth E. Holcombe and Capt. Thomas S. Roberts to shoot down both MiGs. Bernard Nalty in his *Tactics and Techniques of Electronic Warfare* states that the pilots' success was due more to chance—they were in the right place at the right time—than to the information provided by the EC-121D.[20]

To keep track of Air Force planes, the Big Eye crews worked out a system to identify a specific flight by assigning an identification response to one aircraft in the flight. When the IFF operator on the EC-121D interrogated this signal, the plane's transponder automatically replied, giving the position of the flight, provided that the pilots cooperated by keeping the transponder on. The transponder, however, was not universally liked by the pilots. One unproven fear, according to Marshall Michel, "was that the North Vietnamese were able to interrogate and read the U.S. transponders on their radars." The presence of Navy carrier planes over the North also complicated the task of identifying unknown aircraft, as naval aviators did not turn on their IFF during the early part of the war until they were approaching their carrier. This, according to Nalty, "caused some uneasy moments on board the EC-121Ds."[21]

Communication between the EC-121Ds and the friendly aircraft operating over North Vietnam was also a problem because the EC-121Ds were equipped with an unreliable, aging, low-power UHF radio set. The quality of the radio warnings they issued varied from day to day depending on how well the radios functioned, the weather, and the amount of traffic. The communication problem was further complicated by low-level station keeping and the mountainous terrain. The EC-121D radios usually did not have the range to communicate with the Air Force aircraft coming into North Vietnam via Laos from airfields in Thailand, so radio messages had to be broadcast by ground relay stations. Personnel from Big Eye Task Force were so concerned by this situation they started a crusade for assignment of radio relay channels dedicated to their use, so that real-time situation information could be fed through a high-powered radio relay platform to the fighters over North Vietnam.[22]

It took a year before the solution to the communication problem was solved when two KC-130A tankers equipped with the ARC-89 radio relay system arrived in theater under the Combat Lightning program (see chapter 14). This variant of the four-jet Boeing aerial tanker automatically relayed radio messages either to American tactical aircraft or to the control centers that directed them. The Combat Lightning KC-130As arrived at U-Tapao Royal Thai Navy Airfield ninety miles south of Bangkok on September 20 and 22 and flew their first mission on October 5, 1966. Although air refueling was not the normal or main mission of these tankers, they could and often did provide service in emergencies.[23]

These first two aircraft provided only interim coverage. To provide twenty-four-hour coverage required five aircraft. Because of delays in converting additional KC-130s, SAC sent two EC-135L post attack command and control aircraft to U-Tapao in mid-May 1967. By then, the name of the Big Eye Task Force had been officially changed to College Eye. Starting at the end of May, the four Combat Lightning aircraft began twenty-four-hour coverage orbiting in the Gulf of Tonkin. Three additional KC-135As joined the Combat Lightning fleet in the first part of October. In addition to carrying regular radio relay gear, these aircraft were outfitted with secure voice communications equipment.[24]

The long overwater flights required for the Big Eye missions were strenuous on the crews, especially for those flying the low-level Alpha orbits. That type of flying was much more demanding of the pilots, who could no longer rely on the automatic pilot to do most of the work. The EC-121D was not a comfortable aircraft either. Its air conditioning system had been designed for a commercial aircraft flying at medium altitude, not for a low-flying military aircraft. On hot Asian days, "the air conditioning was practically useless, and the heat generated by electronic equipment sent temperatures soaring." George F. Schreader, who flew twenty-eight College Eye missions in various EC-121s, "considered it 'an awful plane to fly.'" The Connies, he explained "were loud as hell inside, were uncomfortable to sit in no matter where you tried to sit or were assigned to sit, and the smell of exhaust fumes and hot oil and fluids that coated every visible mechanical part clung to your clothes and permeated your skin like a coating of aerosol lacquer."[25]

One of the unforeseen duties assigned to the Big Eye Task Force after it began operations was to issue border warnings to friendly aircraft about to trespass over the People's Republic of China. The border warning was unpopular among fighter pilots, who readily admitted that they might become disoriented in a dogfight and stray over China. Some aviators went so far as to shut down their IFF gear when flying near the border.[26]

On May 8, 1966, an incident occurred that revealed major shortcomings in both Hammock and Big Eye's coverage. On that day, four Air Force EB-66s on a SIGINT mission escorted by four F-4Cs strayed into Chinese air space. The SIGINT mission at Da Nang sent seven messages to the control and reporting post at Monkey Mountain warning of an impending border crossing, which in turn relayed the messages to the ARCP mission aircraft, a Navy Big Look EC-121M that was supposedly flying in the Gulf of Tonkin. But the Navy's Big Look mission had been scrubbed, possibly for lack of fighter support, "so the warnings," according to NSA's history, "went nowhere."[27]

The Chinese scrambled four MiG-17s in response to the border intrusion, resulting in a dogfight in which one of the MiGs was shot down, crashing seventeen miles inside the Chinese border. The Chinese complained of a border violation and threatened to widen the war. Although the Air

Force claimed the Americans had never strayed over the border, the Chinese released pictures of the wreckage and of the auxiliary fuel tanks dropped by the F-4Cs.

"The problem," described in NSA's history, "was not the collection or interpretation of the SIGINT. Rather, it lay in getting the intelligence to the pilots where it would be effective. During this incident, the cumbersome, uncoordinated nature of the warning system, plus the last-minute absence of a critical communications relay platform, had precluded any chance of warning the Air Force flight of its navigational error."[28]

This incident led to a full-scale investigation of command-and-control procedures in Southeast Asia. The "Pearl Harbor question" kept coming up: why, if SIGINT was available, wasn't it used? To answer this question, the Joint Chiefs of Staff established a board of inquiry headed by Marine Corps Brig. Gen. Robert C. Owens Jr., assigned to the Joint Chiefs of Staff as director for Command Areas, Operations Directorate. The Owens board discovered that the Big Eye Alpha mission had not been flown that day because of equipment failure. Even if it had been flown, the board concluded, it would not have covered the area where the incursion had taken place, and they recommended establishing an additional Big Eye orbit over Laos. Although a test to determine what coverage could be obtained from an orbit maintained from a point just north of Vientiane, Laos, was conducted in June, the Big Eye force lacked the resources to cover all three orbits until reinforcements arrived in May 1967.[29]

The Owens board also made several recommendations dealing with the "clutter that was clogging the warning system." One of these undoubtedly led to the deployment of the Silver Dawn aircraft mentioned earlier. Other recommendations led to a reorganization of the command-and-control structure that brought all ACRP flights, including Big Look, under the control of the Seventh Air Force Tactical Control Center to ensure that there would always be a Big Eye in orbit when strike missions were flown.[30]

CHAPTER 12
BIG SAFARI, COLLEGE EYE, AND COMBAT APPLE

On November 21, 1966, the Big Eye Task Force received a message from the 552nd Airborne Early Warning and Control Wing that an EC-121D, temporarily configured with special test instrumentation, was on its way to the task force for feasibility testing. The aircraft, identified by the tail number 57-143184, had been modified by Detachment 2 of the Big Safari project within the Air Force Material Command and configured with a top-secret device for identifying hostile aircraft via their onboard IFF system. The objective of the testing, which was to be conducted under the code name Quick Look, was to determine how well this device worked.[1]

Big Safari was a highly secret program established in the early 1950s for the acquisition of special mission aircraft. The program was set up to avoid the labyrinth of bureaucratic regulations regarding procurement in order to provide quick-reaction modifications, system integration, and technical field support for new systems installed on special reconnaissance aircraft. In the early part of 1966, Headquarters U.S. Air Force issued a Southeast Asia Operational Requirement (SEAOR) to the Big Safari program to develop an airborne platform that could determine and locate North Vietnamese surface-to-air missile sites. The project was named Sea Trap and assigned to the liaison office run by Detachment 2 at the Ling-Temco-Vought's Electrosystems Division facility in Greenville, Texas.[2]

A Navy EC-121P, which had previously been modified by the Navy to provide SIGINT capabilities, was selected for the Sea Trap program. To configure the aircraft to Air Force standards, the Sacramento Air Logistics Center procured and installed Wright R-3350 engines. A new Air Force tail number was painted on the aircraft, which was flown to Greenville.

As soon as the aircraft arrived, LTV Electrosystems began to replace the ventrally mounted APS-20's search radar antenna with a high-gain ELINT antenna and install the associated Brigand system electronics to allow the crew to determine accurately the location of enemy early warning radars (it did not work well on Fan Song radars, which did not stay on the air long enough). In May, Sea Trap's mission was augmented to providing warning of hostile aircraft in addition to locating SAM sites. To accomplish this task, LTV Electrosystems added the APX-49 IFF system that was already in use in the Big Eye EC-121Ds as well as an experimental QRC-248 system that was designed to identify enemy aircraft by interrogating the SRO-2 IFF transponder installed in all Soviet fighter aircraft. The QRC-248 was developed in response to a problem with EC-121 Warning Star during the Cuban Missile Crisis when it was discovered that the search radar carried by the EC-121 early warning aircraft could not detect low-flying MiGs because of ground clutter. A prototype installed in the Air Defense Command's Key West radar station successfully interrogated the Russian-built IFF installed on a Cuban MiG at 200 nautical miles.[3]

By the time Sea Trap arrived in Southeast Asia for testing, the project had been renamed Rivet Top. Tests of the QRC-248 were subsequently conducted during twelve missions flown from December 25, 1966, through January 14, 1967. During these missions it was found that with the QRC-248, the EC-121D could locate and identify North Vietnamese MiGs at a range of 175 miles. To protect the secrecy of QRC-248's effectiveness, NSA prohibited operators using the device from actively interrogating each radar return for fear that this would tip off the North Vietnamese GCI controllers. Instead, the USAFSS operators only passively read the IFF interrogations initiated by the North Vietnamese GCIs. Only bearing information was available in this mode, although in some cases this could be correlated with returns from the APS-20 search radar. The success of the tests, according to the report submitted by the Quick Look test team, showed that full implementation and use of the QRC-248 allowed Big Eye aircraft to interrogate electronically and display signals emanating from the Soviet's SRO-2 IFF, thereby providing positive identification of the North Vietnamese MiGs, all of which were equipped with the device.[4]

After the QRC-248 tests were finished, Sea Trap in all likelihood (the record is not clear) returned to Greenville for further modifications that were urgently requested by the Big Eye Task Force. These modifications involved improving the height-finding radar, installation of secure communications, improvements to the plane's air conditioning system, adding an enemy IFF readout capability, providing an improved navigation system, adding a digital processing system, and making improvements to the air-to-ground surveillance radar. Gen. Hunter Harris Jr., commander of the Pacific Air Force, wanted these changes made as rapidly as possible, writing in December that "BIG EYE as an element of Combat Lightning providing data to the TACC (North Sector), should receive, all requested modifications at earliest possible date. Accordingly[,] as equipment becomes available, it should be installed as rapidly as possible, rather than delay its incorporation until a more extensive modification program can be established."

Based on the future equipment listing for this aircraft, it appears that along with the modifications listed above, LTV's engineers added four SIGINT positions to intercept communications between MiG pilots and their GCI station, and three KY8 secure radios they could use individually or simultaneously to inform the Tactical Air Control Center of real-time classified intelligence. LTV quickly completed the modifications and delivered the aircraft for operational test and evaluation in March. The aircraft, now known as "Rivet Top" but designated as an EC-121K, accommodated a crew of thirty: two pilots; one navigator, two equipment maintenance personnel, and twenty-five operators (seventeen to support the electronic consoles and eight relief operators).[5]

From Greenville, Rivet Top was flown to the Air Warfare Center at MacDill Air Force Base, Florida, for ninety days of operational testing by Detachment 2. Hank Maifeld, a Voice Intercept Operator with a top-secret clearance, was assigned to the project. He remembers arriving at MacDill and being "directed to a big hangar on the flight line where a sign was taped to the door reading Detachment 2 Tactical Air Warfare Center Project RIVET TOP. The door was ajar as I walked in. The entire gigantic hangar was empty." He had arrived before the EC-121K. He asked the staff sergeant on duty "what the hell was going on here?" "Well," replied the sergeant, "we are it, we have got to get everything ordered that we are going to need for the next six months and have it ready to load on two C-141s that will be arriving in a couple of weeks." It wasn't long before Maifeld was on his way to Udorn, Royal Thai Air Force Base (RTAFB), Thailand, for a 180-day deployment as one of Rivet Top's EWOs.[6]

Maifeld and Rivet Top arrived at Udorn in July and joined the College Eye Task Force that had recently relocated there from their temporary base at Ubon. The first operational flight to employ the QCR-248 took place over the Gulf of Tonkin on July 21, 1967, under the flight name of Ethan Bravo. By the beginning of September, a second aircraft fitted with the QRC-248 arrived at Udorn. This was probably one of the College Eye EC-121Ds that had received a QRC-248 as part of group "A" modifications recommended by General Harris. By then, restrictions on the use of the QRC-248 had been lifted by the NSA and the JCS and missions began to be flown with QRC-248 in the active mode. When operated in the active mode, the QRC-248 radiated pulses that interrogated the enemy IFF set, which replied with a distinctive pulse train. The range and bearing of the MiG were then plotted using regular radar methods. The active mode provided strike commanders and the Tactical Air Control Center at Monkey Mountain with a much more accurate picture of MiG operations; it showed where their standard orbits were and gave a more precise count of enemy aircraft.[7]

The QRC-248 was originally viewed as an augmentation of the search radar system, the APS-9S. Its ability to effectively detect enemy aircraft, however, was even greater than anticipated, as described by the College Eye Task Force commander, who told Captain Carl Reddel, who was collecting historical data on College Eye, that "With the advent of the QRC-248 we were able to detect aircraft which we had not previously seen. It was somewhat frightening for us to realize that in the past there had been many aircraft that we had not seen.... In fact it so far increased the ability of COLLEGE EYE to detect enemy aircraft, that we later recommended to 7th Air Force, and received approval to discontinue the low altitude radar platform in favor of two, and later three, sorties, making almost exclusive use of the enemy IFF."[8]

In mid-October the College Eye Task Force along with Rivet Top moved to Korat Royal Thai Air Force Base. Rivet Top's assignment was to provide strike force commanders and flight commanders of forces penetrating the North Vietnam air defense system with real-time intelligence concerning MiG activity and the location of SAM sites. Its missions were coordinated with a scheduled air strike so that it would be on station operating at a reasonable stand-off distance from the actual SAM-defended areas in advance of the strike's arrival and staying until the air strikes egressed. Most missions covered two strikes in a day, requiring a turnaround at Da Nang between strikes. The crews flying these fourteen-hour missions were typically on their orbiting station for four hours. The rest of the time was spent getting to and from their orbiting point. In addition to advising the strike force commander or a flight commander as to the location of MiGs within five miles of their aircraft, the equipment on Rivet Top provided information on the location of an active SAM site in azimuth and range three to seven miles from their location. If Rivet Top maintained a lock on the Fan Song radar for five minutes, it might be able to pinpoint the site's location within a half-mile radius.[9]

Difficulties in obtaining an accurate location of the Fan Song radars associated with each SAM site due to the maze of signals emanating from North Vietnam mitigated Rivet Top's ability to perform this mission successfully. But its ability to alert Air Force planes to the presence and position of MiGs was an unmitigated success that was recognized by the commanders of Seventh Air Force and the Pacific Air Force. "Of the twenty MiG kills registered by the Air Force from August 1967," according to an NSA SIGINT history, "thirteen were attributed to Rivet Top." The aircraft had originally been dispatched for a 120-day test, but the Air Force chief of staff ordered it to remain in the area until another suitable replacement could be found. The aircraft stayed until September 1968, five months after President Johnson halted bombing north of the 19th parallel.[10]

Rivet Top was intended to function as an extension of the Tactical Air Control Center at Monkey Mountain. Although it carried COMINT operator stations and an enemy IFF display based on the QRC-248 as well as a display panel for the Fan Song radar, some of the functions were poorly done. Airborne fighter control was hampered by the lack of automated equipment to compute and display the multiple radar/IFF tracks being monitored. Rivet Top's operators were unable to correlate intercepted conversations with specific flights of MiGs and were unable to determine which U.S. aircraft might be under attack.

The need to pass real-time threat warning by call sign directly to the threatened aircraft over discrete control channels, giving hostile positions in relative range and bearing, was achieved through the addition of the GPA-122 IFF/SIF* coder/decoder and improved radio sets. The GPA-122 was a microelectronic IFF subsystem that increased the usefulness of the EC-121D's IFF system. It allowed each weapons controller on board the aircraft to passively track six discrete SIF codes, selectively identify any one of the six in real time, and actively read out the discrete transponder code of any aircraft as fast as he could "gate" the return. The first EC-121D modified for the GPA-122 arrived at Korat on January 29, 1968. By June 15, 1968, all of College Eye's EC-121s had been equipped with the IFF/SIF interrogator.[11]

College Eye's task force had been enhanced by the arrival of six EC-121s that had been modified with the communication intercept equipment installed on Rivet Top, which in combination with the QRC-248, greatly increased its capability to monitor the North Vietnamese MiGs. The modification, given the code name Rivet Gym, consisted of four voice communication intercept stations manned by Vietnamese-speaking intelligence specialists assigned to the USAFSS who monitored all communications between the MiGs and their controllers. The first test flight of a Rivet Gym–equipped aircraft, designated as the EC-121M, was made over the Gulf of Tonkin on May 10, 1968. These six Rivet Gym–modified aircraft, according to Carl Reddel, "provided the most important operational anti-MiG capability in Southeast Asia in mid-1968, a capability shared with Rivet Top."[12]

After President Johnson halted bombing above the 19th parallel on March 31, 1968, College Eye missions were redirected to support fighter-bomber missions over southern North Vietnam. When the bombing throughout North Vietnam ended on November 1, 1968, College Eye's EC-121s changed their orbits from the Gulf of Tonkin to orbits over Laos,

---

* Selective Identification Feature

and their missions shifted from tracking and warning of hostile aircraft to tracking friendly aircraft to prevent border violations. The task force was subsequently withdrawn from Southeast Asia in 1970, when the last of its planes left Korat for Itazuke Air Base, Japan, on June 29, 1970.[13]

The College Eye EC-121Ds were replaced by four EC-121Ts flying out of Korat in missions orbiting over Laos, where they maintained surveillance over air operations in the northern part of that country. The EC-121T was an improved version of the EC-121D that retained all of their capabilities plus additional ones derived from thirty-three enhancements that included

- IFF/SIF beacon tracking/decode through a real-time on-board computer;
- computer "rate-added" tracking of manually initiated hostile directions;
- addition of symbology to display systems;
- computer-assisted intercept control programs;
- computer-formatted air-to-air control datalink message transmission;
- software flexibility to tailor tactics to mission type and geography;
- capability for in-flight reprogramming to adjust to dynamic tactical situations;
- redundant digital data downlink (beyond-line-of-sight and relayed-line-of-sight media);
- secure, high-power, beyond-line-of-sight voice mode;
- new navigation systems and computer interface to increase radar stabilization accuracy in ground reference.[14]

On September 11, 1967, a new SIGINT aircraft, the RC-135M, identified as Rivet Card, began flying ACRP missions in the Gulf of Tonkin to conduct intelligence for the Seventh Air Force and provide early warning to strikes in North Vietnam. The RC-135M was one of six Military Airlift Command C-135 Stratolifter cargo planes converted for peripheral reconnaissance missions. The four-engine, jet-powered RC-135M was an upgraded Boeing 707 airframe with more powerful fan-jet engines. It was faster, could fly higher, and remain on station longer than the much older EC-121s that continued be the main workhorse for College Eye. Rivet Card was an improved version of SAC's RC-135D with an advanced SIGINT suite with thin-line transistorized receivers, small digital displays, interference cancellation equipment, and four-track narrow-band recorders.[15]

The RC-135Ms, which were affectionately named Hognoses due to their distinctive nose radomes, were assigned to SAC's 82nd Strategic Reconnaissance Squadron (SRS) operating out of the Kadena Air Base on Okinawa. Linguists and SIGINT operators were provided by the 6990 Security Squadron established at Kadena specifically to support these aircraft. The primary purpose of the missions flown by the 82nd SRS—called Combat Apple—was to provide MiG-warning and SAM-location communications while gathering additional SIGINT, ELINT, and COMINT intelligence. They were conducted, for the most part, on orbits over the Gulf of Tonkin and typically averaged eighteen hours in the air (three hours ingress, twelve hours orbiting, three hours egress). All Combat Apple missions were scheduled for refueling somewhere off their orbital paths and out of harm's way. It took about a half-hour for the KC-135 aerial tankers to deliver 80,000 pounds of fuel. Pre- and postflight briefings added another two to four hours to the total time the crew spent on the mission. A typical Combat Apple mission was manned by half a dozen SAC personnel (pilot, copilot, two navigators, two EWOs), and anywhere from twenty-five to thirty USAFSS personnel that included the air mission supervisor, tactical air control operators, multichannel operators, the GDRS† operator, the communicator, cryptologists, signal intercept ops, and maintenance personnel. Intelligence relating to gun movements and AAA activity was also a high priority, according to George F. Schreader, a linguist specialist who flew on numerous Combat Apple missions. "We intercepted a lot of communications relating to gun emplacements, and movements," he explained in his

---

† The General Directorate Rear Services operator who listened to the signals generated by Group 559, the NVA's unit responsible for moving troops, weaponry, and material from North Vietnam to South Vietnam.

book *Hognose Silent Warrior*, "and passed along immeasurable amounts of real-time information downlinked to Seventh Air Force."[16]

To counter the MiG threat, the electronic warfare compartment in the back of the Combat Apple RC-135M came equipped with two Tac-Op positions for Voice Intercept Operators: the controller, an advanced linguist, in the Tac-Op 7 position, and his subordinate in the Tac-Op 8 position. Both of these positions were dedicated solely to listening in on all MiG and North Vietnamese ground controller communications to discern what actions were being directed by ground controllers to the MiG pilots. Other information gathered simultaneously via other SIGINT and ELINT systems was all passed along in real time to pinpoint the MiGs' origination point and operating locations.[17]

This method of tracking the MiGs was effective because of the GCI doctrine followed by the North Vietnamese pilots. The Soviet Union had supplied the North Vietnamese with radars and other related equipment for a system of radar control for ground-controlled interception. Soviet doctrine called for strict radar control that guided interceptors to a point behind the target, where they could attack from the most advantageous position. The North Vietnamese pilots, who were trained to follow this doctrine, were almost totally under the management of a ground controller who dictated all their actions.[18]

The North Vietnamese GCI system was directed by a senior controller stationed at the headquarters of the Vietnam People's Air Force and Air Defense headquarters located at Bac Mai military airport in Hanoi. He managed the principal GCI controllers at each of the airfields used by the MiGs. The three major bases established by the beginning of Operation Linebacker I were: Phuc Yen, just north of Hanoi, home base late in the war for the 921st Fighter Regiment's MiG-21s and Mig-17s; Gia Lam airfield, the international airfield in Hanoi, where MiG-21 fighters were also stationed; and Kep Air Base, thirty-five miles northeast of Hanoi, home base for the 923rd Fighter Regiment, which consisted mostly of MiG-17s.[19]

Combat Apple and its ability to listen in to the communications used to control the North Vietnamese MiGs became an important element of Project Teaball (see chapter 16) that provided real-time data necessary to defeat the Vietnam People's Air Force.

CHAPTER 13

COMBAT MARTIN AND AN AGM-45 FRIENDLY FIRE INCIDENT

In the second half of November 1967, Seventh Air Force losses to North Vietnamese MiGs soared when seventeen of its aircraft were shot down. "The [MiG's] controllers," wrote Marshall Michel, "now were able to control two pairs of MiG-21s instead of just one, and the MiG-21s began to coordinate their attacks with the MiG-17s. Additionally, the MiGs now were willing to make multiple attacks instead of single hit-and-run attacks." The losses due to the surge in MiG activity and the improvement in the tactics used by their controllers concerned Air Force leaders. One solution considered by the staff at the headquarters of the Pacific Air Forces was to disrupt the North Vietnamese GCI using communication jamming.[1]

During the past year the EB-66s had been equipped with ALQ-59 communication jammers, but they had found that communication jamming interfered with friendly as well as hostile aircraft. Despite these problems, the EB-66s, along with Navy EAK-3Bs equipped with ALQ-92 communication jammers, occasionally made attempts to disrupt communications between the MiGs and their North Vietnamese GCI controllers. To avoid drowning out friendly messages, electronic warfare officers waited for a "start jamming" code word issued by the strike leader as he neared the target area and another code word to cease jamming when the strike had exited the target area. Both code words were transmitted over a prescribed UHF channel monitored by the jamming aircraft.[2]

The ALQ-59 was a powerful state-of-the-art communication jammer that could jam up to six VHF channels simultaneously. Jamming a radar took relatively low power, since it was a matter of interfering with the faint return echo from a radar pulse. In contrast, jamming communications meant disrupting a direct link from a transmitter to a receiver, and so it took substantially more power to drown out a voice radio broadcast with raw noise. Instead of attempting to disrupt the communication channel with noise, the ALQ-59 intercepted the GCI's voice message and retransmitted it after a short time delay so that the rebroadcast transmission was superimposed on the original. This garbled the sound, making it extremely difficult to understand what was being broadcast. Who came up with the idea of putting the ALQ-59 into an F-105F is not known. But it appears that a version of the ALQ-59, designated the QRC-128, was redesigned to fit into an F-105F. It was too bulky to be placed in a pod, so it was placed in the rear cockpit after the ejection seat was removed to make room for the equipment. Because the QRC-128 responded automatically to a predetermined radio frequency, an EWO was not needed. The "new backseater" was dubbed "Colonel Computer" by flight crews. F-105s fitted with this equipment were referred to as "Combat Martins" and were readily identified by the large square blade antenna installed just behind the cockpit.[3]

When Col. Hyman "Marty" Selmanovitz, assistant to the Commander Electronic Warfare, Seventh Air Force, learned about the proposal to equip F-105Fs with the QRC-128, he became concerned about NSA's response. Selmanovitz must have known about Combat Apple's ability to listen in on North Vietnam's communications since it was the source of a great deal of secret information being provided to Seventh Air Force controllers. An officer of his rank and position within the electronic warfare community would have been briefed on Combat Apple and the need to maintain secrecy. Selmanovitz thought that Combat Martin was a good idea because the MiGs could not operate without it and would be completely defeated. But he wanted to make sure that the Seventh Air Force was not making any waves with regard to the NSA's wishes. "You have to go to the National Security Agency to get their clearance," he told his immediate superiors. "Otherwise, they will not let you use it." Selmanovitz never discussed their response, which remains unknown.[4]

Modification of the five F-105Fs selected for conversion to the Combat Martin configuration began at the Sacramento Air Logistics Center at McClellan Air Force Base in January 1968. All five were assigned to the 388th TFW, at Korat RTAFB, Thailand. The last of the five, bearing tail number 62-4444, arrived on March 25, 1968, just days before bombing of North Vietnam north of the 19th parallel was halted on April 1. Combat Martin aircraft only performed one mission in their role as a communications jammer before the NSA, which was in charge of U.S. strategic intelligence, ordered the Seventh Air Force to cease and desist immediately, since the NSA believed that the intelligence obtained by monitoring the channels outweighed the benefits of jamming them.[5]

On April 16, 1972, the air war escalated when B-52s were used to attack targets in the Hanoi/Haiphong area for the first time during Operation Freedom Porch Bravo. Seventeen of the big bombers struck fuel storage tanks at Haiphong in the predawn hours, setting fires and generating clouds of smoke that could later be seen on the bridge of a Navy carrier a hundred miles away. The strike was preceded by fifteen Navy A-6 Intruders, some armed with AGM-78 Standard ARMs, that struck SAM sites in the Haiphong area. It seems likely one or more of these aircraft launched at least two antiradiation missiles (although some claim that it was an Air Force aircraft) toward the Gulf of Tonkin, striking the guided missile destroyer *Worden* (DLG 18).[6]

The day before the strike was scheduled to take place, the *Worden* was assigned duty as the search and rescue ship for the north section of Yankee Station, a fixed point of the coast of Vietnam, officially Point Yankee, from which U.S. carriers operated. On the morning of the attack Lt. (jg) George Guy Thomas, who had just begun to stand his first watch on the ship, was just entering the starboard wing of the bridge when there was a bright flash and a loud explosion high up on the ship's mainmast. He had just taken another step toward the starboard bridge wing when "the overhead [ceiling to landlubbers], exploded with a very, very loud 'BANG.'" A second explosion, much louder, followed within seconds. At first Thomas believed that the *Worden* was under attack from North Vietnamese PT boats, but the second explosion was much more powerful than anything a PT boat could have fired. What struck the *Worden* were two antiradiation missiles that had gone astray. What aircraft fired the weapon and which antiradiation missile hit the *Worden* has never been accurately determined, although Thomas claims it was a Shrike fired by an Air Force plane.[7]

Thomas' firsthand account of what it was like to be on a vessel struck by one of these missiles is riveting:

> There were pieces of red-orange, extremely hot shrapnel flying all over the place, ricocheting off everything: the floor, the radar scope housing, the chart table, the walls, the lee helm with its engine-order handles, the armor plate windows. The only place where the shrapnel was not thick seemed to be right where I was. It was swarming all around me . . . the shrapnel had hit every other person on the bridge, seriously wounding almost all of them. The explosion knocked all of us down, and at the time, I thought I was the only one not seriously hit. I did take a piece of shrapnel through the thumb, but I was uncut other than that.[8]

The blast, which occurred about forty feet above the port bridge wing, caused 850 holes one-quarter to one-half an inch in diameter. It took out two of the ship's radars and all the communication and ECM antennas. It damaged wire runs and support equipment and injured anyone within a fifty- to seventy-five-yard radius of blast. One sailor was killed and eight others wounded, six of whom were airlifted to the amphibious assault *Tripoli* (LHA 7) for treatment. Fragments of the two missiles, which were scattered all over the ship's top side, were collected by the crew. At least one of them was marked "made in the USA," clearly indicating that the *Worden* had been hit by friendly fire.[9]

CHAPTER 14
AUTOMATING THE TACTICAL AIR CONTROL SYSTEM

COMBAT LIGHTNING, RED CROWN, AND THE NTDS

As previously noted, Hammock, the manual reporting system for disseminating SIGINT data collected by the 6924th Security Squadron on Monkey Mountain, was limited to a maximum of ten tracks, and its quality and timeliness was affected by the number of steps involved. As the tempo of the air war increased, Hammock was flooded with data, which limited its effectiveness. By then a project to develop an automated system to provide the rapid transfer of SIGINT data, called Ironhorse, was already in the works.[1]

The Ironhorse system was initiated in November 1964 when R8, the NSA office responsible for the development of processing and telecommunications for SIGINT, was given the job of developing equipment for visually displaying SIGINT information derived from the tracking of aircraft obtained from communications intercepted from the North Vietnamese air defense system. Data from a maximum number of inputs had to be reduced and fused into a common database and presented to battle commanders. To accomplish this the R8 team utilized two computers that took electronic signals from the field, used it to establish a database, analyzed the data, and presented it on a series of CRT consoles.

It took two years to assemble the necessary equipment and to write and debug the software. By September 1967 the system had been tested and was ready to be installed in a building on the Da Nang Air Base. When personnel from the 6924th Security Squadron and their equipment began arriving at the site, they discovered that the building designated to house the Ironhorse complex had been destroyed by a Viet Cong rocket attack in July. The dilemma of where to establish the Ironhorse complex was solved by modifying four H-1 vans converted to house twenty intercept positions and three more for the computers and communication equipment. The vans were airlifted to Da Nang in November and an engineering team made up from the USAFSS and NSA arrived to complete the startup operations. Testing began in mid-December, when Ironhorse sent its first data to the TACC-NS. Modifications continued to be made to the system as it was being put through its routines. An enhanced voice intercept position was installed to accommodate the growing use of voice communications by the North Vietnamese air surveillance system. Software modifications to the interface cleaned up the garbled tracks sent to the TACC-NS. When it became operational on March 27, 1968, Ironhorse removed one of the communication roadblocks between the USAFSS SIGINT collectors and the air controllers.[2]

At the heart of Ironhorse were the analysts assigned to the twenty intercept consoles installed in the vans. Each analyst, or "intercept operator," sitting in front of the console's CRT display was trained to recognize tactically important North Vietnamese tracking information indicated on the display and to forward it to the TACC-NS. Because all the tracking data were displayed, the analyst had to decide what was tactically important and to which air command element he had to forward the information. The system displayed up to 120 tracks, though, realistically, this number cluttered the screen beyond recognition, so considerably fewer were displayed. The analyst could select a track by typing in its number or else select it directly from the screen with a light pen. The tracks he selected then went into the Seek Dawn System (phase III of the Combat Lightning program) installed within TACC-NS on Monkey Mountain and passed to the tactical air control center at Tan Son Nhut, which was responsible for air defense from Pleiku south.[3]

Project Seek Dawn was a computerized system developed by the Air Force within Combat Lightning initiated in October 1966 to extend and automate the tactical air control system in Vietnam. It was a modified version of the Back Up Interception Control (BUIC) system used as an alternated reserve for the SAGE air defense computers that were operated by the North American Air Defense Command to defend the continental United States against Soviet bombers. The software modifications included in Seek Dawn added operational features and conversion of geographical data to fit Southeast Asia. It also accepted, processed, and displayed data from the Naval Tactical Data System (NTDS) and the Marine Tactical Data System, and it received information from USAF EC-121M Big Look and Navy EA-3B Deep Sea aircraft.[4]

By 1969, Combat Lightning had grown into a comprehensive data collection system that linked USAF, USN, USMC, and USA tactical data systems, forming a comprehensive air picture over North Vietnam. It provided a real-time exchange of tactical air operations information throughout the theater for the first time. While this system received a variety of inputs from intelligence collectors throughout Southeast Asia, classification problems and dissemination concerns by national intelligence agencies that "owned" the data meant that little information was actually passed to aircrews operating over North Vietnam. The USAF also made use of a U.S. Navy radar picket ship operating in the Gulf of Tonkin.

Red Crown was the code name and radio call sign for the Navy ship tasked with monitoring all air activity over North Vietnam and the Gulf of Tonkin. The ships assigned to this duty were equipped with the most advanced radars and communications equipment operated by highly trained personnel. In addition to its shipboard radar, these ships had naval SIGINT, national SIGINT, and an E-248 airborne radar feed linked into its Combat Information Center (CIC), giving it a complete picture of the air situation over Hanoi. Red Crown ships frequently helped Navy and Air Force fighters engage and destroy North Vietnamese MiGs over North Vietnam.[5]

After the *Maddox* incident in August 1964, Yankee Station—the operating area for the Seventh Fleet's carriers in the Gulf of Tonkin—moved closer to the strike targets in North Vietnam. The increased sortie rate and the need for inflight refueling made it difficult for the carrier force commander to obtain a total air picture over the Gulf of Tonkin and the targets in North Vietnam. This, plus the possibility of attack by North Vietnamese aircraft, made it essential to positively identify all aircraft in the area. To accomplish this, a Positive Identification Radar Advisory Zone (PIRAZ) was established using data from Red Crown's radar to identify and track all aircraft within range of the ship's radar as well as that provided by airborne surveillance radars on E-1Bs assigned to the carriers. The PIRAZ concept was introduced in the spring of 1966, and the guided-missile cruiser *Topeka* (CLG 8) was tasked to evaluate the concept in which one ship performed the following functions within a designated zone in the Gulf of Tonkin:

1. positive identification and tracking of all aircraft in the zone;
2. CAP control;
3. flight following;
4. SAR assistance.

The success of the *Topeka*'s mission led to the establishment of PIRAZ on June 15, 1966, inaugurated when the guided missile cruiser *Chicago* (CG 11) took station off the coast of Vietnam and began operations under the Red Crown call signal. In order to get complete air search radar coverage of the skies over North Vietnam, Red Crown was situated twenty-five miles from the mouth of the Red River.[6]

Chicago was one of the first ships in the Navy to receive a first production version of the Naval Tactical Data System. The NTDS was a computerized information system that used data from a variety of sensors on different ships via datalinks to produce a single unified map of the battlespace. This information was displayed within the ship's Combat Information Center and then relayed back to the other ships and their weapon's operators.[7]

The impetus for the creation of the NTDS began with the introduction of jet aircraft at the end of World War II, when officials in the Navy realized that the manual plotting techniques that had proved successful during that war would be unable to handle a massed attack of these high-speed aircraft. As early as October 1945, Adm. Ernest J. King, the wartime Chief of Naval

Operations, wrote a letter to the chiefs of the Bureaus of Ordnance and Aeronautics expressing his concerns regarding the problems associated with the large amount of data that radar was cable of providing and the need to provide a method of automating it:

> The display of information was slow, complicated and incomplete, rendering it difficult for the human mind to grasp the entire situation either rapidly or correctly and resulting in the inability to handle more than a few raids simultaneously. Weak communication prevented information from being properly collected or disseminated either internally aboard ships or externally between ships of a force. . . . A method of presenting radar information, instantaneously and in such a manner that the human mind . . . may receive and act upon the information in the most convenient form; [plus] instantaneous dissemination of information within the ship and force.[8]

In 1954, Lt. Cdr. Irvin McNally, then serving at the Navy Electronics Laboratory (NEL) as the laboratory's radar project engineer, became aware of the radar data-handling problem after he was temporarily assigned to the Office of Naval Research in charge of Project Lamplight, a study group of military and civilian technical personnel convened by the secretary of defense to examine various problems involved in the defense of North America. While serving as the Navy's project officer for Lamplight, McNally studied the Semi-Automatic Ground Environment air defense system (SAGE) that had been developed by MIT. For three months McNally listened to briefings on SAGE and the advanced developments in radar with little, if any, mention of data automation. Concerned about the lack of radar data automation and the absence of any discussion on the potential use of the SAGE concepts at sea, McNally discussed his unease with Nathan Rochester, the IBM engineer who was Lamplight study director. Rochester encouraged McNally to develop a concept paper on the ideas that they had discussed for an anti-air-warfare combat automation system for the Navy.

The fifteen-page typewritten concept paper that McNally prepared as a result of his discussions with Rochester became one of the final reports issued by Project Lamplight. The paper contained a detailed summary of the automated system McNally felt was needed to defend the fleet against the type of saturation air attacks the Navy had experienced in World War II. All combat ships, he believed, from guided missile frigates on up, should have a self-contained, automated radar data plotting and weapons assignment capability that could be shared within a task force so that all the ships in the task force could see the same battle picture. To share the tracking load, McNally envisioned a fully automatic radio data link tying all the task force computers together. Digital computers in each ship would simultaneously process one thousand target tracks and show whether they were friendly, hostile, or unknown. The system would also compute the relative threat of each hostile or unknown target, rank the threats, and recommend the appropriate weapon response. McNally also included the general performance requirements for two new radars that would be needed to interface with the new system. McNally named his proposal the Naval Tactical Data System.[9]

The NTDS in *Chicago*'s Combat Information Center (CIC) consisted of a series of modular consoles each of which was dedicated to a specific function: surface operations, air control, electronic warfare, sonar, underwater battery fire control, detection and tracking, weapons control, and display and decision making, each of which was connected to a central digital computer. The CIC was manned twenty-four hours a day by a crew of thirty-five to forty officers and men. The operator of each console interrogated the computer simply by placing a cursor over a target shown on his CRT display and pushing a button on the console, which instantaneously revealed the course, speed, altitude, range, and bearing on any target. As Capt. Garette Lockee, the former commanding officer of the NTDS-equipped *Wainright* (DLG 28), noted, "The NTDS [did] away with the old problems of grease pencil bookkeeping, vertical plot tracking, and display of enemy and friendly forces. It [gave] the commander a capability of understanding, instantly, the threats and various alternatives to tactical problems or situations."[10]

In order to ensure that unknown aircraft entering the PIRAZ zone could be identified, it was necessary for Red Crown keep track of, and account for, all friendly aircraft in the area and sorties against targets in North Vietnam. All flights and sorties planned by the Air Force, the Navy, and the Marine Corps had to be passed to Red Crown twelve to fourteen hours in advance. This information was recorded and displayed in a way that was immediately accessible to Combat Information Center operators, and the aircraft involved were to check in and identify themselves when entering the zone and check out when leaving.

The NTDS system on the *Chicago* and the other Navy NTDS ships that rotated for Red Crown duty performed a variety of functions. These were listed by Lockee and explained further by David Boslaugh in his history of NTDS in combat as follows:

> Air Control—of two kinds: advisory control and close control. Advisory control means the shipboard air controller sets an area or limits within which the aircraft can fly, after which the controller monitors the plane's location and advises if the pilot has strayed out of the boundaries. Advisory control was used for combat air patrol fighters, airborne tankers, and airborne early warning aircraft. Close control has the PIRAZ air controller sending specific speed, heading and altitude orders such as to an interceptor tasked with identifying a stray, or shooting down a hostile.
>
> Provide Assistance—including vectoring airborne tankers under close control to strike aircraft low on fuel, vectoring search and rescue aircraft to the position of downed airmen, relaying information from strike pilots back to their carriers, and providing numerous kinds of operational information to pilots such as weather reports, locations of other units, and directions back to their carriers.
>
> Hostile Aircraft Warnings—advising pilots of the locations of unidentified or hostile aircraft that might be a threat to friendly aircraft.
>
> Border Warnings—when friendly aircraft were approaching the border of Communist China or the Chinese island of Hainan.

Identify Hostiles and Destroy—with missiles or interceptors under close control, any hostile considered a direct threat to the PIRAZ ship or nearby friendly aircraft.

Barrier Combat Air Patrol—stationing and controlling combat air patrol interceptors.

Flight Following—including tracking and communicating with each strike force from a Yankee Station attack carrier; with one flight follower on the PIRAZ ship assigned to each strike force. It was the duty of the flight follower to positively identify each airplane in the attacking formation and maintain a track on it into the target and back out again to its carrier.

Prosecute or Assist in Search and Rescue Missions—to find and rescue downed pilots. The operators would monitor voice and identification-friend-or-foe (IFF) transmissions from the strike aircraft listening and looking for distress calls or signs. Any plane in distress was closely watched, and if it went down the PIRAZ operators would enter the last known location into NTDS so that rescue helicopters waiting at the raid exit point could be quickly vectored to the crash site. Most of the PIRAZ ships also carried search and rescue helicopters, and were credited with saving many pilots lives, or preventing their captivity.

Relay of Surface to Air Missile Warnings—of potential hostile surface missile activity to friendly aircraft when the PIRAZ ship's electronic countermeasures operators detect radio emanations from hostile surface to air missile systems.[11]

The tasks handed the ships assigned to Red Crown duty were formidable. During her four months of duty in 1967, the guided-missile cruiser *Long Beach* (CGN 9), for example, kept track of approximately 30,000 U.S. aircraft sorties. Although the *Long Beach* was armed with both RIM-8 Talos and RIM-2 Terrier missiles, and her air controllers identified and tracked hundreds of MiGs, rules of engagement emanating from the Joint Chiefs in Washington restricted their use for fear that they might destroy American planes or fall into a North Vietnamese city. In May 1968, Adm. Ulysses Simpson Grant Sharp Jr., the Commander-in-Chief of the United States Pacific Command, convinced the Joint Chiefs to allow *Long Beach* to engage the MiGs with her surface-to-air missiles.[12]

Long Beach's efforts were supported by the EA-6As of VMCJ-1, which began using their ALQ-55 communication disrupters to sever the communications between North Vietnamese ground controllers and any MiGs under their control in a tactic known as "MiG-baiting." The ALQ-55, operating in the 100–210 MHz band, was capable of distinguishing threats from enemy communication bands from those of civilian beacons and civilians. It identified them, evaluated them, and jammed them with a signal that was noncontinuous in order to avoid detection.[13]

As Tom Carter, an EA-6A ECMO recalled, "The North Vietnamese [MiG operational] procedure was iron-clad. They never did anything unless their GCI controller told them to do it. The MiGs would come out of Hanoi every day, fly south of Vinh, turn out towards the water, then do a 180 and fly back, then go back north to Hanoi. Somebody came up with the bright idea of 'what if you turned on an ALQ-55 jammer at the point that they turn and head back inbound?' and they would never hear the missiles coming."[14]

An EA-6A from VMCJ-1 successfully used this tactic on May 23, 1968, when the *Long Beach* launched two ram-jet powered RIM-8 Talos long-range missiles at a pair of enemy aircraft sixty-five miles away. Two minutes and fifty seconds after the first missile left the launching rails, according to the CINCPAC summary, "all of the Talos directors recorded on their [radar] scopes the sudden blooming and the expanding clutter of debris associated with a direct hit. At the same instant, electronics signals from the MiG abruptly stopped. Two minutes later, the second Talos detonated on the falling wreckage." This, according to several sources, was the first time that a ship-launched guided missile had brought down an aircraft in combat. In September, *Long Beach* repeated the feat by hitting a second MiG at sixty-one miles.[15]

Communication jamming was used again on July 9 when an EKA-3B variant of the F3D-2Q and an EA-6A used their ALQ-55s to disrupt an attempt by two MiGs to intercept a Navy RF-8G photoreconnaissance plane north of Vinh. One of the MiGs—according to an evaluation by the Air Force Electronic Warfare Center—intent on pursuing the RF-8G, failed to receive a warning of the photo plane's F-8 fighter escort, which shot it down. On the following day, a Navy F-4 destroyed a MiG-21 under similar circumstances.[16]

Communication jamming by an Air Force EB-66E and a Marine EA-6A was also used on September 22 to intercept two MiGs that were approaching American aircraft conducting strikes in the panhandle of North Vietnam. The two electronic warfare aircraft used their VHF jammers at the request of the *Long Beach* on the Red Crown station. The guided missile cruiser recalled the friendly fighter bombers and launched a Talos missile that downed one of the MiGs. An after-action analysis indicated that the communications jamming might have prevented the MiG from receiving a radioed warning of the missile firing. According to Nalty, "This was the last apparent success of this type of electronic countermeasure until 10 May 1972, when communications jamming figured in the destruction of one MiG-21 and seven MiG-17s."[17]

In November and December 1969, the increase in enemy activity and aircraft losses precipitated a re-evaluation in the Air Force of the MiG threat, which until recently had been regarded as only sporadic. In late December 1971 it became apparent that additional warning and surveillance aircraft were required, so CINCPAC approved the deployment of a College Eye Task Force of four EC-121T aircraft from Kwang Ju Air Base, Korea, to Udorn in Thailand. With the air war now focused in the complex of U.S. air bases in Thailand, Ironhorse was shifted to the 7th Radio Research Field Station at Ramasun, Thailand, and was renamed Ironhorse II. In April 1971, the Da Nang mission closed down and the personnel from the 6924th Security Squadron deployed to Thailand. From then on, Ironhorse became just another input into Combat Lightning.[18]

CHAPTER 15
PROUD DEEP ALPHA, RGM-8H TALOS, AND OPERATION LINEBACKER I

When Rolling Thunder ended in November 1968 and all bombing above the 19th parallel ceased, the North Vietnamese, relieved of the constant necessity of rebuilding bridges and repairing road and rail cuts and the constant hazard of armed reconnaissance overhead, began funneling men and supplies to the South. SAM sites were moved down into Route Package 1, the Demilitarized Zone (DMZ), and Laos. Interdiction efforts against enemy supply lines in the Laotian panhandle did not stop the infiltration that continued under cover of night, weather, and jungle canopy as the North Vietnamese constructed new roads, trails, bypasses, and truck parks to move men and supplies south. Aerial photography uncovered many traces of these roads, along with new pipeline sections, concentrations of heavy artillery guns, and transporters. ELINT traced the progress of Spoon Rest SA-2 acquisition radars and Fan Song tracking radars through the North Vietnamese passes into Laos and the DMZ.[1]

During the three-and-a-half year bombing halt, the North Vietnamese improved and greatly expanded their air defense system, which now fielded two hundred Soviet-supplied surveillance and fire-control radars. There were close to thirty SA-2 SAM sites strategically distributed throughout the country, with the majority located around Hanoi. The number of fighter aircraft in the Vietnamese People's Air Force had also increased substantially and now included 140 MiG-17s, 31 MiG-19s, and 94 MiG-21s. It was considered by many to be one of the finest air defense systems in the world.[2]

Toward the end of 1971, Vietnamese air defense activity involving both SAM firings and MiG attacks had begun to deny B-52s, gunships, and forward air controllers' access to the infiltration areas used to move troops and

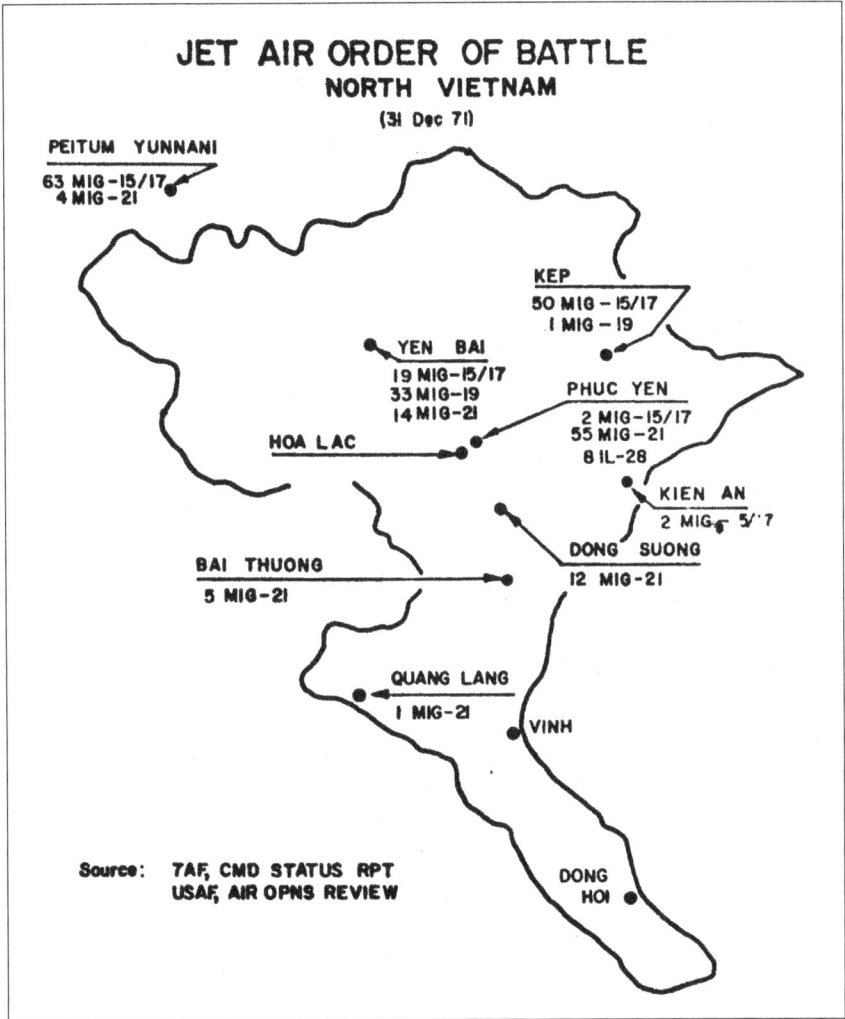

North Vietnamese jet aircraft order of battle *United States Air Force*

equipment from North Vietnam into the South, making interdiction efforts over the Ho Chi Minh Trail increasingly difficult. The North Vietnamese had also begun using their GCI radars to guide MiGs on intercepts of U.S. aircraft and had begun to link their GCI radars with SAM sites in a way that permitted each battery to leave its radars on only a very short time, reducing their vulnerability to U.S. antiradiation missiles. Using this tactic, the North Vietnamese left their Fan Song radars off until the instant of firing.[3]

This new enemy approach to targeting U.S. aircraft threatened Seventh Air Force's aerial reconnaissance mission in Route Package 1 and air operations in the border areas of South Vietnam and Laos. It was not until January 1972 that Gen. Creighton Abrams, commander of the U.S. Military Assistance Command, Vietnam, finally received authority to attack the GCI radars. But in the interim, enemy defensive efforts became a constant hindrance to the interdiction campaign.

By December the increase in enemy activity and aircraft losses led to an Air Force re-evaluation of the MiG threat, which had heretofore been only regarded as sporadic. After a failed MiG attempt against a B-52 on November 20, the Joint Chiefs directed CINCPAC to proceed with Operation Proud Deep, with the objective of

1. Destruction of MiGs on the ground and attainment of a level of damage of Bai Thuong and Quang Lang sufficient to inhibit further use of these bases by the NVAF for MiG operations against B-52s and gunships in Laos.

2. Destruction of logistical and other military targets in NVN south of 18 degrees north, with priority on targets of greatest importance to the enemy as storage and input elements for his logistics system in Laos.[4]

The Proud Deep operation was held up by bad weather. In the meantime, on December 20, North Vietnamese MiGs shot down three USAF F-4s in the same day, eliciting a plea from General Abrams to execute Proud Deep. Adm. Thomas H. Moorer, chairman of the Joint Chiefs, sent an execute authority the next day. It widened the target area to all valid military targets in North Vietnam south of 20 degrees north and restricted the duration of the strike to seventy-two hours rather than the five days provided in the plan, although weather conditions were so bad that the strike had to be extended. The operation, rechristened Proud Deep Alpha, was conducted between December 26 and 30 and involved 1,025 sorties of USAF and Navy strike aircraft. It was the biggest attack and deepest penetration of North Vietnam since the November 1968 bombing halt.

The operation, which was conducted in order to destroy MiGs on the ground and prevent the use of selected enemy airfields, did not achieve the desired results. The destruction of fuel dumps, airfields, transportation points, and military complexes was hampered by bad weather that limited bombing accuracy due to the reliance on radar. During Proud Deep Alpha, EB-66C/Es of Korat's 42nd Tactical Electronic Warfare Squadron flew a total of thirty-three sorties, jamming all enemy frequency bands to neutralize the defensive networks. Nevertheless, the North Vietnamese SAMs managed to bring down three U.S. aircraft.[5]

In late December Wild Weasels, using an AGM-78 Standard ARM, destroyed a Soviet P-35 (Bar Lock) radar site near Barthelemy Pass. The Russian early warning radar was part of the North Vietnamese air defense system guarding the road network in Route Package 4 leading to the pass, which was the supply gateway to the Pathet Lao forces in Laos. The highly mobile trailer-mounted Bar Lock had two large parabolic antennas that rotated at seven revolutions per minute. The radar operated in the E- and F-bands with one megawatt of output and was capable of alternating the pulse rate and frequency to counter attempts to jam it. "Covert personnel on the ground examined the site immediately after it was destroyed and discovered it had been manned by Russian operators."[6]

An additional Bar Lock radar was located near the Mu Gia Pass, another important gateway to a well-camouflaged road network in Laos that the North Vietnamese Army used to supply insurgents in South Vietnam. It was attacked on February 4, 1972, using a new weapon: the RGM-8H Talos. The RGM-8H was an antiradiation version of the Navy's long-range surface-to-air missile developed in the 1950s to defend the Navy's ships from attacks from bombers and guided missiles. The first version of Talos, the

RIM-8A, entered service on board the guided missile light cruiser *Galveston* (CLG 3) in March 1959. Work to develop a version of Talos that passively homed on a radar installation began in 1965.[7]

In the spring of 1971, the Talos armed guided-missile cruiser *Oklahoma City* (CLG 5) took on board one of the new, highly classified RGM-8H antiradiation missiles. The RGM-8H was similar to the RIM-8G but could be fitted with various seekers to cover different frequencies and employed some electronic counter-countermeasures. As Philip Hays, who was serving in the *Oklahoma City* at the time, explains, "We conducted a test firing off Okinawa in July 1971 to train the crew with the ARM missiles. Then we waited for an opportunity to use them."[8]

The *Oklahoma City* steamed into the Gulf of Tonkin in January 1972, as Hays put it, "to do some 'radar hunting.'" Hays was on watch as the Weapons Control Officer on the night of February 4, when the electronic warfare specialists in CIC detected emissions of a Bar Lock radar seventy-four miles away in the Mu Gia Pass near the border between Vietnam and Laos. Hays and some of his buddies had cobbled together a special piece of electronic equipment that accurately determined the pulse-repetition frequency of the radar. This allowed them to keep track the of the bearing and range to the Bar Lock, which was notorious for changing frequencies during operation. Hays remembers continuously passing parameters to the fire-control party as the RGM-8H was being prepared for launch.[9]

As soon as the missile was launched, Hays ran up to the fire-control station above Weapons Control where the SKQ-1 telemetry set was located. When Hays arrived, telemetry data from the RGM-8H indicated that the missile appeared to be functioning correctly. This was quite a relief, since two RGM-8Hs previously fired by the *Chicago* at a North Vietnamese radar site had failed after launching. Hays saw the missile acquire the Bar Lock radar, and he followed the telemetry data showing the wing movements generating the lateral accelerations as the missile acquired and began homing on the target. The signal was lost just before the predicted intercept so fuzing was not visible. The electronics warfare people in CIC told Hays the radar signal had disappeared about the same time the missile arrived. Hays did not know for sure if they had destroyed the target until the next day when his boss, Cdr. Merlin L.

Foreman, showed him post-strike aerial reconnaissance photographs showing what remained of the trailer lying on its side at the edge of a thirty-foot-diameter crater. This appears to have been the one and only time the RGM-8H destroyed a North Vietnamese radar.

On January 20, 1972, after reflecting on the current assessment of North Vietnamese military intentions, General Abrams predicted that "The enemy will use MIGs, SAMs, AAA to complicate our operations. We expect his recently intensified MIG activity to continue and to be directed against our air operations. He is expected to position SAMs and AAA just north of the DMZ, and has already moved these weapons into the Laotian panhandle to counter our operations in these areas." General Abrams' prediction was a harbinger of things to come.[10]

Map showing details of first AGM-8H combat firing *Courtesy Phil Hays*

On March 23, 1972, peace negotiations conducted in Paris with the North Vietnamese were terminated by the United States because of a lack of progress. Seven days later, on March 30, the North Vietnamese Army, having assembled a force of 140,000–150,000 men, along with six hundred tanks and armored vehicles and a substantial amount of ammunition and supplies, moved across the Demilitarized Zone (DMZ) dividing North and South Vietnam. In response to the invasion of South Vietnam, known as the Easter Offensive (the Nguyen Hue Campaign to the Vietnamese), President

Richard M. Nixon initiated Operation Freedom Train. Aircraft from the carriers on Yankee Station attacked Vietnamese supply concentrations south of the 18th parallel and SAM sites defending stockpiles just north of the DMZ. The carrier air strikes, which began on April 5 and continued through April 9, were conducted against vehicles, lines of communication (roads, waterways, bridges, railroad bridges, and railroad tracks), supply dumps, and missile sites. B-52s were brought into the act on April 9 to strike the Vinh railroad yard and Vinh POL (petroleum, oil, and lubricants) supply. It was the first use of B-52s in North Vietnam since October 28, 1968.[11]

As Air Force historian Melvin F. Porter observed, "Once the [North Vietnamese] invasion started and the enemy's intent became clear, the President, his National Security Council, and the JCS prepared for an all-out effort to suppress the North Vietnamese assault and to set the stage for meaningful peace negotiations." Although the United States defeated the invasion, the flow of personnel, supplies, and material did not diminish. To achieve the necessary military result, President Nixon concluded, bombing "would have to be brought to the enemy's heartland around the Hanoi-Haiphong area."[12]

On the weekend of April 15–16, seventeen B-52s, in a test of penetrating the SAM defenses, bombed the fuel storage tanks at Haiphong, setting fires that were visible from 110 miles away. Twenty F-4s, using a new tactic, dropped a chaff forming corridor six to eight miles wide and thirty to thirty-five miles long to protect the bombers, which in conjunction with ECM so degraded the North Vietnamese radar that aircraft flying in the chaff corridor were almost immune to SAMs and radar-guided anti-aircraft fire. The B-52s were also preceded by fifteen Navy A-6s that struck SAM sites around the target area. Although the North Vietnamese launched thirty-five SA-2s, none of the missiles found their targets. Shortly afterward, carrier aircraft joined Air Force fighter-bombers in battering a tank farm and a warehouse complex on the outskirts of Hanoi. When these attacks failed to slow the offensive, naval aircraft began mining the harbors on May 8, and two days later the administration extended the aerial interdiction campaign throughout all of North Vietnam. President Nixon, a big football fan, named it Operation Linebacker.[13]

Chaff Laying

As Marshall Michel pointed out, "Chaff corridors were the single most effective countermeasure to the SA-2 missile during Linebacker." To form a corridor five to six miles wide and eighty nautical miles long required eight to twelve F-4s in a fixed formation flying straight and level at fixed altitude of 20,000 feet while they dropped chaff from their pylon-mounted ALE-38 dispensers. These high-capacity dispensers provide instantaneous bloom for laying chaff corridors and, if necessary, could be automatically activated by the aircraft's threat warning system.

The chaff flight, however, was easily identifiable to North Vietnamese search and height-finding radar, since it was restricted (aerodynamically by the chaff pods) to approximately 480 knots and generally flew in a straight line toward its target. The chaff escort flight was then also easily identified through its faster airspeed and weaving pattern, designed to enable it to stay with the slower chaff flight. Since the chaff had to have time to disperse if it were to be effective, the enemy could count upon fifteen to twenty minutes before the strike force came through. By this time, the chaff itself had outlined a perfect corridor of the ingress, probable target(s), and egress route of the strike force.

To avoid giving the target away too early, the chaff delivery tactics were changed, as described by Michel: "Instead of dropping chaff inbound, the chaff flight would bracket the target and then turn back towards the direction of the flight, join to the normal formation and drop the chaff corridor outbound, with the [CAP] escorts remaining behind to cut off any MiGs. This also allowed the strike flight to delay its entry while waiting for the chaff to deploy, throwing off the timing of the North Vietnamese defenses."

Sources: Michel, Operation Linebacker I 1972, 65; "AN/ALE-38 Chaff/Flare Pod," Command: Modern Operations/Modern Air Naval Operations, accessed August 8, 2023, http://cmano-db.com/pdf/weapon/605/; Porter, "Linebacker: Overview of the First 120 Days," 44.

The establishment of Operation Linebacker increased the burden on the EC-121Ts tasked with supporting the increased air strikes taking place over North Vietnam. Using the call sign Disco, the EC-121Ts, equipped with both QRC-248, Rivet Top electronic suites, and Vietnamese linguists based at Korat, provided limited GCI support. The planes arrived on station—in orbits over Laos or the Gulf of Tonkin—before the day's strike began. As the fighter-bombers and their escorts approached, the airborne controllers on board the EC-121T identified each flight, established radio contact, then followed transponder returns throughout the mission. The EWOs in the electronics compartment observed the replies of North Vietnamese transponders, sometimes triggering the devices, sent MiG warnings, and shifted patrolling F-4s to meet the attack. The EWOs also issued collision warnings and helped pilots find tankers in emergencies. But the EC-121Ts, "with only five radios, eight frequencies, and two controllers simply could not keep track of everything that was taking place." Moreover, the need to turn in orbit and the slow movement of the aircraft's antenna limited its ability to control large raids. The EC-121s using the call signs Big Eye, College Eye, and Disco, however, were not able to reliably detect targets flying below 8,000 feet and, as Steven Fino wrote, "were unable to relieve the Phantoms of the stringent VID [visual identification] requirements" that was needed before they could engage the adversary with their long-range, radar-guided AIM-7 Sparrow missiles.[14]

Most of the air strikes conducted during Operation Freedom Train had been flown south of Route Packages 5 and 6, below the effective range of the GCI-controlled MiGs based in the northern part of the country. The SA-2 continued to be a major threat, due in large part to a change in the tactics employed by the North Vietnamese missile men. In October 1966, the Soviets dispatched a team of experts from the PVO-Starny (Soviet National Air Defense Force) to examine the results of the air campaign. To counter the homing capabilities of the Wild Weasels and the Shrike missiles, the technical experts instructed the Vietnamese to reduce the time the Fan Song radar was on. Once the radar was turned off, however, it took more than a minute to warm up, by which time the U.S. aircraft were gone. Steven

Zaloga, the best authority on the SA-2 in the West, skillfully recounts the Soviet's attempts to overcome this problem:

> Instead of switching the radar off, Soviet engineers modified existing maintenance equipment to allow the Fan Song operator to switch the transmitter into a "dummy load," which left the radar at full power but not transmitting. As a result, the radar could wait in ambush until the last possible moment, and then switch from dummy load to transmit mode to ambush an unsuspecting fighter. The 236th Missile regiment also pioneered the technique of switching from automatic to manual tracking, called "three point" guidance, when under intense electronic jamming, scoring a first victory using this tactic on August 12, 1967.[15]

Despite these new tactics, U.S. electronic countermeasures continued to reduce the effectiveness of the SA-2. "In the August 1967 fighting over Hanoi, 66 percent of all SA-2 missiles lost control, and more than 8 percent lost control so near the ground that they impacted near the city, causing considerable death and destruction." The North Vietnamese became so desperate that they began using untested track-on-jam tactics in an attempt to steer the missiles using the target's own jamming signal.[16]

Soviet engineers continued to work on these problems and to make improvements in the SA-75MK Dina (SA-2) system. By the end of 1969 they had issued Technical Bulletin-1 that introduced the following modifications:

- The addition of a switch that changed the missile guidance, reducing the minimum effective altitude to five hundred meters.

- The addition of a switch to use in case of attempts to jam the proximity fuze (feared by the Soviets but which never happened).

- The addition of a "bird house" over the horizontal antenna that housed an electro-optical system that allowed two operators to track the target visually when electronic jamming prevented normal operation. It only worked in daylight, however, and was vulnerable to cloudy conditions.

- The addition of a device that made the launch calculations automatic.
- A change in the missile's guidance beacon that increased its output from twenty to eighty watts.

The latter defeated the beacon jammer pod, and its use was discontinued once this change was discovered by Seventh Air Force. The Strategic Air Command never received this information, however, so the B-52 bombers continued to try and jam the SA-2s using this outdated technique. By the summer of 1972, every SA-2 system in North Vietnam had been retrofitted with these modifications.[17]

The North Vietnamese missile crews also introduced a new firing tactic that was designed to counter the defensive maneuver used by U.S. fighters, which consisted of quickly diving away and then pulling up sharply, thus avoiding the missile, which could not turn fast enough to engage the aircraft. Instead of firing one missile at a target, the Vietnamese fired three: the first high, the second low a few seconds later, and the third in the middle. The second missile closed on the U.S aircraft as it dived; then, when the aircraft pulled up, it flew into the third missile. Another tactic, used when jamming or chaff blanked out the NVA guidance systems, was to fire a barrage of missiles hoping for a lucky hit, or at least causing the attackers to drop their ordnance and break away. As Operation Linebacker proceeded, the North Vietnamese adopted a new strategy of using their MiGs to attack the B-52s. Eschewing the SAMs, which proved ineffective in the face of American countermeasures, the MiG-21s homed in on one of the flights of B-52s, executed a single high-speed pass, launched an Atoll air-to-air missile, and retired.[18]

EC-121M Warning Star, circa 1969 *United States Navy*

Douglas RB-66C. The first electronic warfare B-66 version sent to Southeast Asia was the RB-66C (later called the EB-66C). *United States Air Force*

An AD-5Q (later redesignated EA-1F) Skyraider, assigned to Carrier Tactical Electronics Warfare Squadron (VAW) 33 on board the carrier *Independence* (CVA 62), pictured in flight on May 1, 1962 *United States Navy*

EKA-3B Skywarrior of VAQ-135 in 1971 *United States Navy*

F3D-2Q Skynight of VMCJ-2 *Public domain*

An EA-6A Electric Intruder in flight, date unknown *United States Navy*

A US Navy A-4E Skyhawk of VA-164, from the USS *Oriskany*, en route to attack a target in North Vietnam during November 1967. *Public domain*

Of the 143 F-105F trainers built, 86 were converted into Wild Weasels like the one pictured here. Because of the high losses attributed to such a dangerous mission, though, there were typically fewer than a dozen aircraft available for missions at any one time. *United States Air Force*

F-105D on the tarmac *United States Air Force*

The elements of a "hunter-killer" team: F-105F Wild Weasel with Shrikes and F-105D with bombs *United States Air Force via Aviation Geek Club*

An F-105G Wild Weasel about to be refueled in flight. Note the ALQ-105 pod near the fuselage, the inboard wing-mounted AGM-78 Standard ARM, and the outboard AGM-45 Shrike. *United States Air Force*

A Boeing RC-135M Rivet Card at an airport in Southeast Asia, circa 1968 *Source TK*

CHAPTER 16

TEABALL, NTDS, AND RED CROWN

In the late spring of 1972, Gen. John Vogt, Seventh Air Force commander, sent an eyes-only message to the Air Force chief of staff, Gen. John Ryan, describing his analysis of the air war situation. The problem, he explained, was the proficient North Vietnamese pilots using single high-speed passes firing Atoll missiles, and inexperienced U.S. pilots rotating into the combat zone every year. Vogt demanded a better system to provide MiG warnings to U.S. aircraft operating outside existing radar coverage. Classification problems and dissemination concerns by national intelligence agencies had prevented aircrews from receiving MiG warnings on a timely basis. Although the Air Force had been obtaining this information for some time, all warnings had to pass through the filter of TACC-NS, unless extraordinary circumstances intervened. But, as NSA historian Thomas R. Johnson wrote in his Cold War history of cryptography: "Every request to pass warnings directly to operations people had encountered the implacability of the director of Air Force Intelligence, Maj. Gen. George J. Keegan Jr." Compounding the problem was the "Bull's Eye" system in which all intelligence information was broadcast to pilots on the guard channel in terms of distance and bearing from the center of Hanoi, known as the Bull's Eye.[1]

Between May and June, the U.S. Air Force shot down five MiGs, but lost seven F-4s in air-to-air combat. This negative kill-to-loss ratio was anathema to Air Force leadership. In early July General Vogt appealed to General Ryan again for a new approach to the intelligence warning system. Ryan called Vice Adm. Noel Gayler, the director of NSA, asking that he reactivate the "Y" service, a form of COMINT the USAFSS used during the Korean War to support the Tactical Air Command. In response, Gayler sent

FIG. 16.1

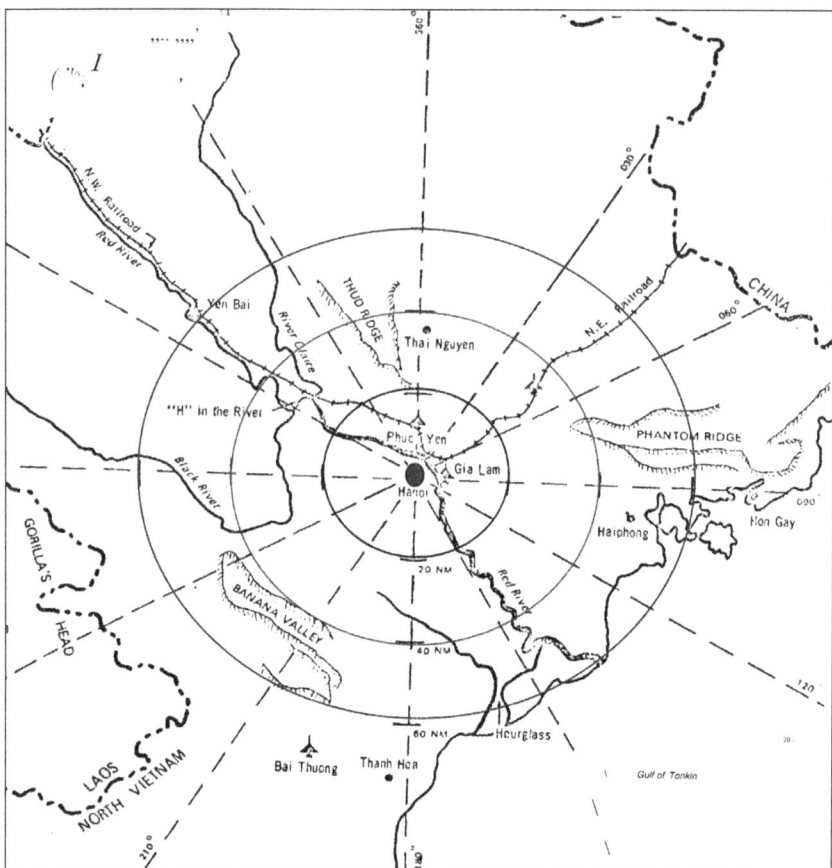

The bull's-eye map used to identify and provide the location in real time of air and land threats to U.S. aircraft operating over the north part of North Vietnam. This map was in every squadron, and crews memorized it. *United States Air Force, as reproduced in Michel*, Operation Linebacker I 1972

a team of SIGINT experts to Saigon headed by Delmar Lang, a retired Air Force lieutenant colonel and civilian employee of NSA who had been in charge of the COMINT team in Korea that provided real-time SIGINT to the TACC. After several days of research, Lang rediscovered an untapped source of COMINT that could be exploited and used to satisfy Seventh Air Force needs. "This unique source of intelligence," as Lang later recalled, "was North Vietnamese azimuth and range ground-controlled intercept (GCI) radar position reports on North Vietnamese MiGs." Although these NVA communication links were identified and picked up by U.S. Army intercept operators as early as 1965, they were not monitored regularly until two U.S. Air Force operators at the 6908th Security Squadron (SS), at Nakhon Phanom (NKP) Royal Thai Air Force Base, recognized the significance of this intercept and began to copy it daily. The North Vietnamese radar operators used their Bar Lock and Big Bar B radars to provide the azimuth and range of U.S. aircraft used to plot interception vectors that was passed to the MiGs over HF and VHF communication channels. These line-of-sight communications could be reliably copied only from a high-altitude airborne collection platform such as the SIGNIT U-2, called Olympic Torch, which captured the communication signals that were downlinked to the 6908th at Nakhon Phanom.[2]

Lang believed that if the information captured from the North Vietnamese GCI controllers could be processed in real time, it could be used by an Air Force weapons controller for both offensive and defensive purposes. He was sure that the data, "if properly presented, could provide a U.S. weapons controller with basically the same air picture that was available to the North Vietnamese GCI controller, thus giving the U.S. controller the capability to know the enemy's intent as well as the exact location of his aircraft." Lang briefed General Vogt on July 8, 1972, on the possibility of exploiting the North Vietnamese azimuth and range GCI tracking in conjunction with tactical air communications to support U.S. air operations. Vogt approved the concept and sent Lang to the 6908th to determine the feasibility of processing and correlating the tactical air and GCI tracking communications in real time.[3]

During the Korean War, Lang recommended placing a Chinese linguist next to the ground controller in the TACC. The Chinese linguist heard the information at the time of intercept and quickly told the ground controller, who immediately used the information to instruct the pilots on the fighter to take appropriate action. "He had repeatedly offered to do the same in Vietnam," according to Gilles van Nederveen, "but commanders in Southeast Asia had turned him down."[4]

Lang knew that Vietnamese voice communications revealed the takeoff of the MiGs and which B-52 sortie would be targeted. He also knew that Olympic Torch was intercepting those communications and that the intercept operators were sitting at the 6908th Security Squadron at Nakhon Phanom. He recommended that the takeoff and targeting information be passed to a collocated Seventh Air Force controller, who would alert the Air Force defensive patrol in the gulf. When the MiGs arrived, theoretically the F-4s would be waiting for them.

Once in Saigon, Lang's group ran into opposition from Seventh Air Force intelligence, who opposed the idea of giving raw data directly to the pilots. Larson, in his article on what became Project Teaball, maintains that "Despite this opposition, General Vogt decided to proceed with this new direct support concept and gave the Air Staff Action Group the priority needed to make it happen." With Vogt's blessing the Air Staff Team set up

FIG. 16.2

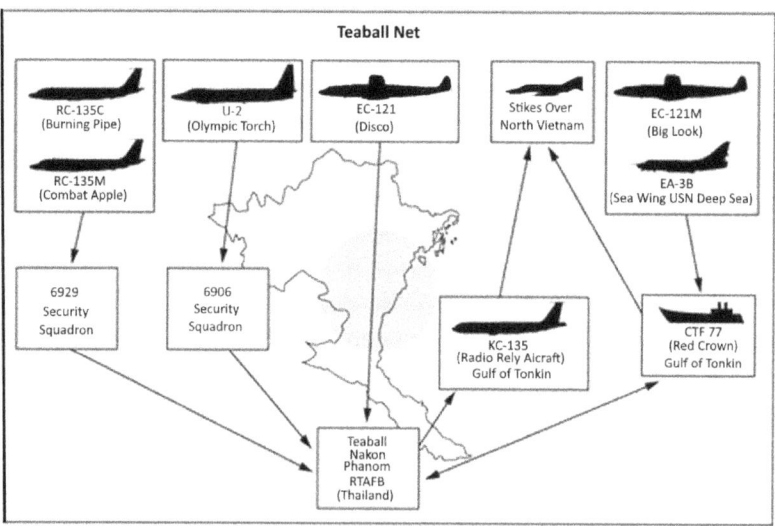

Graphical representation of Teaball elements and communication system. Note the absence of communication links from Combat Apple RC-135Ms or Olympic Torch U-2s. Data from these sources was first sent to the USAFSS facility at Nakhon Phanom before being passed to the Teaball center. *Van Nederveen, "Wizardry for Air Campaigns: Signals Intelligence Support to the Cockpit," 2012*

operations at NKP where they would have online access to radios monitoring North Vietnamese air operations.⁵

When operations began on July 26, 1972, COMINT and ELINT information collected from Combat Apple RC-135Ms orbiting over Laos and the Gulf of Tonkin as well as that collected from Olympic Torch was passed, via a USAFSS squadron at NKP, to the Teaball operations center installed in a 11-foot by 17-foot expandable M-292 van mounted on a 2 1/2-ton military truck adjacent to the 6908th SS intercept area. The operations van also had access to radar data from EC-121Ks, code-named Disco, orbiting over Laos and the Gulf of Tonkin as well as radar and SIGINT data sent by datalink from Red Crown. The control van was packed with radios and intelligence personnel who tracked, plotted, and recorded air movements. This information was forwarded to the Ironhorse air controllers, who passed the information to the F-4 combat air patrols via Combat Lightning KC-135 radio relay aircraft, code named Luzon, operating over Laos and the Gulf of Tonkin.⁶

While the control van was being set up, Lt. Col. William L. Kirk, of the Air Staff Action Group, visited each fighter wing in the Seventh Air Force to explain Teaball procedures in order to overcome the "Green Door" syndrome* that prevailed among the pilots, that is, the belief they were not getting enough timely intelligence.

Van Nederveen claims that "'Teaball' provided critical GCI assistance in sixteen of the fifty-nine air-to-air engagements during this period." He asserts that its success was based on the Ironhorse system for displaying collected SIGINT data to a few cleared weapons controllers in the Teaball operations room. Maj. Gen. Doyle Larson, USAF (Ret.), in his article on Teaball, went even further and declared that Teaball was an instant success and was responsible for the much-improved kill-to-loss ratio that saw U.S. aircraft shoot down thirty MiGs while losing only ten aircraft in air-to-air engagements from July 29 to the end of the war.⁷

---

* The "Green Door" syndrome, coined by Ronald T. Smetek, referred to valuable intelligence data that was kept locked away in a vault in the squadron or wing headquarters building behind a green door.

Marshall Michel, who flew combat missions in Vietnam and has written several books on the air war, had a less positive view of Teaball. As he wrote in his short book on Operation Linebacker I:

> *Teaball* was very cumbersome and slow. There were delays in *Teaball* receiving information, and delays from the time *Teaball* acquired the information and relayed through *Disco* (which often had to use an unreliable radio relay KC-135A Combat Lightning aircraft operating under the callsign *Luzon*) which cancelled out its value for use in "real time." General Vogt was able to dictate by decree that *Teaball* was to be used as the primary control agency despite its unreliability, but this order was only honored in the breach. Flights that had a choice of controlling agencies, like the MIGCAP flights, still used Red Crown.[8]

Michel went on to state that "General Vogt gave his brainchild *Teaball* the credit [for the greatly improved kill ratio], but few agree with him and after the war in a formal review, *Teaball* was excoriated and considered of 'limited usefulness.'"

Red Crown had several advantages over Teaball that made for more effective and timelier GCI. The guided missile cruisers assigned to Red Crown duty were now equipped with the SPS-48E, a long-range, three-dimensional, air search radar system that provided contact range, bearing, and height information that was displayed on the ship's NTDS consoles. The SPS-48E had a contact range in excess of two hundred nautical miles and had a much greater ability to pick up low-flying aircraft. It extended Red Crown's radar coverage over North Vietnam and allowed it to keep track of all air traffic in the PIRAZ zone, especially around Hanoi and Haiphong. It was the closest radar to North Vietnamese airfields and was in the best position to offer radar information to U.S. aircraft in the area.[9]

The NTDS was also a more robust system that had been in operation for a much longer time. Unlike Ironhorse, which was a tactical SIGINT system only, the NTDS operated as the heart of Red Crown's CIC. NTDS was a self-contained system that provided the air intercept controller all the information he needed as well as the ability to monitor the SPS-48E's radar screen. It was also linked with other NTDS-equipped ships operating in the Gulf of Tonkin. This extended its overall radar picture beyond the ship's own radar. The detachment from the Naval Security Group (NSG) assigned to the ship worked from a space called Supplemental Radio (SupRad) and typically consisted of an Officer-in-Charge, a Division Chief, a Leading Cryptologic Technician who was the analyst and report writer, six Cryptologic Technicians, one or two Vietnamese linguists, three Cryptologic Technician Communication Operators, and one Cryptologic Maintenance Technician. The NSG operatives, according to a former member of that elite group, "were very, very busy providing the location and status of the bad guys to the air intercept controllers, in as near to real time as possible for a minute. . . . This was especially critical to the MiGs flying below the radar coverage," which they had several ways of detecting and tracking.[10]

As Michel aptly noted, "Red Crown's greatest asset was its staff of exceptionally good GCI controllers." Men such as OSCS Larry Nowell were radar experts trained in air-to-air combat with years of practical experience conducting GCIs. Nowell, who had enlisted in the Navy in 1958, was the Air Intercept Controller Supervisor on the guided missile cruiser *Chicago* when the ship departed from Subic Bay on May 2, 1972, having been urgently recalled to PIRAZ duty after a four-month tour in the Gulf of Tonkin in response to the North Vietnamese army's invasion of the South. Air operations increased dramatically as U.S. aircraft conducted strike and interdiction missions throughout North Vietnam to stem the movement of men and supplies. For the next two and a half months, *Chicago* monitored all aircraft flying over the gulf, directed friendly CAP, and coordinated fighters escorting B-52 raids. The ship vectored damaged aircraft around enemy missile sites, set up tanker rendezvous points for planes low on fuel, and directed helicopters on rescue missions.[11]

Chief Nowell was in the thick of the air actions. After enlisting in the Navy in 1958, he was sent to radar school, which was the first step in a process that led to his duties as an Air Intercept Controller (AIC). From radar school, Nowell was assigned to the guided missile frigate *Mahan* (DLG 11), one of three ships outfitted with the newly developed NTDS for testing and evaluation. During the six years that Nowell served on the *Mahan* he learned all aspects of the NTDS and qualified as an NTDS watch stander, which was invaluable requisite for his next assignment to the Pacific Fleet's Air Intercept Controllers School in San Diego. Nowell served on the guided-missile cruiser *Horne* (DLG 30) before being posted to the *Chicago*.[12]

Nowell began providing critical information to Air Force and Navy fighters threatened by enemy MiGs in North Vietnam and Laos when the *Chicago* first took over Red Crown duties in December. His job, as he later explained, was "to draw the pilot a continuous mental picture of where he is in relation to the overall tactical situation." The Air Force pilots flying MiG CAP favored him over their own GCIs, because he could guide them to the MiGs. One pilot described his talents as "an instinct for aerial tactics and flying. He isn't just good. He's uncanny." He was "the quarterback, calling the plays." While Nowell was an excellent intercept controller, he often relied on help from the NSG operators in order to confirm that the blips showing up on his radar screen were hostiles, one of the requirements under the rules of engagement. Getting this timely information was one of the bottlenecks in the Air Force's system.[13]

This was a problem on the *Chicago* too, as the ELINT operator or his supervisor had to get someone's attention to report anything. As Guy Thomas, then a lieutenant in charge of the ship's NSG Detachment at the time, explains, "There was a lot of peripheral information to be gleaned from it that was not being fully fused into the calculation as to what is happening. To make matters worse, from SupRad we did not know when the coordination team was very busy (as was usual during hostile activity) or could be interrupted to report information of widely varying timeliness requirements." Thomas solved the problem by getting a sound powered phone installed in CIC next to the AICs so that he could act as a liaison between SupRad and the AICs in the CIC.[14]

While the *Chicago* was deployed to Vietnam on her fifth cruise, Larry Nowell was involved in more than one hundred air-to-air engagements, and with the help of the "radio guys" in SupRad, he was credited with directing three Navy and nine Air Force pilots—all flying F-4 Phantoms—to MiG kills. In mid-May, Nowell started using the call sign "ACE" after he became the first AIC to register five successful intercepts. He was flown to an aircraft carrier and told he was being nominated for the Distinguished Service Medal and promoted to senior chief. When he returned to *Chicago* that evening, he went straight to SupRad with the news and said, "I owe you guys at least half of this medal. I could not have done it without you."[15]

For his actions in directing twelve MiG interceptions as an Air Intercept Controller on the *Chicago*, Larry (ACE) Nowell became the first enlisted man awarded the Navy Distinguished Service Medal for combat action. His accomplishments were lauded throughout the fleet and his accomplishments became legendary.[16]

CHAPTER 17
COMBAT TREE PHANTOMS, PROWLERS, AND WILD WEASEL IVs

The first MiG CAP missions over North Vietnam at the start of Linebacker I were flown by the F-4Ds of the 432nd TFW home-based at Udorn, the closest air base to Hanoi. In December the wing received eight F-4Ds equipped with the APX-80 IFF interrogator that was similar to the QRC-248. The success of the QRC-248, which made a profound difference in the U.S. picture of the air war, led the Air Force to begin development of a more advanced version that could be deployed in U.S. fighters. It had to be reduced in size in order to fit into the cockpit of an F-4 and needed to be automated so it could easily be used by the Weapons System Officer (WSO) who operated the F-4D's radar. The APX-80 and the F-4Ds equipped with it both bore the code name Combat Tree. Like its predecessor, it could be used in a passive mode where it received and processed IFF replies sent to MiGs in response to their own GCI, or it could be used in an active mode to trigger the MiG's responses.[1]

The WSO in the rear seat of the F-4D used the APX-80 interrogator to search for North Vietnamese MiGs. Combat Tree "challenged" or "interrogated" each transponder it came across and analyzed the return to determine whether or not it was coming from a hostile aircraft. The returns were plotted on a display that enabled the WSOs to keep track of all the aircraft in the vicinity. Combat Tree detected and identified enemy aircraft in either mode up to sixty nautical miles away, three times as far as the F-4's radar, which lacked the ability to determine whether the radar contact was friend or foe. That latter factor was a significant detriment to the tactics available to the F-4s due to the U.S. rules of engagement, which mandated that targets could not be engaged beyond visual recognition. Before the introduction of APX-80, the F-4s had to identify the enemy visually before opening fire, which for all intents and purposes eliminated the use of the AIM-7 Sparrow air-to-air missile, which negated one of the F-4s biggest advantages over the MiG.[2]

While Combat Tree was a vast improvement over the F-4's radar, it did not eliminate the rules of engagement, which were modified to account for the capabilities of the APX-80. Combat Tree F-4Ds had strict rules of engagement with regard to how close friendly aircraft were before they fired a missile. If the MiG was coming head on, the friendly aircraft had to be about fifteen miles away; if the F-4 attacked from behind the MiG, it could fire if the friendly aircraft was five miles away. Any closer, or if there was any misgiving about the range or identity of the target, the F-4 had to identify the target visually before firing. "Even when the F-4D could not fire early, [Combat] Tree was still useful because it told the F-4 crew that a MiG was in the area and allowed them to set up their visual identification pass knowing that they were probably attacking a MiG." Combat Tree allowed the F-4Ds to locate MiGs without the help of the GCI controllers in Red Crown or Disco, which was a tremendous advantage given the difficulties encountered with radio communications. It also cut down the time and fuel needed to chase down radar targets that turned out to be friendly.[3]

Because of the small number of Combat Tree aircraft available, they were typically assigned as the flight leader (01 position) and second element leaders (03 position). The (02) and (04) positions in the flight were usually filled with the F-4Es, which now carried the General Electric M61 gun that provided the flight with some gun capability without forcing the F-4D Combat Trees to carry gun pods.[4]

Combat Tree had several positive effects. First, it allowed U.S. fighters to acquire MiGs at much greater ranges than they could from a radar return. Second, it allowed them to make contact looking down in many cases. This meant that free-roving MiG CAPs were much more effective than they had been previously. Unfortunately, the combat advantage offered by the APX-80 was compromised after Combat Tree crews called out over the radio that they were locking onto MiGs at long ranges. The North Vietnamese monitoring the MiG CAP radio calls must have figured, or possibly learned from an intelligence leak, that their IFF was the problem, for the MiGs began to cut down on the use of their IFF until the critical moments of an engagement.[5]

When President Nixon ordered that the bombing of North Vietnam be renewed in April 1972, only one Wild Weasel Squadron (WWS), the 17th WWS, remained in Southeast Asia, the others having been withdrawn after the bombing was stopped in 1968. The 17th WWS deployed, with their newly configured F-105Gs, to Korat Royal Thai Air Force Base on December 1, 1971, where they were assigned to the 366th TFW. The F-105Gs were some of the sixteen F-105F Wild Weasels that had been modified at the beginning of 1968 so they could fire the new AGM-78B Standard antiradiation missile. Key features that differentiated the F-105G from the F-105F were the side fairings that had been added to the fuselage to hold the Westinghouse Electric Company's QRC-380/AlQ-155 electronic countermeasures system. Because of space limitations, the jamming equipment, equivalent to the QRC-380, could not be placed within the fuselage. The fairings eliminated the need to use one of the pylons for jamming, allowing more ordnance to be carried. The radar homing and warning gear installed on the F-105Gs included the APR-35 receiver used to target the Standard ARM, the APR-36, and APR-37, which were updated versions of the APR-25/26 that were more reliable and effective, and the ALR-31 "SEE SAM" that alerted the crew if they were being targeted by a Fan Song radar.[6]

As Larry Davis noted, all available air power in Southeast Asia was being employed in an effort to stop the North Vietnamese offensive in the South. "At the same time, virtually all of North Vietnam was now an open target and for the first time in the war B-52 Stratofortresses flew missions north of the DMZ." In April, to beef up the support for the South Vietnamese army, the Air Force began a series of deployments under Operation Constant Guard. To take the load off the 17th WWS, which was bearing the brunt of the effort to suppress SAMs, the Air Force sent a full squadron detachment of new F-105Gs to Korat. The detachment, which was assigned to the 561st TFS, arrived on June 4, 1972, and immediately began combat operations.[7]

As missions over North Vietnam increased, the number of U.S. aircraft downed by North Vietnamese SAMs began to increase. In response, the 366th TFW increased the number of Iron Hand hunter-killer teams sent to seek out and destroy SAM sites. The Iron Hand strikes were typically composed of two F-105G Wild Weasel and two F-4E Phantoms. For the Iron Hand missions, the F-4Es were configured to carry SUU-30 submunition dispensers loaded with 220 softball-sized CBU-52 submunitions, better known as "cluster bombs." These small bomblets were a new weapon introduced in 1970. The CBU-52 bomblet, which contained a 0.65-lb. high-explosive charge that spewed shrapnel when initiated, was designed for use against personnel and thin-skinned vehicles. When the pylon-mounted SUU-30 was released, ram air forced its two halves open, dispensing the bomblets over a wide area.[8]

During the hunter missions, the F-105Gs were sent ahead to locate the targeted SAM site using their onboard sensors. When the site was located, they launched their antiradiation missiles (most often the Shrike) to take out or shut down the site's Fan Song radar, neutralizing the SAMs. This allowed F-4Es to attack the site without worrying about being hit with an SA-2. They inundated the site with large numbers of cluster bombs spread over a wide area, destroying the site's radar vans, missiles, and equipment and killing its personnel.

In July 1972, a new aircraft—the EA-6B Prowler—arrived in Vietnam when a four-plane contingent from VAQ-132 flew on board the *America* (CV 66) and started conducting operations in the Gulf of Tonkin. The EA-6B was the first aircraft designed from the ground up for electronic warfare. The plane was designed around the A-6 airframe, lengthened by adding a 4-foot by 6-foot section in the nose to provide space for a crew of four (pilot and three EWOs). It had a strengthened airframe, increased internal fuel capacity, and a bulbous receiver antenna fairing at the top of the vertical stabilizer. These changes increased the airplane's weight, which required more powerful engines and a beefed up landing gear. The most striking characteristic was its canopies that were coated with a micro-thin layer of gold to prevent stray electromagnetic radiation from interfering with the onboard instruments.[9]

The heart of the of the EA-6B's ECM suite was the ALQ-99 Tactical Jamming System (TJS), the first fully integrated computer-controlled jamming system. The on-board portion of this system consisted of a computerized CPU that processed threat signals, drove the operator displays, and controlled up to four external jamming pods carried on the aircraft's four wing racks; a Systems Integrated Receiver (SIR) designed to sample the threat environment; and the operator stations. The SIR group, made up of antennae and receivers, was mounted in the Prowler's distinctive fin-top fairing and the blisters on either side of the tail fin. The 950-lb. jamming pods contained two high-powered noise-jamming transmitters, an exciter/processor, a tracking receiver, and a ram air turbine to provide electric power.[10]

Also on board was the highly sophisticated ALR-42 radar warning receiver, which was able to detect and prioritize radar threats. Data from the ALR-42's signal analyzer was passed to the ALQ-99's central processor for use in programming the selected jammers. The power to each jammer was managed by the ALQ-99's computer, ensuring that the greatest threats received the most power. The optimization of the jamming transmitters and their modularity allowed the Prowler's ECM equipment to be tailored to the needs of specific missions.

The ALQ-99 TJS could be operated automatically, semiautomatically, or manually. The automatic mode, which was the simplest to use, did all the work, evaluating the threat and initiating the response. The semiautomatic mode reduced the operator's workload by providing and analyzing threat data that the operators could then select in order to provide the optimized jamming response. The manual mode created the most work for the operators. It required the operator to manually call up the information they needed to evaluate the threat, which could then be used to select and initiate a suitable jamming response.

VAQ-132 flew its first combat support mission with the EA-6B on July 11, 1972. Four more of the squadron's EA-6Bs joined the *Enterprise* (CVN 65) in September. The EA-6Bs averaged five to six missions each day from the carriers supporting Navy and Air Force strikes against the North Vietnamese.[11]

The role of the EA-6B during these missions was to provide stand-off jamming to deny precise location information to the NVA's GCI, search, and acquisition radars and to disrupt its communication links. As Capt. Albert A. Gallotta Jr. explained to the Senate's Ad Hoc Subcommittee on Tactical Air Power of the Committee on Armed Services during the defense appropriation hearings for fiscal year 1974, there were three principal reasons why the EA-6Bs operated "from orbits just off the North Vietnam coast" and were not used "in the penetration mode for any penetration in Vietnam." Unfortunately, the reasons given during Captain Gallotta's testimony before the subcommittee were deleted for security reasons. But the reason for keeping the EA-6Bs off the coast can be deduced from other sources of information related to the SAM threat. The primary reasons for keeping the EA-6B away from the SAM sites over North Vietnam, according to these sources, was its lack of maneuverability due to the characteristics of the airframe, its lack of power, and the heavy load and drag associated with the ECM pods it normally carried. As USAF Maj. Gen. John J. Burns revealed in his testimony to the subcommittee, neither deceptive jamming nor noise jamming provided enough protection to eliminate the need to maneuver. In order to avoid a SAM, Vice Adm. William D. Houser explained, "You had to be able to maneuver your airplane and outmaneuver the telephone pole chasing you." It was something the EA-6B Prowler was not designed to do.[12]

There may have been other reasons too. The tactical jamming pods on the original ALQ-99 initially covered only four of the seven radar bands used by the Soviets, a problem alluded to by Captain Gallotta but not clarified in the public record, which contained numerous redactions. Another issue may have been the lack of a missile warning system, which was not delivered with the first production EA-6Bs. Although the APR-27* was later added to the EA-6Bs in Southeast Asia, it proved difficult to integrate into the aircraft's ECM suite due to the high radiated power of

---

* ALQ-100 countermeasure sets may also have been added, although this system was not used very much, as it interfered with the TJS and was unreliable (see Whitten, "MCARA Aircraft > EA-6B ICAP II").

EA-6B's jammers. Despite these issues a number of studies indicated that the EA-6B[†] proved effective in reducing the number of losses due to SAMs and played a significant role in supporting Air Force operations during Linebacker II (see chapter 16). One such report produced by the Air Force Special Communications Center concluded that the EA-6B was the most effective tactical electronics jamming system employed against the North Vietnamese air defenses at any time during the war. The ALQ-99 TJS was so superior to the ECM jammers carried by the Air Force's EB-66s that the Air Force considered acquiring the Prowler as a replacement for the aging EB-66. Although the advantages of the ALQ-99 were recognized by the Air Force, its leadership concluded that the EA-6B lacked the flight performance to meet the Air Force's needs. Instead, the Air Force selected the high-performance F-111 Aardvark to house a version of the EA-6B's jamming system.[13]

In September 1972, six F-4C Wild Weasel IVs from the 67th TFS assigned to the 18th TFW arrived in Korat to augment the F-105Gs already there. The F-4C Wild Weasels, identified officially as EF-4Cs, were designed in parallel with the F-105F Wild Weasel III program that began in January 1966. Because of attrition, aging airframes, and the ceasing of production of the F-105F in January 1965, the Air Force realized that the definitive answer to the Wild Weasel problem was the development of another airframe. The two-seat F-4C appeared to be the ideal choice to replace the F-105s, and in 1966 the decision was made to develop the F-4C Wild Weasel IV in parallel with the Wild Weasel III program, and plans were made to install the same electronics going into the F-105s. Although the two aircraft were comparable in size, the F-4C had a second engine, which required twice as much wiring, control cables, and electronics just to run the aircraft, not to mention the additional wiring needed to handle the greater variety of ordnance and missiles. "The Phantom was a jam-packed aircraft and simply could not handle the added electronics and wiring required to properly install Wild Weasel equipment."[14]

---

† In addition to the EA-6B, Marine EA-6As and Navy EKA-3Bs and EP-3Bs provided electronic warfare support during Operation Linebacker II.

While McDonnell-Douglas engineers, who were converting the F-4C under Project Wild Weasel IV-A, mounted the APR-25 radar homing and warning system and the APR-26 launch warning receiver internally, they could not find room for the IR-133 panoramic receiver's electronics. Instead, it was installed on a pylon in a pod mounted in place of one of the Sparrow missiles normally carried by the F-4. These modifications, according to Wild Weasel expert Larry Davis, "did not work well. First, the Weasel gear was not compatible with the McDonnell F-4 wiring, causing erratic displays, high interference, or no display at all." It took more than a year to figure out what the problem was and to replace the high capacitance cable they attempted to use in place of the low capacitance coaxial cable needed. As a result, the first F-4C Wild Weasel was not finished until the summer of 1969. The F-4C was a step backward for the Wild Weasel program, as it lacked the room in the airframe for the electronic interface needed for the AGM-78 Standard ARM, which limited it to the use of the AGM-45 Shrike.[15]

CHAPTER 18
LINEBACKER II

THE PRELIMINARIES

The air strikes during Linebacker I were effective in bringing the North Vietnamese leadership back to the bargaining table, and negotiations to end the war began once more on September 15, 1972. The North Vietnamese thought they could pressure the United States into a peace treaty ending the war, thinking correctly that the American electorate had tired of the war and would support a candidate who promised peace. President Nixon, who was re-elected in November and who wanted credit for ending the war, had to take action before the new Democratic Congress cut off funds to continue the fighting. At the end of November, after foot dragging by the Vietnamese negotiators had stalled the peace talks, Nixon ordered the Joint Chiefs to prepare plans for a three-to-six-day strategic bombing campaign using massive flights of B-52 bombers. The president believed that the psychological effect of the B-52s, which dropped their bombs from 30,000 feet and were seldom seen or heard until their bombs exploded, would have a greater effect than the destructive power they wrought. "Nixon," explained William P. Head, "wanted northern civilians to feel the sheer terror U.S. airpower could elicit."[1]

The Eighth Air Force, headquartered in Guam, was in charge of bombing and refueling operations for Southeast Asia. In August 1972, Gen. John C. Meyer, SAC commander, ordered Eighth Air Force planners to prepare an operations plan for using B-52s over Vietnam. The plan they prepared in response to Meyer's command called for extensive attacks against Hanoi and Haiphong using multiple-bomber formations simultaneously attacking from different directions in an attempt to confuse and defeat the North Vietnamese defenses. Meyer, concerned that the collateral bomb damage in this plan might cause a large number of casualties—which Nixon wanted to avoid—rejected the Eighth Air Force plan. Instead, he detailed his staff to create a new plan, which they hurried to accomplish in just three days.[2]

The B-52s were assigned to SAC as part of the U.S. nuclear deterrent, and their tactics and training were geared to this mission. SIOP, the Single Integrated Operational Plan for war with the Soviet Union, called for the B-52s to penetrate Soviet airspace at low level. The crews of the nuclear armed bombers did not have to perform precision attack; all they had to do was to drop one bomb in the general area of the target. As Peter Grier wrote, "The situation in North Vietnam was very different. The B-52s would fly at high altitudes and use radar guided systems to drop conventional munitions. Destruction of a rail yard or power plant would entail placing weapons directly on target." To ensure target destruction, the B-52s needed to attack several times in concentrated groups. With little time for extensive planning, General Meyer's staff formulated what William Head called "an inflexible scenario that sent three waves of bombers on the same route at the same altitude and at the same times for the first three days." To avoid civilian casualties from collateral bomb damage, crews were instructed to adhere strictly to the planned course and fly in a trail formation, with three cells of three aircraft each striking the same target area with up to ten minutes separation between cells. To ensure the bombing was accurate, they had to maintain a straight and level flight for four minutes before the bomb release to allow the bombing computers to stabilize. According to Head, "Staffers at Eighth Air Force were alarmed by the repetitive routing, and some feared casualties as high as 16 to 18 percent." General Meyer, using figures based on the planned nuclear attacks against the Soviet Union, which had individual bombers penetrating at low altitudes and using short-range air missiles (SRAM) to suppress SAM defenses, estimated losses at 3 percent.[3]

The operational plan prepared by SAC's staff had to take into account the differences between the B-52D and B-52G models, both of which were needed to fill out the order of battle to achieve the objectives of Linebacker II. In December 1965, less than six months after the first B-52s became involved in the Vietnam War, the Air Force initiated a special modification program known as the Big Belly project to convert most of the B-52Ds to

FIG. 18.1

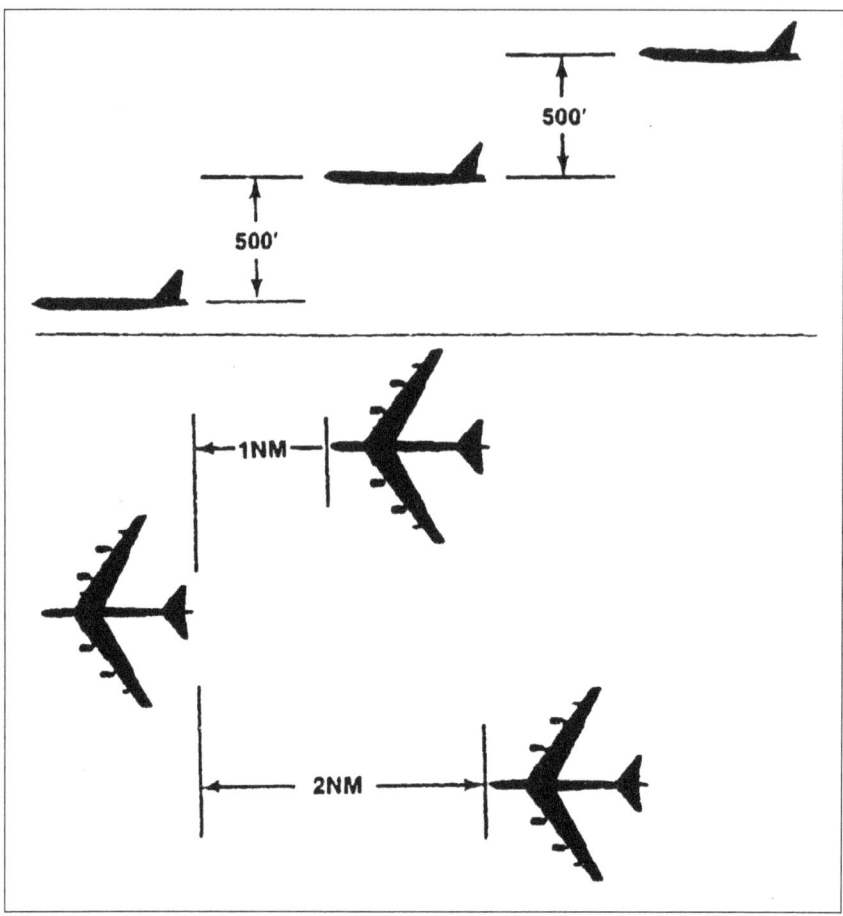

B-52 cell formation *Nalty,* Tactics and Techniques of Electronic Warfare, *2012*

carry more bombs for the conventional warfare configuration needed for operations in Southeast Asia. In addition to the twenty-four 500-lb. or 750-lb. bombs that the B-52Ds carried externally on wing-mounted pylons, the bomb bay was reconfigured to allow the aircraft to carry eighty-four instead of twenty-seven bombs before the Big Belly modifications. This upped the B-52D's maximum bomb load to 60,000 pounds.[4]

The B-52G was SAC's primary weapons system to be used against the Soviet Union in the event of nuclear war. It featured a "wet" wing that increased the plane's range by increasing the amount of fuel that could be carried through the elimination of the space taken up by the rubber bladders in the wings of previous models. The increased range allowed the B-52G to fly more missions than the D models, because they did not require in-flight refueling as often. The new wing also saved weight by dispensing with the ailerons, relying entirely on the wing's spoilers for lateral control. The removal of the ailerons did create some control problems, including the tendency for the B-52G to experience Dutch roll and for the nose to pitch up. The B-52G also had a squared-off vertical stabilizer that was eight feet shorter than the Model D's "shark fin." The B-52's small, twenty-seven-bomb load was a problem because it took four B-52G cells (twelve aircraft) to drop the same number of bombs as a single three-plane B-52D cell. The G model differed from the D model in one other aspect: the gunner in the tail of the D model was moved to the flight deck, where he remotely controlled the radar-directed quad .50-caliber machine gun in the tail.[5]

FIG. 18.2

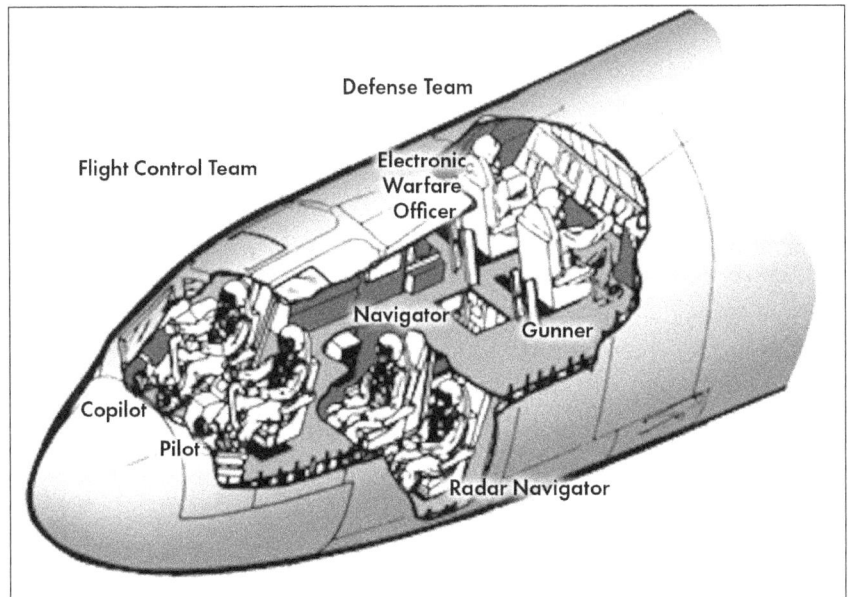

B-52G flight deck *United States Air Force*

Between 1967 and 1969, a program known as Rivet Rambler was begun to modify the B-52Ds with the latest electronic warfare equipment, known as the Phase V ECM suite. By 1971, all the B-52Ds had been so modified under this program, which involved the installation of one ALR-18 automated receiving set, one ALR-20 panoramic receiver set, one APR-25 radar homing and warning system, four ALT-6B and six ALT-22 continuous wave jamming transmitters, two ALT-32H high-band jamming systems, one ALT-32L low-band jamming set, two ALT-16 barrage-jamming systems, six to eight ALE-24 chaff dispensers, two ALE-25 chaff dispensers, and six ALE-20 flare dispensers.[6]

As the first "Special" B-52 bombing missions into the Hanoi-Haiphong area were being planned, there had been a great deal of discussion between the Eighth Air Force, which supplied the bombers, and SAC's staff concerning the tactics used by the B-52s. Instead of the steep, 45-degree posttarget banked turn that SAC insisted the B-52s make immediately after they released their bombs—a routine tactic taught by SAC to avoid the blast from a nuclear bomb—the Eighth wanted the B-52s to make two shallow turns of fifteen degrees. The planners in the Eighth Air Force rightly believed that the steep turn would direct the bombers' jamming signals, sent from antennas mounted on the bottom of the B-52's fuselage, away from the enemy radars during the turn, making the B-52s more visible to the Fan Song radars that guided the SA-2s. Although SAC agreed that the steeper banked turn would blank the jammers, they still insisted that the B-52s use it. As Marshall Michel noted, "the seemingly innocuous 45-degree post-target turn had never been tested against the captured Fan Song SAM radars available to the United States, an oversight that was to have tragic consequences later."[7]

Michel chastised the SAC planners for using the standard B-52 tactics in their operational plan. SAC had earlier tested the B-52s against the Fan Song surrogate at Eglin Air Force Base during the Giant Stride program aimed at developing penetration tactics for B-52s attacking missile-defended targets, but the tests only revealed the burn-through problem and did not address the use of a wide-spaced jamming formation. The latter was an anathema to SAC because of their priority to deliver bombs accurately. Notwithstanding whatever tests may have occurred during the Giant Stride program, which was probably conducted in 1968, Michel was convinced that SAC should have answered the following questions:

- What was the effect of a steep posttarget turn on the B-52's jamming capability?
- How effective was mutual, overlapping cell jamming? How far apart could the B-52s be in a cell—spread laterally, horizontally, and vertically—before the distance started to impact on their mutual jamming?
- How did SAC's standard anti-SAM maneuver, a series of gentle turns, climbs, and descents, known as the "TTR maneuver," affect the Fan Song's ability to track a jamming B-52?
- Did opening the bomb doors increase the B-52's radar return?

Michel thinks that simulated testing was not considered because it would have taken funds from SAC's budget and would have raised questions of why it had not been done before. It seems more likely that the SAC planners felt that the Phase V ECM suite would take care of the Fan Song radar problem. They knew that the ECM suite carried by the B-52s had more jamming equipment than any other aircraft in the Air Force's inventory. Since the B-52 was designed to penetrate the Soviet Union's defended air space, the people at SAC were convinced "that their electronic countermeasures would surely overpower the defenses of a third-world country."[8]

As the first messages containing plans for Operation Linebacker II began to be received at Eighth Air Force Headquarters, the staff, according to Michel, "saw [that] the SAC plan contained the seeds of disaster." The errors in SAC's plan were made strikingly clear by Michel:

> The single file of B-52s would allow the North Vietnamese to easily track one cell after the other and engage them one at a time. The 100-knt jet stream wind would move the B-52s through the target area quickly, but it also would give the radar navigators less time to get lined up on their targets. But the major problem was that,

after bomb release, the SAC plan called for the B-52s to make their post-target turn west, back *into* [author's emphasis] the jet stream wind that would slow the bombers dramatically over the heaviest part of the NVN defenses.[9]

While the staff of the Eighth Air Force prepared the target lists for the B-52 raids on Hanoi and Haiphong, the North Vietnamese General Staff formed a special headquarters group, designated Division H61, to investigate the problem of shooting down the American B-52s that heretofore had seemed to be immune to their SA-2s. Division H61 immediately gathered a number of radar operators from missile regiments around Hanoi and sent them to the middle of the country to observe and document the effects of B-52 jamming. Photographs of the radar scopes while jamming took place were taken to be studied further by the operators. On October 6, 1972, Division H61 convened a meeting with the radar operators sent to observe the B-52 jamming to see what they had learned. Using the information collected during this meeting, the North Vietnamese Air Defense Command produced a manual called "How to Fight the B-52," complete with several meters of film taken of radar scopes full of B-52 jamming.[10]

Before November 5, 1972, B-52s striking supply routes and staging areas above the DMZ broke off or diverted if the EWOs on board detected a high SAM threat level. On that date, a B-52D bearing the call sign Copper 3 was damaged by an SA-2. "The SAM [reportedly] was so close that the [rear] gunner heard the motor of the missile prior to detonation." This incident—which was the first time a B-52 had been damaged by an SA-2—had a chilling effect on the B-52s, so much so that they began to divert more regularly. General Vogt became tired of supplying large escorts to the B-52s only to find out that they had broken off their attacks in the face of a potential SAM threat. He sent a strongly worded message to General Meyer at SAC Headquarters complaining about the situation. General Meyer responded by designating most missions into North Vietnam as "press on" missions, which James McCarthy described as "one in which the aircraft continued to the target despite SAM or MIG activities in particular and aircraft systems degradations in general."[11]

The North Vietnamese missileers shot down their first B-52 on the night of November 22, when an SA-2 exploded beneath one of the bombers assigned to the 96th Bomb Wing during a "press on" mission against a storage area near the city of Vinh, a focal point for supplies and equipment moving south along the coastal supply network 150 miles north of the DMZ. Olive 2, piloted by Capt. Norbert J. Ostrozny, sustained lethal damage when the SA-2 that had visually been sighted seconds before bomb release exploded beneath the B-52D, riddling the airplane with shrapnel that punctured the underbelly, starting fires in both wings and in an engine, which knocked out all power. Captain Ostrozny was able to reach the Mekong River and the border with Thailand before the right wing "folded up over the fuselage," causing an uncontrollable roll to the right. Ostrozny hit the bail-out alarm button and all six crewmen successfully escaped the dying aircraft and parachuted to safety (the B-52D only had five ejection seats; the gunner had to bail out through a hole left when the gun capsule was automatically jettisoned). The position of Olive 2's bailout was reported to Search and Rescue by Olive 3's navigator, and all six of the crew were rescued within minutes by helicopter.[12]

Olive 2 was lost despite the extensive support package that accompanied the strike. Four F-4s laid a chaff corridor across the target area, and three EB-66 aircraft provided stand-off jamming and electronic surveillance. Five additional F-4s were operating in two separate MiG CAP orbits while a pair of F-105 Iron Hand aircraft provided SAM suppression support. Other support aircraft included tanker aircraft and search-and-rescue teams.[13]

The downing of Olive 2—the first B-52 lost after the bomber had flown more than 112,000 combat sorties since June 1965—was a milestone in the Vietnam War. Although five B-52s had been damaged by SA-2s and 286 SAMs sighted by B-52 crews, it was SAC's first combat loss in Southeast Asia. An analysis of the factors causing the loss of Olive 2 revealed "several unfortunate occurrences," according to the CHECO report later produced by Maj. Calvin R. Johnson. Numerous malfunctions in Olive 2's ECM equipment prevented the Electronic Warfare Officer from ascertaining the exact position of his jamming transmitters. When he received the call of a SAM sighting by the crew, he made the most appropriate response, selecting the preset modes on his jamming transmitters and jamming the track-while-scan and downlink beacon signals in the blind. A second factor was the location of the chaff corridor with respect to the cell and the location of the SAM sites. Chaff corridors were laid so that the strike force flew through or along it for protection. It was well known that aircraft close to, but outside, the chaff corridor would be highlighted by enemy radars. In the case of Olive 2's cell, severe winds at altitude or one of several possible miscalculations by strike or support aircraft caused the chaff cloud to be of little protection that day.[14]

The first of the bombing strikes ordered by President Nixon under Operation Linebacker II were scheduled to begin on the night of December 18, 1972. In preparation for the first night's mission, Brig. Gen. Andrew B. Anderson Jr., commander of the 57th Air Division, and the staff of the Eighth Air Force, briefed the wing commanders leading the B-52s. Details were provided of the target routes to be flown inbound and outbound, and there was an in-depth description of radar aiming points for bomb release. There was considerable discussion of enemy defenses, supporting forces, and target area tactics. In accordance with the instructions issued by the Joint Chiefs, the operation would involve three days of around-the-clock bombing of the Hanoi-Haiphong area. McCarthy, then a colonel, who was charged with briefing the first wave, described the forces involved as follows:

> Tactical fighter and fighter-bomber forces from Seventh Air Force and comparable aircraft from the Seventh Fleet would strike during the day and the B-52s would strike at night. The first night's sorties would consist of 54 B52Gs and 33 Ds from Andersen, plus 42 Ds from U-Tapao. These 129 planes would fly in three different waves, with approximately four to five hours between waves. Strikes would first be made by the U-Tapao crews against Hoa Lac, Kep, and Phuc Yen Airfields, suppressing MIG-21 "Fishbed" operations, while the rest of the force attacked Hanoi area targets, coming in generally from the northwest. This cone of attack was selected to insure [sic] ready identification of radar aiming points and give minimum exposure time to the lethal SAM defenses deployed in

the target area. After the aircraft had dropped their bombs on target they would turn and exit the target area on tracks heading back to the west and northwest. The second and third days were to be repeats of the first day, except the total number of aircraft would be 93 for Day Two and 99 for Day Three. The axis of attack and withdrawal routings were to be essentially the same for all three days, but comparison of the mixture of targets spread over the three days shows that there was considerably more variety to ingress and egress routing than there was first believed to be. The one constant in all strikes during the three days would be the post-target turn (PTT).[15]

The target list for the first night's attack established by the Joint Chiefs included the Kinh No vehicle repair facility, the Yen Vien railroad yard, the Hanoi railroad repair shop, the Hanoi International Radio Station, and the MiG bases at Hoa Loc, Kep, Yen Bai, and Phuc Yen. All, except for the radio station, were selected as suitable for the wide bombing patterns expected to be delivered by the B-52s. No SAM sites were selected, even though they were the greatest threat to the bombers. Also not targeted were any of the SAM storage facilities where SA-2s were removed from their shipping containers, assembled, and placed on trucks to be transported to the SAM battalions. These were important facilities for the North Vietnamese defenses, as each SAM site had just twelve missiles available for use. Once these were fired, the site needed to be replenished. The storage facilities, which were inside large warehouse complexes, would have made excellent targets for the B-52s, but may have been eliminated from the target list for fear of creating large numbers of civilian casualties.[16]

The briefing for the B-52 crews scheduled to take off from Andersen Air Force Base on Guam began at eleven o'clock in the morning (another group of B-52s would take off from U-Tapao Royal Thai Navy Airfield in Thailand). The projection showing the route to the targets in Hanoi was highly unsettling to the men who were about to attack what was considered to be the most heavily defended city in the world. Capt. Steven Brown, a B-52 copilot, stated that it looked like "a line of ants going to a picnic."

Even more disturbing to aircraft commander Capt. Dwight Moore was "The map of the Hanoi area on a poster board in front of the briefing room showing known active SAM sites—50 plus dots (it looked like it had measles) with the flight path line going right down the middle of them, then turning right about 150 degrees and coming back out through the dots." Captain Moore did not think that looked good, nor was he thrilled when the weather guy told them they would be turning into a headwind for the outbound leg from the target.[17]

CHAPTER 19
LINEBACKER II

INTO THE DRAGON'S TEETH

During the night of December 18, 1972, and the early hours of the 19th, 129 B-52 Stratofortress bombers from U-Tapao Royal Thai Air Force Base, Thailand, and Andersen Air Force Base on Guam began Operation Linebacker II, striking a variety of targets in three waves hitting the Kinh No storage complex, the Yen Vien rail yard and repair facility, the Hanoi radio station, and three airfields around Hanoi. The B-52s approached their targets in a trail formation of three-ship cells, separated by ten-minute intervals later known as an "elephant walk." While the B-52s were attacking their targets, fourteen F-111 Aardvarks bombed airfields, port facilities, transportation centers, and the radio communication center at Lang Truoc airfield. The F-111s, which arrived in Thailand on September 25, 1972, were an all-weather, swing-wing fighter capable of flying at night, at low altitude, and at supersonic speed.[1]

The strike force was supported by a multitude of other aircraft. Sixty-three F-4 Phantoms were used as escorts and provided a combat air patrol over MiG airfields. Seventeen Wild Weasels (thirteen F-105Gs and four F-4s) aided by thirty-four Navy A-6s and nine A-7s were assigned the Iron Hand mission of suppressing the SAMs. Three Air Force EB-66Es from the 42nd Tactical Electronic Warfare Squadron accompanied the first and second waves to provide electronic countermeasures. Five Navy EKA-3Bs did the same for the third wave. Assigned to orbit just forty miles north of Hanoi, the EB-66Es attempted to jam both GCI and SAM radars that were employing electronic countermeasures such as band switching and frequency changes to make it more difficult for the U.S. aircraft to jam them.[2]

To hide the B-52s from Fan Song radars, twenty-two F-4s (four first wave, ten second wave, eight third wave) carrying ALE-48 bulk chaff dispensers or M-129 chaff bombs preceded each wave in an attempt to create a chaff corridor three miles wide and fifty miles long. In order to create the corridor, the chaff-dispensing F-4s had to fly through the defended area at 35,000 feet. To maintain this altitude with the added drag of the chaff dispensers, the F-4s had to fly with one engine in afterburner and the other at military power. This left only a small margin of performance to maneuver if they encountered SAMs. "The F-4 guys scheduled to fly the first Chaff sorties," according to the account later given by Rudy Smart, "figured they had been given a suicide mission." The chaff corridor, they believed, was going to look like the tip of an arrow pointing directly at them. The North Vietnamese would see this and launch their missiles right at the point. The F-4 pilots, according to Smart's recollection, thought they "were good as dead."[3]

Because of unexpected high winds, the chaff corridor was blown off course, leaving the B-52s dependent on their own electronic countermeasures. The most important jammers carried by the B-52s were those covering the E/F-bands of the Fan Song's acquisition radar and the downlink channels used to control the missiles. The B-52s had a lot of jamming power, especially those modified with the Phase V ECM suite (all the D models and most of the Gs had ten jammers, including six of the more powerful ALT-28s). The few B-52Gs that had not been upgraded still had seven jammers, although the older ALT-6Bs and ALT-13s provided marginal jamming signals. For mutual support it was necessary for all the aircraft in a cell to maintain precise spacing, altitude, and airspeed within a cell. The three airplanes in a cell flew almost in trail with a one-mile separation between airplanes, with each succeeding aircraft five hundred feet higher than the one in front. If one of the aircraft moved out of the formation, one-third of the cell's mutual protection was lost. An aircraft separated from the cell stood out "like a sore thumb because his jammers highlight his position." This allowed the North Vietnamese missileers to target the B-52 using a track-on-jam mode of operation.[4]

As the first wave of B-52s approached North Vietnamese air space from the southwest, the Spoon Rest early warning radar operated by the 45th Radar Company, 291st Radar Regiment, Vietnamese People's Army Air Defense Corps, in the village of Nghe An in the western edge of North Vietnam, began to pick up radar returns from the bombers. The Soviet-made

Spoon Rest radar was one of thirty-one radar stations that made up the first line of North Vietnam's early warning system. The operators watching their Spoon Rest radar scope knew from the jamming pattern—which they had observed many times before—that they were watching radar returns from B-52s. As he watched the returns "proceeding north in a stately procession," Dinh Huu, the commander of the 45th, sent a message to his regimental headquarters advising that a large number of B-52s appeared to be headed for Hanoi, which in turn was forwarded to the Air Defense Headquarters in Hanoi. The position coordinates of the B-52s provided by the 45th Radar Company were entered in wax pencil on a large Plexiglas map that was used to keep track of the American airplanes (in red) and their own MiGs (in blue).[5]

The Air Defense Headquarters monitored the overall picture of the emerging battle, but after identifying the raid it turned control over to the regimental headquarters, which first picked up the U.S. aircraft on its Soviet-supplied P-15 (Flat Face) long-range search radar and associated (Side Net) height finder. The incoming targets were then tracked using the regiment's P-14 (Tall King) and P-12 (Spoon Rest) early warning/tracking radars. When a raid was detected, the regimental headquarters picked out a target, gave it a number, then passed the range, bearing, and altitude information to the SAM battalion most likely to engage the target.[6]

As soon as the missile crews were alerted that the B-52s were inbound, trucks carrying the missile control vans started up their noisy diesel engines to provide power to the units. In the command van, the site's commander turned the power switch to standby mode to warm up the system, which took four minutes. Once it was warmed up, it took just four seconds for the Fan Song to go from standby to full power.[7]

As the B-52s approached the various missile batteries in the target area, the range and bearing information provided by headquarters allowed the battalion commander to locate the target and start a manual plot that was maintained on a plotting board in the control van. The missile crews tried to track the B-52s passively by following the jamming strobe instead of turning on their Fan Song radars to prevent the risk of an antiradiation missile attack, but as Michel indicated, "passive tracking was not working—jamming was too strong."[8]

When the B-52 being tracked came within range of the battalion's SA-2 missiles, the commander switched on the Fan Song radar. If the commander had enough experience and familiarity with the effects of jamming, he could probably locate the jamming strobe produced by the B-52's jammer, which could then be used to track the target by placing the crosshairs on the jamming strobe indicated on the radar's display. The jamming strobe was too unstable for use in automatic tracking, so the crew had to rely on manual tracking in an attempt to shoot down the B-52. This was accomplished by the guidance officers, who tried to keep the target return centered in their radar scopes using small steering wheels under the scope while twisting the gain control knobs in an effort to sharpen the strobe into a useable target.[9]

As each of the B-52s entered the area defended by SAMs, the Electronic Warfare Officer (EWO) in the narrow compartment behind the pilots began the arduous task of identifying and then jamming the Fan Song radars targeting their aircraft. He had two screens to work with: the APR-20 receiver's rectangular 8-inch by 12-inch display that showed all of the North Vietnamese electronic signals, and the APR-25's smaller circular display. The length of the strobe on the APR-20 showed the approximate distance to radar. The EWO selected the right frequency for jamming by centering the cursor over the highest strobe. The direction of the strobe emanating from the center of the APR-25 display indicated the direction of the radar and the length of the strobe its approximate range.[10]

Andrew N. Vittoria Jr., an EWO on one of the B-52Ds in the first wave, explained how he went about this task:

> I used my manual dot to scan from 2800 to 3100 MHz, back and forth, looking for Fan Song radar signals. When I found them[,] I put them up on my expanded trace and marked the point with a grease pencil. Then I narrowed down the bandwidth of four E/F band transmitters [ALT-6 s and Alt-28s], and set two on the Fan Song azimuth beam and two on the elevation beam. Then as instructed, we were to aim the missile downlink signal in the blind. So I estimated where they would be and put a grease pencil mark there. Then I brought my remaining E/F band transmitters and set them on the mark.

I picked up A-band signals from a Spoon Rest acquisition radar and put a jammer [the ALT-32] on it. Once the jammers were transmitting, I monitored them the whole time, to see if the radars were still painting me. And always, out the corner of my eye, I kept watch on Tract Two in case there as a SAM uplink signal.[11]

The B-52G gunner on a later raid told the copilot of his aircraft, Capt. Edward Wildeboor, that while they were nearing the bomb release point, he had observed the plane's Electronic Warfare Officer "on the edge of his seat 'playing his receivers and jammers like a pinball machine' while [the crew] concentrated on the bomb run."[12]

Charlie 01, piloted by Lt. Col. Donald Rissi, was in the lead of a nine-plane formation of three cells of B-52Gs headed for the Yen Vien railroad complex. As Charlie 01 approached the target, it was assigned a number by the North Vietnamese Air Defense Headquarters, which passed its range and bearing information to the 59th Battalion of the 161st Regiment. The battalion, commanded by Nguyen Thang, had already launched a series of SA-2s at an F-111 and a B-52 without success. Now, as he turned the antenna of his Fan Song radar toward the assigned target, he saw heavy jamming on the radar scope, indicating a cell of three B-52s. Thang could not make out the range because of the jamming, but he continued to track the movement of the jamming strobe until the return stabilized, at which point he turned it over to the guidance officers for manual tracking.[13]

"As the B-52 approached Thuan saw the three crewmen were tracking the target well," unknowingly aided by the fact that Charlie 01 was one of the unmodified B-52Gs. The lower-powered ALT-6B jammer enabled the Fan Song radar to burn through the jamming signal at a greater range, giving Thuan's crew just enough time to establish an accurate target track. Noise jamming, as utilized in the ALT-6B, worked by emitting a signal that buried the receiver in noise. The farther away the transmitting radar was, the smaller the return. All the jammer signal needed to override the return was to be stronger than the return. As the target got closer to the transmitting radar, however, the radar return to the receiver got stronger. At some point the return became stronger than the jammer signal, overriding the jamming. A template over Thuan's radar scope marked with engraved lines showed the range of the SA-2 missiles in his battery. When Charlie 01's return was well within the missile's range, he checked the lights on his control board to make sure two missiles were ready to fire and pressed the fire button. Six seconds later he fired a second missile.[14]

Michel described what happened next:

> The first SA-2 missile, 35 feet long and weighing almost 5,000 pounds, was started on its way by its solid fuel booster in the tail, which burned with a very large, bright flame for five seconds. As the fuel in the booster burned out the booster motor dropped off, the liquid-fueled sustainer engine cut in and the Fan Song radar acquired the transponder located in the rear of the missile.
>
> As each of the guidance controllers turned their wheel to keep the strobe centered, the Fan Song's computer began to direct the missile's onboard guidance system to follow the radar beam. This guidance information was sent to the missile by means of a radar beam.[15]

This guidance information sent to the missile, known as the uplink, was detectable by U.S. aircraft equipped with radar warning receivers that indicated that a missile had been fired. The B-52s, however, were under orders not to maneuver during the run-in to the target and the four-minute bomb run, which negated the possibility of using the early warnings to avoid the missiles. All the B-52s were also equipped with chaff dispensers that could have been used to counter the ground-launched missiles, but the EWOs were instructed to release chaff only if their plane came under attack from fighters. Price claims that "Dropping Chaff injudiciously would have betrayed the plane's range and assisted those trying engage it." I tend to believe Michel, who believes that SAC forbid its release "because chaff was difficult to load into the B-52." The latter fits into SAC's culture and their faith in jamming. Price's explanation does not make much sense, since chaff defeated the SAMs when used.[16]

Charlie 01 was about to release its bombs when the first of the two SA-2s launched by the 59th Battalion detonated close beside it, followed shortly by the second. Capt. Arthur "Rex" Rivolo, flying as an escort in his F-4E below, "watched in amazement as the giant airplane cracked open like an egg. . . . It seemed like the wing was blown off and, with fuel streaming out, it rolled slowly over on its back, raining fire as the fuel burned and slowly fell toward the ground in a sheet of flames." Only three of the crew survived the attack. Colonel Rissi; his copilot, 1st Lt. Robert Thomas; and their gunner, MSgt. Walter Ferguson, were killed. It was the first B-52 brought down by the North Vietnamese missileers, but not the only B-52 lost that night to SA-2s. Also lost were Peach 02, another B-52G, and Rose 01, a B-52D.[17]

The loss of Rose 01, which had the latest jammers, was a harbinger of things to come, as the 77th Missile Battalion had discovered a critical weakness in the tactics used by the B-52s that enabled them to shoot down more of the big bombers. During the second wave on the first night of the attack, the battalion's commander, Dinh The Van, had been unable to break out a B-52 from the jamming on his radar screen, but as he watched the radar returns he noticed the point at which the jamming dropped off. Unknowingly, he had discovered what the Eighth Air Force had feared would happen when bombers executed their high-angle post-target turns into the wind.[18]

While he waited for the next attack to develop, Dinh discussed with his fire-control officer how they were going to modify their firing procedure for the next raid. They would manually track a B-52 until the moment the return appeared clearly on the radar screen. At that point they would go into the automatic tracking mode, which was much more accurate than manual control. "It is difficult enough to guide the missiles under normal conditions when the targets are clearly seen," explained Nguyen Thang. "It was even more difficult looking at the silky crepe jamming of the B-52 aircraft on the radar screen. An uneven rotation or a mere jerky movement of the control wheel could cause the missile to deviate from the target by thousands of meters or even detonate in the air."[19]

During the next raid, Dinh watched carefully as the three officers tracked the jamming of the lead B-52 they had been assigned to target. He called for automatic tracking when the return cleared and pushed the button to launch two missiles in close succession.[20]

Rose 01, a U-Tapao B-52D, led the third wave of twenty-one B-52s assigned to hit the Hanoi radio station. As the B-52 approached the target, Hal "Red" Wilson and his copilot, Charles "Charley" Brown, saw a large number of SAMs. As they rolled into the posttarget turn, one of the missiles passed between the wing and the stabilizer without detonating. As the second missile launched by Thang closed on Rose 01, the SA-2's 400-lb. warhead detonated, spewing hundreds of pieces of shrapnel in a circular pattern, blowing a hole in the fuselage big enough for the navigator-bombardier to see the external bomb racks, and setting the cockpit on fire. Bob Steffen, flying behind in Rose 02, saw a fireball as the missile exploded. "For a second," he later told Marshall Michel, "it didn't register, but then I did an almost instantaneous double take as we flew through billowing black smoke, like an oil fire, and then we heard the beepers [emergency radio beacons carried by the aircrew]." Only four of the six Rose 01 crewmen survived. All were picked up by the North Vietnamese and remained in captivity for 103 days.[21]

During the night, the North Vietnamese missileers had shot down three B-52s and damaged two more while also downing a Navy A-7. Although these losses were within the range expected by U.S. commanders, it was a tremendous morale and propaganda victory for the North Vietnamese, "who lost respect for the B-52s, and the idea of them being 'super flying fortresses' became a bit of a joke." But the North Vietnamese forces had paid a high price for this achievement, launching as many as 164 SA-2 missiles. "Each battalion had begun the first night with twelve missiles, six on their launchers and six more on trucks that quickly moved in and reloaded the launchers after the missiles were fired. But," as Michel revealed, "because of a lack of planning by both the General Staff and the Air Defense Headquarters, there were very few pre-assembled missiles in storage warehouses." During the next day the North Vietnamese worked feverously to rearm the missile batteries, but they were hindered by the fact that there were only two technical battalions to supply missiles for three regiments.[22]

While the North Vietnamese were rushing to replenish their stocks of missiles, Air Force leaders were reassessing the results of the night's attacks. Of the five B-52s lost, three were hit just prior to the target, when the bombers were believed most vulnerable due to burn-through, and two were struck in the posttarget turn. Two of the three B-52s had been "G" models, which was somewhat disconcerting because while they represented one-quarter of the force, their losses amounted to two-thirds of those lost. This contributed to the Eighth Air Force doubts about SAC's claim that the unmodified B-52G was just as effective as the modified version.[23]

At Seventh Air Force Headquarters, the problem that needed to be addressed was the failure of the Iron Hand missions to suppress the SAMs. The suppression force fired twelve AGM-78 Standard ARMs and forty-seven AGM-45 Shrikes, but were credited with damaging only two missile sites. Seventh Air Force attributed this to the late arrival of the mission plan, which did not even include the B-52s' entry and exit headings. Without the B-52s' headings, the Wild Weasels could not fly parallel to the bomber stream so they would be pointed at the SAM sites in order to be in a position to launch their missiles as soon as a Fan Song radar came up. In addition, they were unable to meet the B-52s at their scheduled time over the target, forcing them to launch most of their antiradiation missiles pre-emptively, hoping that a Fan Song would be turned on while the missile was in flight.[24]

After interviewing the flight crews, recommendations were made to SAC Headquarters that permission be granted for the bombers to maneuver until just prior to the bomb release point. Suggestions were also made that the ingress and egress routes be changed so that a pattern was not set, which would make it easy for the enemy to preposition his forces.[25]

Partly because the losses on day one were not greater than expected, and because of long lead time from planning to execution, SAC Headquarters made the decision to continue with the same attack plan on the second night. As with the first night's mission, the B-52s were not allowed to maneuver from the initial aiming point to the target. Neither of these conditions were favored by either the flight crews or the staff of the Eighth Air Force, which asked SAC to change both the maneuvering restriction and to change the ingress and egress routes. "Time compression," according to McCarthy, who led the first wave of B-52s from Guam the night before, "made it impossible to clear changes in tactics through the higher [SAC] headquarters. . . . Day Two was to be Day One all over again."[26]

CHAPTER 20

LINEBACKER II

LOSSES LEAD TO NEW TACTICS

The attacks on the night of December 19, 1972, were almost a mirror image of those conducted the night before. In addition to the Bac Giang transshipment center and the Thai Nguyen power plant, which were struck for the first time, the B-52s returned to the Kinh No vehicle repair complex, the international radio station, and the Yen Vien railroad complex. Despite the failure to change routes, timings, or tactics, the B-52s managed to avoid most of the missiles fired by the North Vietnamese, except for Ivory 01, the only B-52 seriously damaged that night. Ivory 01, a B-52D piloted by Maj. John Dalton, was in the lead position of the first cell attacking the Hanoi International Radio Station when it was struck during its post-target turn. The exploding SA-2's 400-lb. warhead caused near-fatal damage to the bomber. "One engine was on fire and the other engine in the pod had flamed out, an alternator was running away and over speeding, cables for the rudder and right elevator were severed and both drop tanks had fuel pouring out of holes the missile had blown in them." Dalton headed for the nearest emergency base, the Marine airfield at Nam Phong, and managed to land the severely damaged aircraft in a spectacular exhibition of superb flying. Because the B-52 did not crash within their borders, the North Vietnamese were not aware of how close they had come to downing another of the "invincible" Stratofortresses.[1]

The Hanoi International Radio Station was frequently used by the North Vietnamese as a propaganda tool to castigate Nixon for prolonging the war and refusing to settle it. Nixon insisted that it be included on Linebacker II's target list even though it was a poor target (it was small and well-protected by blast walls). Nevertheless, during the first two days of the Linebacker II campaign, thirty-six B-52s were assigned to bomb it. Although they dropped 2,900 bombs, they managed to shut the station down for a total of just nine minutes.[2]

The staff and leadership of the North Vietnamese Air Defense Command was furious at the failure to bring down any B-52s that night, despite the number of SA-2s launched by the various missile battalions defending Hanoi. While American aircrews were preparing for yet another night of bombing, all the battalion commanders were called to the Air Defense Command Headquarters for self-criticism and to explain the techniques they had used during the missile launching process and why it had failed.[3]

After examining and analyzing the results of each battalion's engagement, the SA-2 technical experts and training officers decided that the best chance of success was to standardize tactics based on a variation of the automatic tracking technique pioneered by the 77th Battalion. The experts recommended using manual guidance at the jamming strobe until the bomber's return became clear, at which point they should switch to automatic tracking. If they were unable to switch to automatic tracking, they needed to continue in the manual mode. They were also told not to fire until the bombers were close, going "face to face with the B-52s," as it was later described.[4]

The Air Defense Command, having observed that the B-52s had used the same northwest to southeast route to Hanoi during the previous two nights, correctly guessed that the bombers would fly the same route again. To bolster the defenses around Hanoi they moved two battalions from the south to the north of the city so that their missile sites formed a triangle around Hanoi. To protect the missile sites from the dreaded antiradiation missiles, the staff at the Air Defense Headquarters prepared a cunning plan to trick the Wild Weasels into firing their missiles prematurely. When headquarters received word from the early warning system that Wild Weasels were approaching, they planned to order two missile battalions to briefly turn on their Fan Song radars, using the "side pointing technique" to lure the missile away from the site. Shrike was easy to defeat because of the distinctive flight profile used to fire the missile that could easily be identified on the radar display, as well the launch itself. As soon as this was recognized, the North Vietnamese radar operator would briefly turn the Fan Song radar to the side before turning it off. The Shrike followed the beam away from the radar and would then "go dumb" and crash away from the site when it lost the signal. The ruse worked in part because of the overabundance of Shrikes. Only one in five of the antiradiation missiles employed in Linebacker II were the more advanced Standard ARMs. Only one SAM battalion was put out of action by an antiradiation missile, the 76th, when it stayed on the air too long.[5]

After two nights of bombing, B-52 losses at 1.5 percent were well within the acceptable limits previously established by SAC. Although no B-52s had been shot down the night before, a large number of SAMs had still been seen by the B-52 aircrews. As Marshall Michel notes, "SAC did not understand how important feedback was in combat and there were no mechanisms in place to take crew inputs." SAC's leadership did not consider the North Vietnamese defenses to be much of a challenge, which contributed to their initial lack of interest in looking into the problems experienced by the bomber crews while attacking targets in and around Hanoi. This soon changed.[6]

Ninety-nine B-52s took part in the third night's operation that began on December 20, 1972, using the same routes and tactics as the night before along with a chaff cloud that pointed the way to Hanoi before strong winds blew it away. The first wave consisted of sixteen cells of three B-52s, nine of which were tasked against the Yen Vien railroad yard and the adjacent Ai Mo warehouse area. The attack was led by Quilt 01, a B-52G whose ECM suite was degraded by the failure of two of its E/F jammers. The last bomber in the cell, Quilt 03, a B-52G with the older electronic suite, had also lost two of its E/F jammers. Nevertheless, the three planes maintained their integrity and completed their bomb runs unscathed.[7]

As the B-52s led by Quilt cell headed for Hanoi, they were picked up on the North Vietnamese early warning net, which reported their range and bearing to the Air Defense Command Headquarters, which assigned it target T621 and ordered the 93rd Battalion to engage it. When the three B-52s of Quilt cell were within twelve miles of the battalion's site, its commander launched two SA-2s using strobe tracking and manual control until the return broke out, at which point he ordered the automatic control to be turned on. Quilt 03 had just begun its posttarget turn when one of the missiles exploded close by, damaging the hydraulic control system and setting the aircraft on fire. Four of the crewmen managed to eject safely before the B-52 crashed in a ball of flame. The next two cells in line, including the B-52D piloted by Glenn A. Russell, made it through the target area without being hit. As he began his posttarget turn, two SA-2s began to close on the aircraft. His Electronic Warfare Officer, a combat veteran with almost 500 Arc Light* missions, in direct violation of a SAC directive, released chaff, which deflected the missile away from the B-52. A few seconds later, Russell turned sharply to avoid another missile, allowing his aircraft to return to base unharmed. But Brass 02, a B-52G with a degraded ECM suite in the following cell, was also hit during the post-target turn. The stricken aircraft made it back to Thailand before the crew had to eject and were safety recovered. A minute later, Orange 03, three cells behind Brass, was struck during its posttarget turn, went into a spin, and exploded in midair. Only two of the crew escaped.[8]

The loss of the three B-52s—about 10 percent of the first wave, including two of the twelve G models—sent shock waves through the SAC Headquarters staff. They now realized the vulnerability of the unmodified Gs. In what Michel termed "an incredible example of poor planning," they had assigned six B-52Gs in the second wave to attack the Yen Vien railroad yards in the heart of Hanoi's missile defenses for the third time. General Meyer, fearful of losing more of the precious B-52Gs that were the backbone of SAC's nuclear deterrent, recalled the six B-52Gs in the second wave while they were on their way to the target and had yet to bomb. As Michel makes clear, "the North Vietnamese defenses had done something that the Germans, Japanese, Soviets, Chinese, and North Koreans had never been able to achieve. They had made an American bombing raid abort a mission for fear of losses." The remaining B-52Ds of the second wave, which were assigned targets away from Hanoi, struck the Bac Giang transshipment point and the Thai Nguyen power plant without loss.[9]

The third wave that night was made up of twelve B-52Ds assigned to strike the Hanoi petroleum storage area and twelve B-52Gs assigned to hit the Kinh No complex a few miles north of Hanoi. This took them over the

---

* Arc Light missions were carried out by B-52Ds flying from Guam to provide close-air support to ground operations in Vietnam.

heaviest concentration of SA-2s during their posttarget turns. By then "it was apparent after the first raid losses that the unmodified Gs were neither protecting themselves nor the formations adequately and were bearing the brunt of the losses inflicted by Hanoi's SAM sites." The three cells of B-52Gs would drop only 324 bombs, which was the same number of bombs dropped by a single cell of B-52Ds. So recalling the B-52Gs, as McCarthy and Allison note, "would not significantly alter the tonnage delivered on the targets" in North Vietnam.[10]

Michel's account of what occurred at SAC Headquarters at this time is so important that it is quoted verbatim:

> But the SAC staff was clearly disturbed that General Meyer had listened to General Johnson and worse, Seventh Air Force, and canceled the second raid [B-52Gs]. General Harry Cordes, SAC's Chief of Intelligence said, "We wanted to prove that SAC could do the job." And the SAC staff pressed General Meyer to send the G models to attack. . . .
>
> The staff's implication was clear—if the B-52s were recalled, there would be a further diminution of SAC's credibility and reputation in and out of the Air Force. Then, according to Cordes, the staff offered a final, seemingly incredible rationale, telling Meyer, "never in the history of the United States Air Force had a bomber attack been turned back by enemy action." This completely ignored what happened just a few hours before when Meyer had turned back the second raid's G models, and also ignored SAC's policy for most of the war, to turn the B-52s back when they were threatened by North Vietnamese defenses. But the selective memory lapse served its purpose—General Meyer ordered the G models in the third raid to continue.[11]

The results were disastrous. After leading the attack, Olive 01 became separated from the rest of the cell and was hit by an SA-2 during its post-target turn. It "went down like a torch about nine miles north of Hanoi and only two of the six crew members survived." Tan 03, a three-plane cell of unmodified B-52Gs, followed eight minutes later. Tan 03 experienced a complete failure of its bombing radar and had drifted far from the rest of the Tan cell when it was stuck in the forward fuselage by a missile fired by the North Vietnamese 57th Battalion. The gunner, the only survivor, bailed out just before a missile from another site hit the aircraft, which disintegrated.[12]

Five minutes after the B-52Gs attacked the Kinh No complex, the twelve B-52Gs targeting the Hanoi petroleum storage area began their bomb run. An SA-2 detonated close to Brick 02, piloted by Capt. John Miz, during his posttarget turn. It put nineteen holes in the bomber without causing serious damage, and Miz was able to bring the aircraft back to their bases at U-Tapao. The crew of Straw 02, a B-52D that had earlier attacked the Gia Lam railroad yard, was not as lucky. The explosion of an SA-2 ripped through the bottom of the bomber just after the last of its bombs were released. It knocked out all of the power, holed the fuel tanks, and seriously wounded two of the crew. The pilot tried to make the Laos border, but the crew had to bail out over the western part of North Vietnam, where some of the crew were rescued the next day.[13]

Based on the interviews of the aircrews conducted after the action, Air Force intelligence officers estimated the North Vietnamese had fired more than 220 SAMs during the night, resulting in the loss of six B-52s along with a Navy A-6. Another B-52 was severely damaged. Of the twelve B-52Gs of the third wave that were sent to Kinh No, two were shot down and nine of the twelve crewmen were killed. None of the B-52Gs would be sent to Hanoi again. The after-action reports also revealed the continuing difficulties in trying to suppress the SAM sites through the use of the Wild Weasel Iron Hand missions. The Weasels were unable to confirm damage to any missile sites despite having launched sixty-one antiradiation missiles.[14]

Upon returning to base, several of the EWOs reported intercepting signals on their frequency-agile receivers from an unfamiliar radar operating in the I-band. It came from an unknown unfriendly source dubbed T-8209. Most likely it was the signal from a Soviet-supplied SNR-125 (Low Blow) fire-control radar developed in conjunction with the S-125 Neva/ Pechcora (SA-3 Gao) surface-to-air missile system. The Air Force Communications

Center, which was responsible for monitoring and classifying North Vietnamese radar signals, concluded that a new, unjammable North Vietnamese radar was guiding missiles from the SAM site that had been credited for downing three of the B-52s shot down so far. This created a crisis, since the jammers carried by the B-52s could not jam the new system. The Marine Corps EA-6As, which did have a system that could jam the new radar, were flown to bases in Thailand in an emergency airlift in the hope that operating new versions of AGM-45 Shrike and AGM-78 Standard ARMs tuned to the new signal would knock out the phantom North Vietnamese "killer site."[15]

Meanwhile, back at SAC Headquarters in Omaha, the "staff was completely befuddled by the losses." In one night, the North Vietnamese had shot down 3 percent of all the B-52s in Southeast Asia. In meeting after meeting, they discussed ways to bring down the losses. Maj. David Sjolund, an Air Force intelligence officer who had previously calculated the estimated B-52 loss rate, was sent to Eglin Air Force Base to see what could be done to improve the effectiveness of B-52 jamming.[16]

On the fourth night, December 21, 1972, thirty B-52Ds from the 17th Air Division at U-Tapao, were scheduled to hit three targets in the vicinity of Hanoi: the Quang Te airfield, the Bac Mai storage area, and the Van Dien supply depot. The original operation plan that SAC prepared had the same single-file, northwest to southeast routes and tactics that had been used the previous three nights. Earlier that morning, Brig. Gen. Glenn R. Sullivan, the 17th's commanding officer, unhappy with the casualties his force was incurring, sent a message directly to General Meyer in SAC Headquarters recommending new tactics. Although Sullivan had sent an information copy to his immediate superior, the Commander-in-Chief of the Eighth Air Force, General Johnson, he committed a major breach of military procedure that would negatively affect his future prospects for promotion.[17]

Both generals heeded Sullivan's recommendations and saw to it that changes were made to the tactics used on the fourth night's attack. Instead of their posttarget turn back into the jet stream, the bombers would make smaller turns and fly out of North Vietnam over the Gulf of Tonkin. In addition, the bomber streams would be more concentrated than previously with the interval between cells reduced to two minutes instead of the previous four minutes. This reduced the time over target, making it more difficult for the North Vietnamese to pick up succeeding cells.[18]

The attack on Quang Te and Van Dien took place without losing any more B-52s, but Scarlet 01, in the lead cell attacking the Bac Mai airfield, ran into trouble. It lost its bombing and navigation radar, forcing the crew to execute a complicated maneuver so that it could drop back in the formation, which provided the 57th Battalion's Fan Song radar a clear return. Two of its SA-2 missiles exploded under the plane's right wing, which was immediately engulfed in fire. As the bomber began to roll right, the burning wing folded over the top of the B-52 and it plunged toward the ground. Four minutes later, Blue 01, about to drop its bombs on the same target, was hit and exploded after the crew bailed out. No more B-52s were shot down that night, but eight B-52s had been lost in two days, a loss rate twice what SAC had predicted.[19]

Despite recommendations from Adm. Noel A. Gayler, Commander-in-Chief Pacific, and Secretary of Defense Melvin Laird to halt the bombing, President Nixon ordered the bombing to continue on December 22, the fifth consecutive night of the bombing. The loss of two B-52s on December 21 caused SAC to shift targets from Hanoi to the port city of Haiphong, with twelve B-52Ds assigned to strike the Haiphong railroad siding and eighteen B-52Ds to hit the Haiphong petroleum storage area. The two groups were to approach from the south over the Gulf of Tonkin, conduct their bomb run, and then turn and exit over the Gulf of Tonkin. This was good news to the aircrews, who knew that if they were hit, they could make it over the gulf, where they could be rescued by the Navy. Instead of flying in trail, the cells fanned out in three different tracks, and as they approached the targets they abruptly split again and attacked on six different tracks staggered in time, distance, and altitude. The B-52s were supported by fifteen F-111s, thirty-four F-4s, and twenty-four Navy A-7s to disrupt and distract enemy defenses. In another change in tactics, Seventh Air Force directed two flights of F-4s to lay a chaff blanket designed to drift over Haiphong before the bombers arrived, while Navy Iron Hand aircraft attacked seven SAM sites in the Haiphong area.[20]

The attack caught the North Vietnamese off guard. "The chaff blanket confused their tracking radars, and the rapidity with which the B-52 force struck and departed disrupted their manual tracking process." No B-52s were downed, the first time since the Linebacker II attacks began.[21]

On December 23 (the sixth night of attack) twenty-four B-52Ds were sent to bomb the storage area and railroad repair shops near the village of Lang Dang, fifty miles northeast of Hanoi. Six B-52Ds were also sent in flights of two to bomb three SAM sites north of Haiphong. Twelve F-111s and seventy other tactical aircraft supported the heavy bombers. "The bombers were told to delay their post-target turn to clear the target defenses, then make two small turns rather than one large one, and to fly in a significantly larger altitude 'block,' from 31,000 to 38,000 feet, varying their altitudes as they approached the target, then changing altitudes again after they dropped their bombs." A feint attack on Hanoi and a last-minute turn toward the targets caught the North Vietnamese missileers off-guard and only five SAMs were fired. None of the missiles struck any of the B-52s, all of which successfully returned to U-Tapao.[22]

The attack on December 24 was similar to the previous night, with thirty B-52Ds from U-Tapao striking railroad yards at Kep and Thai Nguyen well away from Hanoi. On this night a new tactic was added to the B-52s' playbook: they were now authorized to drop chaff during their now smaller post-target turns. This useful tactic had previously been forbidden by SAC because the chaff was difficult to load.[23]

While the bombing continued over North Vietnam, the B-52 jamming tests conducted against Eglin's SADS-1 surrogate Fan Song radar got underway. The tests confirmed that the posttarget turn did blank the B-52s' jamming antennas, but by then the operational tactics had been significantly changed to reduce the severity and length of the post-target turn. But testing at the Rome Air Development Center in Rome, New York, revealed another problem: the jamming antenna pattern on the B-52s was optimized for the low-level penetration and bombing approach that SAC preferred for their nuclear strike role against the Soviet Union. According to Sjolund, "the antenna pattern beneath the airplane was very narrow in elevation, very broad in azimuth. . . . We just weren't getting the energy into the radars from high altitude."[24]

The Eglin tests also revealed that the B-52's jammers were ineffective against the beacon transponders of the model 11D version of the SA-2 missile that the North Vietnamese had been using since 1971. The transponder had been modified in response to the recommendations of the Soviet technical experts sent to Vietnam in 1966 to examine the performance of the SA-2s. Although the staff at Seventh Air Force had known that the beacons on the new missile were unjammable for the past year, either the air staff at SAC had not read the Seventh Air Force reports or they had not paid attention to them. In any case, the B-52 Electronic Warfare Officers were quick to use their jammers on other parts of the SA-2 system.[25]

Following the missions flown on the December 24, Nixon ordered a thirty-six-hour halt in the bombing. This gave the North Vietnamese breathing space and allowed them to resupply the missile battalions that had expended their immediate supply of SA-2s. Hanoi was worried about resupply. About eight hundred missiles were in storage, but they needed assembly and delivery to the SAM battalions. It also gave SAC a chance to review events so far and give the crews some rest. During the interlude, in a major change in policy, SAC transferred planning and operational authority from Omaha to the Eighth Air Force on Guam, tightening the operational structure and improving coordination. But SAC still retained the prerogative of selecting the targets and the number of B-52s assigned to each target.[26]

With planning now in their hands, the staff at Eighth Air Force prepared an operational plan for the next phase of the air campaign, incorporating many of the suggestions made by the aircrews. The bombing, which resumed on the night of December 26 (day nine of the bombing), was a maximum effort involving 120 B-52s. Instead of dividing the bombers into widely spaced waves, a single massed attack hit all targets simultaneously. The B-52s used four different routes to the target: two raids coming in from Laos and exiting over the Gulf of Tonkin, and two raids entering from the Gulf of Tonkin and split into seven target groups. Seventy-two B-52Ds attacked targets in the Hanoi area: Gia Lam, Doc Noi, and Giap Nhi rail yards; the Kinh No storage complex; the Van Dien vehicle repair shops; and the Hanoi petroleum storage area. Forty-five B-52Gs attacked targets in less defended areas away from Hanoi including the Haiphong and Giap Nhi railroad yards and the Haiphong transfer station. They ran in from all directions at different altitudes and were protected by a ten-thousand-foot chaff corridor twenty miles wide by thirty miles long offset from the target area by an amount proportional to the wind. Four F-111s bombed the airfields fifteen minutes before the B-52s entered their bomb runs to eliminate the MiG threat. Five other F-111s bombed selective targets after the main attack had ended.[27]

To limit SAM losses, SAC planners had asked the Joint Chiefs of Staff for permission to attack the Quinh Loi SAM storage and assembly facility in Hanoi with B-52s. The JCS authorized the attack, but the target, which was in the heart of a Hanoi neighborhood, was not considered suitable for the B-52s because of the potential for a large number of civilian casualties. The weather was overcast, which precluded the use of precision-guided munitions or visual bombing by tactical air. Unpersuaded by flawed claims about the accuracy of LORAN bombing, the Joint Chiefs of Staff gave permission to strike the SAM warehouse using F-4 fighter-bombers led by specially equipped F-4D pathfinders that had the ARN-92 LORAN-D long-range navigation system installed. "LORAN," as defined by Michel, "involved triangulation between three stations that gave extremely precise navigation and position data when the system functioned properly. It had been in combat use in Vietnam since 1970, but had major problems with range and reliability."[28]

Retired Marine Reserve Col. Hays Parks' description of the attack, which he called "one of the more remarkable feats of the air campaign," claimed "the target was destroyed by 16 LORAN-guided F-4s bombing through solid overcast from 20,000 feet." This is at odds with Michel's rendition of the action: "Given the inaccuracies of LORAN and the location of the target, the strike could have caused a large number of civilian casualties, but fortunately as the U.S. aircraft approached the [*sic*] Hanoi the clouds parted and most of the flights bombed visually and accurately. The next day another large tactical strike bombed the Trai Ca SAM storage area using LORAN and missed badly, but Trai Ca was thirty miles north of Hanoi in a relatively uninhabited area."[29]

Despite the loss of two B-52Ds belonging to cells that reached the target with only two aircraft, which left them with insufficient countermeasures protection, the attack conducted that night was considered a tremendous success. "In a little over twenty minutes the B-52s dropped more than two thousand tons of bombs and all indications were that the attack from different directions had befuddled the North Vietnamese defenses."[30]

The highly successful attack conducted during the night of December 26/27 achieved what Nixon had intended. The next morning, the leader of North Vietnam, Lê Duẩn, first secretary of the Vietnamese Workers Party, realizing that North Vietnam's air defenses were declining rapidly and worried about his own support within the Politburo if future raids proved equally successful, sent a message to the United States offering to start negotiations on January 8. Nixon replied that while he was willing to begin formal negotiations on January 8, Henry Kissinger, his national security advisor, wanted preliminary negotiations to begin on January 2.[31]

Despite the North's apparent willingness to negotiate, Nixon did not halt the bombing. Sixty B-52s (thirty from U-Tapao and thirty from Guam) attacked the railyards at Lang Dang, Doc Noi, and Tung Quant, the Van Dien supply center, and three SAM sites on December 27 (day ten). Except for the deletion of targets in Haiphong, the attack was a small-scale version of the previous night's raid. As in the night before, two B-52Ds were lost to SAMs. More ominous was the destruction of two F-4Es by MiGs. Although only two B-52s had been lost, "the ferocity of the North Vietnamese defenses surprised and disturbed the American commanders and dispelled any feeling that they had turned the tide. U.S. crews reported that the North Vietnamese had fired about ninety missiles, that night, more missiles per sortie than they had fired on any other raid."[32]

The bombing continued on December 29 (day eleven) but moved away from the heavily defended capital of Hanoi. Sixty B-52s attacked the Lang Dang railroad yard, the SAM assembly facilities at Phuc Yen, Doc Noi, Sama, and two SAM sites. The big bombers were supplemented by nineteen all-weather F-111 fighter-bombers that struck four airfields, eight SAM sites, and three other targets, along with twenty F-4s and sixty-four Navy A-7s. Few SAM launches were reported and the B-52s suffered no losses.

Having exhausted their supply of SAMs, which made continued defense impossible, the North Vietnamese succumbed to Nixon's demands and agreed to resume the Paris talks on January 2. Nixon halted all bombing north of the 20th parallel, officially ending Linebacker II at 6:59 a.m. Hanoi time on December 30.[33]

CHAPTER 21
LOOKING BACK

Linebacker II was the most intensive bombing campaign in history. During the eleven days of the campaign, B-52s dropped 15,237 tons of bombs in 721 sorties, with the fighter-bombers in 1,274 sorties delivering an additional 5,000 tons of ordnance (see table 21.1). The B-52s accomplished this mission despite the presence of twenty-six SAM sites, twenty-one of which were situated in the Hanoi area. Most American sources claim North Vietnamese missileers fired some 1,240* SA-2s during Linebacker II, which shot down sixteen of the Air Force's bombers at a loss rate less than predicted at under 3 percent. The SA-2s were also responsible for downing two F-111As, three F-4s, two A-7s, two A-6s, one EB-66, and one RA-5C, an insignificant number of losses considering the large number of sorties involved. The North Vietnamese, however, claimed to have fired only 239 missiles while engaging B-52s 143 times. Michel suggests the number provided by the Vietnamese is low for propaganda purposes. He also explains why the U.S. figure is too high (Michel provides a detailed explanation of these numbers on pages 248–49 of *The 11 Days of Christmas*). The truth probably lies somewhere in between but cannot be determined accurately at this point.[1]

The difficulty in accurately determining the number of SA-2s actually fired brings to mind a similar situation during World War II with regard to the claims made by the aerial gunners (and fighter pilots too). The second Schweinfurt raid of October 14, 1943, conducted by the Eighth Air Force, provides an example of the widely inflated values given for the fighters downed by bomber aircrews. The B-17 gunners on that raid claimed to have shot down 186 Luftwaffe fighters, 27 "probables," and 89 damaged.

---

* Alfred W. Price's estimate, based on the Comfy Coat Evaluation, gives the number as 850, three quarters of which targeted the B-52s.

TABLE 21.1
LINEBACKER II: TOTAL SORTIES CONDUCTED (NIGHT AND DAY)[1]

Day			SAM Suppression		CAP/ Escort	Chaff	USAF Other	USMC/Navy		
	B-52	F-111	F-105	F-4				A-6	A-7	F-4
Night 1	129	15	13	4	63	72	0	34	9	0
Night 2	93	28	15	14	61	24	0	41	26	10
Day			6	6	24	12	6			
Night 3	99	12	14	4	61	36	0	21	19	0
Day			6	6	31	8	0			
Night 4	30	12	4	9	23	9		9	6	0
Day			8	8	36	8				
Night 5	30	15	4	11	5	15		14	4	0
Day			8	8	34	12				
Night 6	30	12	7	6	12	16		14	0	0
Day			8	8		18	8			
Night 7	30	3	7	9	22	4		3	2	4
Day			5	4	24	8				
Xmas Standdown										
Night 9	120	10	9	9	34	23		10	11	0
Day			6	5	24	8	9			
Night 10	60	14	14	9	32	23		2	1	0
Day			6	5	34	8	24			
Night 11	60	14	6	1	28	23		14	6	2
Day			5	6	26	8				
Night 12	60	12	6	33	25			9	3	2
Day			6	6	22	9	5			
Total	741	147	163	171	621	344	52	171	87	18

Source: Calvin R. Johnson, "Linebacker Operations September–December 1972 (U)," December 31, 1978, Project CHECO, Office of History, Headquarters Pacific Air Force, 96–102.

[1] Excluding day sorties of U.S. Marine Corps and Navy aircraft that had not been located.

As military historian John M. Curatola noted, "Many men often claimed credit for the same fighter passing through a formation with other claims mere wishful thinking. For that date, German records revealed a loss of only 31 fighters destroyed, 12 written off and 34 damaged—hardly what the Americans claimed." Determining the number of SA-2s launched was affected by many factors, including the methodology used, none of which is indicated for any of the numbers shown above. Was it based on the visual observations derived from the airmen recorded during their after-action briefings (which were plagued by multiple sightings of the same missile by other aircraft in a cell)? Or was it was based on the number of radar lock-ons observed by the EWOs, which missed those launched in barrage, or perhaps by the U.S. electronic intelligence, which Michel claims "is highly suspect"?[2]

The relatively small loss rate was due mostly to the massive use of the B-52's electronic countermeasure suite and the large clouds of chaff that spread over the target area, even though some of these were blown away by the wind. The Air Force also flew 284 suppression sorties in which 421 AGM-45 Shrike and 49 AGM-78 Standard ARM missiles were expended. What effect this had on the defenses remains to be seen. It is clear that the use of these weapons forced the North Vietnamese to shut down their radars as soon as the missiles were fired, but it does not appear that this prevented them from turning the Fan Song radars back on after the missiles had passed and then tracking the B-52s via their jamming signal. Price indicates that "that the evidence suggests that only two Standard ARMs, and no Shrikes, inflicted damage to the control radars at missile sites." The Shrike missed for many reasons. It was deceived by the North Vietnamese, who turned the Fan Song radars on to instigate a launch and then turned it off so the missile would go ballistic, and it became confused when confronted with multiple radar signals. Why the AGM-78 failed to live up to its potential is unknown. Any information regarding data on its effectiveness was most likely classified at the time and may never have been declassified, if in fact it still exists.[3]

Fifty-nine U.S. airmen died supporting Operation Linebacker II, according to the detailed list provided by Gary Joiner and Ashley Dean in their retrospective report published in December 2017. Thirty-two of them died in the fifteen B-52s shot down by the North Vietnamese missiles. Brig. Gen.

James McCarthy, who served as a wing commander in Operation Linebacker II, argues, like other authors (civilian and military†), that "the losses were morally worth it" because the B-52 attacks on Hanoi won the war. It is true that "airpower was the decisive factor leading to the peace agreement," but it did not win the war. The agreement, signed in Paris on January 15, 1973, as Adm. Ulysses S. G. Sharp noted, "was not a formula for peace." The agreement allowed the 210,00–220,000 North Vietnamese troops in South Vietnam to remain, while giving "the North Vietnamese what they had wanted more than anything else—*the complete withdrawal* of allied troops and air forces."[4]

"The ink," explained Sharp, "was not dry on the agreement's pages before the North Vietnamese . . . [began] to violate every provision." They started infiltrating additional troops into South Vietnam, improved their logistics system, increased their stockpiles of supplies, and introduced new weapons including armor, anti-aircraft weapons and SA-2s below the DMZ. "The North Vietnamese, resupplied and sensing a final victory, were eager to fight. In December 1974 they tested whether the United States would resume bombing if they blatantly violated the peace by invading Phuoc Long province." There was no response as Congress rejected appeals for increased aid for South Vietnam. In March 1975 the North Vietnamese launched offensives in the Central Highlands and Quang Tri province that led to the collapse of the South Vietnamese army. On April 30, 1975, the South Vietnamese capital of Saigon fell to the North Vietnamese army, effectively ending the Vietnam War.[5]

When the United States initiated the air war against Vietnam it was already equipped with a number of highly specialized aircraft that were able to capture SIGINT and ELINT data on the rapidly growing North Vietnamese air defense system. These aircraft, which turned out to be highly useful and important assets, had not been designed or acquired for conventional warfare but had emerged in response to the requirements of the Peacetime Aerial Reconnaissance Program (PARPRO) that had been established for peripheral reconnaissance missions around the Soviet Union and People's Republic of China. The Air Force, the Navy, and the Marine Corps had a small number of ECM-equipped aircraft for use against

---

† See Clodfelter, *The Limits of Air Power*, 201.

radar-controlled anti-aircraft and surface-to-air missiles that would prove to be unsuitable in the hotly contested air space over Vietnam.

As noted in chapter 2, when Air Force planes began flying strike missions as part of Operation Rolling Thunder, there were no tactical electronic warfare assets in the Pacific Air Force. Yet despite the knowledge of Soviet SA-2 missile systems gained during the Cuban Missile Crisis, both the Air Force and the Navy were lax in equipping their strike forces with the ECM needed to protect them from SAMs. As I discussed in *Fighting in the Electromagnetic Spectrum*, the interest in electromagnetic warfare, such as that which emerged during the Cuban Missile Crisis, often wanes in the peaceful interlude that follows such crises or wars. This phenomenon can be attributed to a number of factors concerning the nature of electronic warfare. First is the secrecy surrounding its operations and equipment, which limits the dissemination of this information to the public and within the military itself. Since the mission of such aircraft is to collect information or suppress enemy air defenses, it lacks the glamour associated with missions flown by attacking aircraft. This discrepancy is reflected in the attitude and elan of the pilots in the attack community who have great influence over what type of aircraft need to be procured or developed. Then there is the question of funding. When budgets become tight, leadership within the aviation community prefers to spend their limited resources on obtaining more fighter, bomber, or attack aircraft.[6]

Although the Air Force had the foresight to develop the QRC-160 jamming pod before the air strikes on North Vietnam, tactical air's lack of interest relegated this device to storage, where the fragile electronics of that era deteriorated to the point that they were no longer reliable. Fortunately, American industry was quickly able to adapt to the ECM needs of the military community, although the rush to provide the needed ECM devices, such as the case with the APR-24, frequently resulted in equipment that had not been adequately tested, was temperamental, did not take kindly to being mounted on wing pods, or lacked the personnel and parts to maintain them.

As the U.S. military had to relearn during the Vietnam War, the introduction of new weapons inevitably causes the enemy to change tactics, which requires a readjustment or evaluation in the use of new weapons. No jamming method was infallible, as was discovered when the NVA adopted a SAM firing technique using the jamming signals to target flights of strike aircraft with missile barrages, which at the very least caused the fighters to drop their ordnance in order to outmaneuver the SAMs or, in the case of the B-52s, switch to automatic control upon burn through. The North Vietnamese missileers also discovered that they could disrupt attempts to attack their missile sites by turning off or limiting the use of their Fan Song radars. As the North Vietnamese tactics evolved to counter the Wild Weasels, the Weasel aircrews found they could no longer cruise about and locate radars with their own detection gear. Instead, they had to rely upon intelligence reports on specific transmitting sites in order to engage the target. Sometimes the radar signals were so brief that the crew did not have time to use the range tables and had to estimate how sharply to pull up. If the enemy had already launched an SA-2, the crew had to make a best guess of the range and fire the Shrike in the hope that the firing would cause the Fan Song operators to shut down the radar, which would cause the SA-2 to go ballistic and miss them.

Perhaps the biggest obstacle, and the one that took the longest to overcome, was the difficulty in coordinating information among the various commands in Southeast Asia obtained from the multitude of SIGINT assets available. Collating, evaluating, and disseminating this information proved to be a formidable task complicated by the lack of a unified command structure and overly protective concerns of the highly secretive intelligence agencies involved.

The classification problem and dissemination concerns that prevented pilots from receiving MiG warnings on a timely basis was symptomatic of what Michael Herman called intelligence's Achilles' heel, "because the best intelligence in the world is wasted if it is not passed in time to those who can use it." As a result of the Vietnam War, SIGINT was freed from behind the "Green Door" by "creating incentive to push such information quickly to U.S. forces for operational use."[7]

As the outcome of Linebacker II illustrates, the bombing of North Vietnam, in the face of what some considered one of the best air defense systems in the world, was considered a success, both from the number of losses that were "considerably less" than those encountered during prior mass bombardment missions over heavily defended targets and as a means of bringing the North Vietnamese to the bargaining table.[8]

The Vietnam War was a wake-up call for the U.S. military with regard to the need to prepare for electronic warfare. The war fostered the introduction of the EA-6B, the EF-111A, Wild Weasels, antiradiation missiles, jamming pods/ECM pods, and sophisticated radar warning and homing radars. Improvements and further developments in these weapons continued after the war ended and were destined to play an important role in all future conflicts involving American air power.

APPENDIX I

MILITARY AND ECM RADAR BANDS

MILITARY RADAR BANDS

Band	Frequency	Notes
HF	3–30 MHz	High Frequency
VHF	30–300 MHz	Very High Frequency
UHF	300–1000 MHz	Ultra High Frequency
L	1–2 GHz	
S	2–4 GHz	
C	4–8 GHz	
X	8–12 GHz	
Ku	12–18 GHz	
K	18–27 GHz	
Ka	27–40 GHz	
mm	40–300 GHz	Millimeter Wavelength

ECM BANDS

Band	Frequency
A	30–250 MHz
B	250–500 MHz
C	500–1000 MHz
D	1–2 GHz
E	2–3 GHZ
F	3–4 GHZ
G	4–6 GHz
H	6–8 GHz
I	8–10 GHz
J	10–20 GHz
K	20–40 GHz
L	40–60 GHz
M	60–100 GHz

APPENDIX II
SOVIET EQUIPMENT DEPLOYED BY THE NORTH VIETNAMESE

NATO Name	*Soviet Model/Name*	*Band*	*Use*
SA-2 Guideline	S-75 Dina/ V-750 missile		Surface-to-Air Missile
Fan Song	RSNA-75	E/F	SA-2 Fire-Control Radar
Flat Face	P-15	C	Search and Tracking Radar
Spoon Rest	P-12	VHF	Early Warning/ Tracking Radar
Tall King	P-14	VHF	Early Warning Radar
Side Net	PRV-11	E	Height Finder Radar
Whiff	SON-4	S	Fire-Control Radar
Fire Can	SON-9	S	Fire-Control Radar
Bar Lock	P-35/P-37	E/F	Early Warning Radar
Big Bar B	P-30	S	Early Warning/ GCI Radar
Low Blow	SNR-125	I	Fire-Control Radar
Back Net	P-80	E/low S	Early Warning Radar

APPENDIX III

AGM-45 SHRIKE MISSILES

Designation	Guidance Section	Seeker Type
AGM-45A-1	MK 23 MOD 0	E/F-Band
AGM-45A-2 AGM-45B-2	MK 22 MOD 0, 1, or 2	G-Band
AGM-45A-3 AGM-45B-3	MK 24 MOD 0, 1, or 4	Broad E/F-Band with Angle Gating
AGM-45A-3A AGM-45B-3A	MK 24 MOD 2 or 5	Narrow E/F-Band with Angle Gating
AGM-45A-3B AGM-45B-3B	MK 24 MOD 3	E/F-Band with Angle Gating
AGM-45A-4 AGM-45B-4	MK 25 MOD 0 or 1	G-Band with Angle Gating
AGM-45A-5 AGM-45B-5	Cancelled	
AGM-45A-6 AGM-45B-6	MK 36 MOD 1	I-Band with Angle Gating

AGM-45A-7 AGM-45B-7	MK 37 MOD 0	E/F-Band with Angle Gating
AGM-45A-9 AGM-45B-9	MK 49 MOD 0	I-Band with Angle Gating
AGM-45A-9A AGM-45B-9A	MK 49 MOD 1	I-Band with Angle Gating, G-Bias
AGM-45A-10 AGM-45B-10	MK 50 MOD 0	Broad E- to I-Band with Angle Gating

APPENDIX IV

F-4 PHANTOM EW UPGRADES

1965–1966
- Singer CMR-312 "Little Ears"

1966–1967
- Melpar AN/APR-24 C/S/X-band Homing and Warning System

Installed on 4–5 aircraft for VF-151 and VF-161 on CVA-64 USS *Constellation*, 4–5 aircraft for VF-14 and VF-32 on CVA-42 USS *Franklin D. Roosevelt*, and 4–5 aircraft for VF-96 on CVAN-65 USS *Enterprise* during their 1966–1967 Vietnam cruises. This equipment was removed after their cruise.

- Magnavox APR-27 Missile Warning System

Sent as kits and installed on-board during the 1966–1967 Vietnam cruises of CVA-64 USS *Constellation*, CVA-42 USS *Franklin D. Roosevelt*, and CVAN-65 USS *Enterprise*. Future deployments had the APR-27 installed prior to departure.

1967–1968
- Applied Technology APR-25 S/C/X-band Radar Homing System

This was a pre-Shoehorn fit as seen only in aircraft of VG-142 and VF-143 on *Constellation* during its 1967 cruise.

Project Shoehorn Mod 1

F-4 PHANTOM EW UPGRADES 221

Project Shoehorn was a program providing a complete suite of EW equipment to the F-4B. It was considered "state of the art" at the time.

- Melpar APR-30 S/C/X-band Radar Homing System
- Sanders ALQ-51 S/E/F-band Deception Jammer and Lock Breaker
- Magnavox APR-27 Missile Warning System
- Magnavox AlQ-91 IFF Countermeasures System
- Tracor ALE-29 Flare and Chaff Countermeasures Set

The only carrier to bring a full complement of Shoehorn Mod 1–equipped F-4Bs into the Vietnam War was the *Coral Sea* (CVA 23) during her 1967–1968 cruise. Her air group at the time included the F-4Bs assigned to VF-151 and VF-161.

Project Shoehorn Mod 2

- Magnavox APR-27 Missile Warning System
- Applied Technology APR-25 S/C/X-band Radar Homing System

The Navy was not happy with the APR-30 and replaced it with the proven APR-25.

- Sanders ALQ-51/100 S/E/F-band Deception Jammer and Lock Breaker
- Tracor ALE-29 Flare and Chaff Countermeasures Set

1968–1969   Project Shoehorn Mod 3

- Magnavox AlQ-91 IFF Countermeasures System
- Applied Technology APR-25 S/C/X-band Radar Homing System
- Magnavox AlQ-91 IFF Countermeasures System
- Tracor ALE-29 Flare and Chaff Countermeasures Set

The first aircraft to be modified to Shoehorn Mod 3 in 1968 were those on the *Constellation* with VF-142 and VF-143.

Project Shoehorn Mod 4

- Hazeltine AN/APX-76 air-to-air IFF System
- Applied Technology Inc. AN/APR-25 S/C/X-band Radar Homing and Warning System
- Sanders AN/ALQ-51A/100 S/E/F-band Deception Jammer and Lock Breaker
- Tracor An/ALE-29A Flare and Chaff Countermeasures Set
- Magnavox AN/ALQ-91 IFF Countermeasures System

Source: "US Navy F-4 EW development – Revisited," Phantom Phacts, accessed April 23, 2023, https://phantomphacts.blogspot.com/2014/03/us-navy-f-4-ew-development-revisited.html.

NOTES

CHAPTER 1. FIRST STRIKES

1. Malcolm W. Cagle, "Task Force 77 in Action off Vietnam," U.S. Naval Institute *Proceedings* 98, no. 5 (May 1972), https://www.usni.org/magazines/proceedings/1972/may/task-force-77-action-vietnam; U.S. Navy, Bureau of Personnel, "On Guard in the Pacific," *All Hands* no. 572 (September 1964), 4.

2. Pat Paterson, "The Truth about Tonkin," *Naval History Magazine* 22, no. 1 (February 2008), https://www.usni.org/magazines/naval-history-magazine/2008/february/truth-about-tonkin; Delmar Lang, "Memorandum for the Record," October 14, 1964, Subject: Chronology of Events of 2–5 August 1964 in the Gulf of Tonkin, U.S. Department of Defense, accessed September 27, 2023, https://media.defense.gov/2021/Jul/14/2002762861/-1/-1/0/REL1_LANG.PDF, 2; David Sears, "A 'Piercing Arrow' Strikes North Vietnam," *VFW Magazine*, August 2014, http://digitaledition.qwinc.com/publication/?i=215523&article_id=1749279&view=articleBrowser; CTG Seven Seven Pt Six to RUHLHL/CINCPCFLT, 5 August 1964, 0830Z, Folder, "*Operation Pierce Arrow*, 8/1964, 1 of 2," Country Files, Vietnam, NSF, box 228, LBJ Presidential Library, accessed December 21, 2022, http://www.lbjf.org/txt/nsf/cf-vietnam/591194-nsf-co-vn-b228-f5.pdf.

3. Robert J. Hanyok, "Skunks, Bogies, Silent Hounds, and Flying Fish: The Gulf of Tonkin Mystery, 2–4 August 1964," Naval History and Heritage Command, accessed December 12, 2022, https://www.history.navy.mil/research/library/online-reading-room/title-list-alphabetically/s/skunks-bogies-silent-hounds-flying-fish.html, 14–22; Lang, "Memorandum for the Record," 2. For more on the Marine detachment at Phu Bai, see National Security Agency, "Gulf of Tonkin Incidents Additional SEAHAG Inputs (U)," NSA,

accessed September 27, 2023, https://www.nsa.gov/portals/75/documents/news-features/declassified-documents/gulf-of-tonkin/history-of-southeast-asia/release-2/rel2_gulf_tonkin_incidents.pdf.

4. Joseph C. Goulden, *Truth Is the First Casualty* (New York: James B. Adler Inc., 1969), 146; Hanyok, "Skunks, Bogies, Silent Hounds, and Flying Fish," 23.
5. Jim and Sybil Stockdale, *In Love and War* (New York: Harper and Row, 1984), 23, as cited by Paterson, "The Truth about Tonkin."
6. Hanyok, "Skunks, Bogies, Silent Hounds, and Flying Fish," 23.
7. Paterson, "The Truth about Tonkin"; Hanyok, "Skunks, Bogies, Silent Hounds, and Flying Fish," 24; "Tonkin Gulf Resolution (1964)," National Archives, accessed December 25, 2022, https://www.archives.gov/milestone-documents/tonkin-gulf-resolution; President Johnson quoted by Stanley Karnow, *Vietnam: A History* (New York: Penguin Books, 1983), 372, as cited by Paterson, "The Truth about Tonkin."
8. "Tonkin Gulf Resolution," Public Law 88–408, 88th Congress, August 7, 1964, General Records of the United States Government, Record Group 11, National Archives, https://usapoliticaldatabase.weebly.com/tonkin-gulf-resolution-1964.htm; "Tonkin Gulf Resolution (1964)," accessed July 28, 2023, https://www.archives.gov/milestone-documents/tonkin-gulf-resolution; Ronald B. Frankum Jr., *Like Rolling Thunder: The Air War in Vietnam, 1964–1975* (Lanham, MD: Rowman & Littlefield, 2005), 14.
9. William Cahill, "Strategic Air Command SIGINT Support to the Vietnam War," *Air Power History* 66, no. 4 (2019): 31; Bruce M. Bailey, "55th Strategic Reconnaissance Wing," in Philip A. St. John, *50 Years USAF: A Look at the Air Force, Air Force Association and Commemorative Las Vegas Reunion* (Paducah, KY: Turner, 1998), 61.
10. Jeanette Remak, *Boeing B-52: Warrior Queen of the USAF* (Stroud, UK: Fonthill, 2016), Google Books, https://www.google.com/books/edition/Boeing_B_52_Stratofortress/zrZUDgAAQBAJ?hl=en&gbpv=1&dq=RB-47H&pg=PT80&printsec=frontcover; "B-47 Stratojet: Historical Snapshot," Boeing Corporation, accessed February 16, 2023, https://www.boeing.com/history/products/b-47-stratojet.page; U.S. Air Force, "Characteristics Summary—Reconnaissance RB-47H," accessed February 16, 2023, https://www.avialogs.com/aircraft-b/boeing/item/3633-2751rb-47hstratojetcharacteristicssummary-october1963; Forrest L. Marion, "A Hot Day in a Cold War: An RB-47 vs. MiG-17s, April 28, 1965," *Air Power History* 53, no. 3 (Fall 2006): 28; Dario Leone, "RB-47H Down: When a Soviet MiG-19 Downed a USAF

Elint Plane over the Barents Sea," The Aviation Geek Club, August 19, 2018, https://theaviationgeekclub.com/rb-47h-shot-down-when-a-soviet-mig-19-downed-a-usaf-elint-plane-over-the-barents-sea/. Note that while several sources refer to the ten- to twelve-hour-long missions endured by the EWOs, the official USAF flight characteristics based on flight tests of the Silver Dawn RH-47B show a maximum combat radius flight time of 6.4 hours.

11. Roger P. Fox, *Air Base Defense in the Republic of Vietnam 1961–1973* (Washington, DC: Office of Air Force History, 1979), 1; Lee March, "Going to War in Korea and Vietnam: The Decisions of Harry Truman and Lyndon Johnson," chap. 4 in *Between Memory and Mythology: The Construction of Memory of Modern Wars*, ed. Natalia Starostina (Newcastle upon Tyne, UK: Cambridge Scholars, 2015), 59; Theo van Geffen, "The Air War against North Vietnam: Hoa Railroad and Highway Bridge (Part 1)," *Air Power History* 65, no. 2 (Summer 2018): 8; Frankum, *Like Rolling Thunder*, 15.
12. Jacob Van Staaveren, *Gradual Failure: The Air War over North Vietnam 1965–66* (Washington, DC: Air Force History and Museums Program, 2002), 9.
13. Van Staaveren, *Gradual Failure*, 16; van Geffen, "The Air War against North Vietnam," 8.
14. Van Staaveren, *Gradual Failure*, 17; "Kosygin Hailed in Hanoi: He Praises the Vietcong," *New York Times*, February 7, 1965, 1, 6; Joseph Seltzer [Executive Officer National Estimates, CIA], Memorandum for the United States Intelligence Board, Subject: SNIE 10-65: Communist Military Capabilities and Near-term Intentions in Laos and South Vietnam, February 1, 1965, Central Intelligence Agency Collection, box 7, folder 111, The Vietnam Center and Sam Johnson Vietnam Archive, Texas Tech University, accessed December 28, 2022, https://vva.vietnam.ttu.edu/images.php?img=/images/041/04107111001.pdf, 8.
15. Van Staaveren, *Gradual Failure*, 22; van Geffen, "The Air War against North Vietnam," 8.
16. Van Staaveren, *Gradual Failure*, 25.

CHAPTER 2. OPERATION ROLLING THUNDER

1. CINCPACAF to 5AF/13AF, Subj: TEW, 14/0239Z December 68, as cited in Robert M. Burch, "Tactical Electronic Warfare Operations in SEA, 1962–1968," Contemporary Historical Evaluation of Compat Operations (CHECO) Report, February 10, 1969, HQ PACAF, Directorate, Tactical Evaluation, CHECO Division, 24.

2. Frankum, *Like Rolling Thunder*, 19–21; Van Staaveren, *Gradual Failure*, 78, 84.
3. Scott Duncan, "The Combat History of the F-105," *Aerospace Historian* 22, no. 3 (1975): 122–23, http://www.jstor.org/stable/44523414.
4. Van Staaveren, *Gradual Failure*, 70, 86.
5. Van Staaveren, *Gradual Failure*, 86.
6. William S. Borgiaz, *The Strategic Air Command: Evolution and Consolidation of Nuclear Forces, 1945–1955* (Westport, CT: Greenwood Publishers, 1996), xii; Edward G. Longacre, *Strategic Air Command: The Formative Years (1944–1949)* (Offutt Air Force Base, NE: Office of the Historian, Headquarters Strategic Air Command, 1990), 1, 21–23; Craig C. Hannah, *Striving for Air Superiority: The Tactical Air Command in Vietnam* (College Station: Texas A&M University Press, 2002), 21; Marshal L. Michel III, *The 11 Days of Christmas: America's Last Vietnam Battle* (San Francisco: Encounter Books, 2002), 3, 25–26; Mike Worden, *Rise of the Fighter Generals: The Problem of Air Force Leadership 1945–1982* (Maxwell Air Force Base, AL: Air University Press, 1998), 85.
7. Michel, *The 11 Days of Christmas*, 3; Hannah, *Striving for Air Superiority*, 23, 25, 30.
8. Dennis W. Drew, "Rolling Thunder 1965: Anatomy of a Failure," Report No. AU-ARI-CF-86-3 (Maxwell Air Force Base, AL: Air University Press, 1986), 37; Worden, *Rise of the Fighter Generals*, 77.
9. Donald R. Baucom, "Editorial Note," *Air University Review* 33, no. 2 (January–February 1982): 33, https://babel.hathitrust.org/cgi/pt?id=mdp.39015081905674&view=1up&seq=195&size=125&q1=face%20defenses%20consisting.
10. Theodore van Geffen Jr. and Gerald C. Arruda, "Thunderchief," *Air University Review* 33, no. 2 (January–February 1982): 46, https://babel.hathitrust.org/cgi/pt?id=mdp.39015081905674&view=1up&seq=189&size=125&q1=face%20defenses%20consisting; Hannah, *Striving for Air Superiority*, 49, 52.
11. John P. Piowaty, "Reflections of a Thud Driver," *Air University Review* 33, no. 2 (January–February 1982): 52–53, https://www.google.com/books/edition/Air_University_Review/vFTD2gFd99QC?hl=en&gbpv=1&dq=thunderstick+II&pg=PA57&printsec=frontcover; Jack Broughton, *Thud Ridge: F-105 Thunderchief Missions over Vietnam* (Manchester, UK: Crecy Publishing, 2006), 37.
12. Frankum, *Like Rolling Thunder*, 22; Robert E. Morrison, "How VQ-1 Supported Military Actions with SIGINT during Vietnam Part 1 of 7," Station HYPO, accessed January 7, 2023, https://stationhypo.com/2018/04/22/how-vq-1-supported-military-actions-with-sigint-during-vietnam-part-1-of-7-guest-post/; Angelo Romano and John D. Herndon, *From Bats to Rangers:*

A Pictorial History of Electronic Countermeasures Squadron Two (ECMRON2-2) Fleet Air Reconnaissance Squadron Two (VQ-2) (Simi Valley, CA: Ginter Books, 2017), 62; Alfred W. Price, *The History of US Electronic Warfare*, vol. III, *Rolling Thunder through Allied Force, 1964–2000* (Arlington, VA: Association of Old Crows, 2000), 22–23; John Pike, "S-60 Anti-Aircraft Artillery," Federation of American Scientists, accessed January 7, 2023, https://man.fas.org/dod-101/sys/land/row/s-60.htm.

13. Robert E. Morrison, *Naval Communications Station Philippines Fleet Support Detachment Da Nang, Republic of Vietnam (Det Bravo) Command History* (published by the author, 2017), 9–10; Larry D. Brosh, "EC-121 Layout," VQ Association, accessed February 15, 2023, https://vqassociation.org/history/.

14. Brosh, "EC-121 Layout"; Alfred W. Price, *The History of US Electronic Warfare*, vol. II, *The Renaissance Years, 1946 to 1964* (Arlington, VA: Association of Old Crows, 1989), 281–84; Price, *The History of US Electronic Warfare*, 3:15–16.

15. Theodore J. Turner, "Seventh Air Force History," Seventh Air Force, accessed January 28, 2021, https://www.7af.pacaf.af.mil/About-Us/Fact-Sheets/Display/Article/408382/seventh-air-force-history/; Bernard C. Nalty, *Tactics and Techniques of Electronic Warfare: Electronic Countermeasures in the Air War against North Vietnam 1965–1973* (Washington, DC: Office of Air Force History, 1977), 19; Courtland C. Moore, "EB-66C Out-Country Electronic Reconnaissance, 1965–67, A Case Study," Report No. 3655 (Maxwell Air Force Base, AL, 1968), 13, 19–20; Gilles Van Nederveen, "Sparks over Vietnam: The EB-66 and the Early Struggle of Tactical Electronic Warfare," ARI Paper 2000-03, College of Aerospace Doctrine, Research and Education, Air University, Airpower Research Institute (Maxwell Air Force Base, AL, 2000), 34.

16. Greg Goebel, "Douglas A3D Skywarrior & B-66 Destroyer," AirVectors, accessed January 9, 2023, http://www.airvectors.net/avskywar.html; Moore, "EB-66C Out-Country Electronic Reconnaissance, 1965–67," 6; 15; Price, *The History of US Electronic Warfare*, 2:179–80; Van Nederveen, "Sparks over Vietnam," 11.

17. Van Nederveen, "Sparks over Vietnam, 11.

18. Price, *The History of US Electronic Warfare*, 2:179–80.

19. Price, *The History of US Electronic Warfare*, 2:211, 248–49; Van Nederveen, "Sparks over Vietnam," 14–15.

20. Moore, "EB-66C Out-Country Electronic Reconnaissance, 1965–67," 12–17; Peter E. Davies, *B/EB-66 Destroyer Units in Combat* (Oxford: Osprey Publishing, 2021), 32.

21. Moore, "EB-66C Out-Country Electronic Reconnaissance, 1965–67," 22–23.
22. Van Nederveen, "Sparks over Vietnam," 41, 55.
23. Burch, "Tactical Electronic Warfare Operations in SEA, 1962–1968," 34.
24. Van Nederveen, "Sparks over Vietnam," 42; Chris Hobson and David Lovelady, "The B-66 Variants in Southeast Asia (Oct 1965)," Vietnam Air Losses, accessed October 2, 2023, https://www.vietnamairlosses.com/index.php/sidelines/1965/oct65; Merle L. Pribbenow, "The 'Ology War: Technology and Ideology in the Vietnamese Defense of Hanoi," *Journal of Military History* 67, no. 1 (January 2003): 181.
25. Van Nederveen, "Sparks over Vietnam," 43–44.

CHAPTER 3. ELECTRIC SPADS, VOODOOS, SKYNIGHTS, AND ELECTRIC INTRUDERS

1. Richard R. Burgess and Rosario M. Rausa, *U.S. Navy A-1 Skyraider Units of the Vietnam War* (London: Osprey, 2013), 17; U.S. Navy, Naval Air Command, "Standard Aircraft Characteristics: Navy Model EA-1F Skyraider," American Aviation Historical Society, accessed July 28, 2023, https://www.aahs-online.org/images/Navy_SAC/EA-1F.pdf; John B. Nichols and Barrett Tillman, *On Yankee Station: The Naval Air War over Vietnam* (Annapolis, MD: Naval Institute Press, 1987), 88.
2. Burgess and Rausa, *U.S. Navy A-1 Skyraider Units of the Vietnam War*, 18; Nichols and Tillman, *On Yankee Station: The Naval Air War over Vietnam*, 89–90.
3. Julian Lake, quoted in Price, *The History of US Electronic Warfare*, 3:51. Lake's recollection was wrong; Lacouture retired as a captain.
4. Wersaja Polska, "Douglas EKA-3B 'Skywarrior/Whale,'" Virtual Museum of the Vietnam War, accessed May 16, 2023, https://vietnam.net.pl/EKA3Ben.htm; Rick Morgan, *A-3 Skywarrior Units of the Vietnam War* (Oxford: Osprey, 2015), 53.
5. Morgan, *A-3 Skywarrior Units of the Vietnam War*, 54; Martin Streetly, *Airborne Electronic Warfare: History, Techniques and Tactics* (London: Jane's, 1988), 40.
6. Morgan, *A-3 Skywarrior Units of the Vietnam War*, 54; Streetly, *Airborne Electronic Warfare*, 40.
7. John T. Correll, "Take It Down! The Wild Weasels in Vietnam," *Air Force Magazine*, July 1, 2010, https://www.airforcemag.com/article/0710weasels/; Cagle, "Task Force 77 in Action off Vietnam," 75; Robert W. Love Jr., *History of the U.S. Navy: Volume Two 1942–1991* (Lanham, MD: Stockpole Books,

1992), 517; William Momyer, *Airpower in Three Wars* (Washington, DC: Government Printing Office, 1978), 118.
8. Nalty, *Tactics and Techniques of Electronic Warfare*, 2; Van Staaveren, *Gradual Failure*, 114; William A. Hewitt, "Planting the Seeds of SEAD: The Wild Weasel in Vietnam (Thesis, Air University, Maxwell Air Force Base, 1992), 2; Chris Pocock, "Commentary by British U-2 Historian Chris Pocock on the CIA History," National Security Archive, accessed January 20, 2023, https://nsarchive2.gwu.edu/NSAEBB/NSAEBB434/docs/W-P%20History%20Notes%20by%20CP%20on%20redactions%20lifted%20Aug13.pdf, see reference to page 335; Gregory W. Pedlow and Donald E. Welzenbach, *The Central Intelligence Agency and Overhead Reconnaissance: The U-2 and OXCART Programs, 1954–1974* (Washington, DC: Central Intelligence Agency, 1992), 234; Price, *The History of US Electronic Warfare*, 3:44.
9. Van Staaveren, *Gradual Failure*, 116.
10. Price, *The History of US Electronic Warfare*, 2:263–64.
11. Gerald Sensabaugh, as quoted by Price, *The History of US Electronic Warfare*, 2:264; "Nowadays, the Indispensable Jamming Pod for Fighters, at the Beginning of Its Birth, Why Did the U.S. Air Force Not Like It?" iNews, September 30, 2023, https://inf.news/en/military/da8d44cfb3d64a5860b-2de35641f0437.html.
12. Bud Voland, quoted in Price, *The History of US Electronic Warfare*, 2:298; Price, *The History of US Electronic Warfare*, 3:26.
13. Anton D. Brees, quoted in Price, *The History of US Electronic Warfare*, 3:27.
14. Price, *The History of US Electronic Warfare*, 3:26–27.
15. Price, *The History of US Electronic Warfare*, 3:27; Burch, "Tactical Electronic Warfare Operations in SEA, 1962–1968," 118.
16. H. Wayne Whitten, *Silent Heroes: U.S. Marines and Airborne Electronic Warfare 1950–2012* (Lutz, FL: Colonel H. Wayne Whitten and Associates, 2011), 55; J. T. O'Brien, *Top Secret: An Informal History of Electronic & Photographic Reconnaissance in Marine Corps Aviation 1940–2000* (Anaheim, CA: Equidata Publishing, 2004), 114, 136, 257; Gordon Swanborough and Peter M. Bowers, *United States Naval Aircraft since 1911* (London: Putnam Aeronautical Books, 1990), 198–99; Joe Coles, "Enter the Skyknight: Hornet Pilot Shares the Dark History of the Douglas F3D 'Night Killer,'" *Hush-Kit: The Alternative Aviation Magazine*, accessed November 9, 2021, https://hushkit.net/2020/12/29/enter-the-skyknight-hornet-pilot-shares-the-dark-history-of-the-douglas-f3d-night-killer/.

17. Whitten, *Silent Heroes*, 22–23; H. Wayne Whitten, "MCARA Aircraft > F3D-2Q/EF-10B Skyknight- History," Marine Corps Aviation Reconnaissance Association, accessed July 28, 2023, https://www.mcara.us/EF-10B_history.html; O'Brien, *Top Secret*, 80, 122–23.
18. Whitten, "MCARA Aircraft > F3D-2Q/EF-10B Skyknight—History"; Price, *The History of US Electronic Warfare*, 2:198–99.
19. Whitten, *Silent Heroes*, 55; O'Brien, *Top Secret*, 226; Jack Shulimson and Charles M. Johnson, *U.S. Marines in Vietnam: The Landing and the Buildup 1965* (Washington, DC: History and Museums Division, Headquarters, U.S. Marine Corps, 1978), 27.
20. Whitten, *Silent Heroes*, 59–61; Nalty, *Tactics and Techniques of Electronic Warfare*, 21.
21. Whitten, *Silent Heroes*, 62; O'Brien, *Top Secret*, 255.
22. Whitten, *Silent Heroes*, 63–64.
23. Nalty, *Tactics and Techniques of Electronic Warfare*, 20–22.
24. Hewitt, "Planting the Seeds of SEAD," 2; Steven J. Zaloga, *Red SAM: The SA-2 Guideline Anti-Aircraft Missile* (Botley, UK: Osprey, 2007), 16–17; Pribbenow, "The 'Ology War," 177; Patrick K. Barker, "The SA-2 and Wild Weasel: The Nature of Technological Change in Military Systems" (Thesis, Lehigh University, 1994), 10–13; Burch, "Tactical Electronic Warfare Operations in SEA, 1962–1968," 19.
25. Joe Copalman, "Glory Days// EA-6A in Vietnam." *Combat Aircraft Journal*, accessed June 6, 2019, https://www.keymilitary.com/article/electric-intruder-war; H. Wayne Whitten, "Marine Composite Reconnaissance Squadron One (VMCJ-1) History," Marine Corps Aviation Reconnaissance Association, accessed June 6, 2019, https://www.mcara.us/VMCJ-1.html.
26. Whitten, *Silent Heroes*, 64; Nalty, *Tactics and Techniques of Electronic Warfare*, 21.
27. Price, *The History of US Electronic Warfare*, 2:267; H. Wayne Whitten, "MCARA Aircraft > Grumman EA-6A Intruder—History," Marine Corps Aviation Reconnaissance Association, accessed July 8, 2023, https://www.mcara.us/EA-6A_history.html.
28. Whitten, "MCARA Aircraft > Grumman EA-6A Intruder—History"; Whitten, *Silent Heroes*, 50; see also statement of Gen. William H. Green, commandant of the Marine Corps, *Department of Defense Appropriations for 1964. Hearings before a Subcommittee of the Department of Defense of the Committee on Appropriations and the Committee on Armed Services, U.S. Senate*, 88th Cong., 2nd Sess. on H.R. 10939, Part 1, 579.

CHAPTER 4. THE SAM THREAT

1. Intelligence Advisory Committee, "Soviet Capabilities and Probable Programs in the Guided Missile Field, *National Intelligence Estimate 11–5-57*, TOP SECRET, 12 March 1957," 9, 14–15, National Security Archive, accessed January 23, 2023, https://nsarchive.gwu.edu/document/19840-national-security-archive-doc-14-intelligence; Steven J. Zaloga, "Defending the Kremlin: The First Generation of Soviet Strategic Air Defense Systems 1950–60," New York Military Affairs Symposium, accessed January 23, 2023, https://nymas.org/defendingthekremlin.htm; Eric Hehs, "Carmine Vito: U-2 Pilot," *Code One* 17, no. 1 (2002), accessed October 27, 2021, https://www.codeonemagazine.com/article.html?item_id=167. See also the Secret "Surface-to-Air Guided Missile Sites in the Moscow Area," Wikimedia, https://commons.wikimedia.org/wiki/File:SAM_rings_of_Moscow,_1957_CIA_estimate.png.
2. Zaloga, "Defending the Kremlin"; "Boris Vasilievich Bunkin," Global Security, accessed January 17, 2023, https://www.globalsecurity.org/wmd/world/russia/bunkin.htm; Nalty, *Tactics and Techniques of Electronic Warfare*, 2; "V-75 SA-2 Guideline," Federation of American Scientists, accessed January 25, 2023, https://nuke.fas.org/guide/russia/airdef/v-75.htm.
3. Nalty, *Tactics and Techniques of Electronic Warfare*, 4; Carlo Kopp, "Engagement and Fire Control Radars," Air Power Australia, accessed October 27, 2021, http://www.ausairpower.net/APA-Engagement-Fire-Control.html#mozTocId517282.
4. Carlo Kopp, "Almaz S-75 Diva/Desna/Volkov Air Defense System/HQ-2A/B /CSA-1 /SA-2 Guideline," Air Power Australia, accessed October 27, 2021, http://www.ausairpower.net/APA-S-75-Volkhov.html; Michel, *The 11 Days of Christmas*, 38; Carlo Kopp, "SNR-75M3 Fan Song E Engagement Radar," Air Power Australia, accessed January 25, 2023, https://www.ausairpower.net/APA-SNR-75-Fan-Song.html.
5. Richard P. Hallion, *Rolling Thunder 1965–68: Johnson's Air War over Vietnam* (Oxford: Osprey, 2018), 27; "Soviet Aid to North Vietnam," Global Security, accessed January 10, 2023, https://www.globalsecurity.org/military/world/vietnam/hist-2nd-indochina-ussr.htm.
6. Pribbenow, "The 'Ology War," 177.
7. "Surface-to-Air Missiles (SAMs)," Global Security, accessed February 21, 2023, https://www.globalsecurity.org/military/world/vietnam/nva-ad-sam.htm; Hallion, *Rolling Thunder 1965–68*, 27.
8. Burch, "Tactical Electronic Warfare Operations in SEA, 1962–1968, 28; Melvin Porter, "Air Tactics against NVN Air/Ground Defenses 27 February

NOTES TO PAGES 41–46

1967," Project CHECO 7th AF, Directorate of Operations Analysis, 11–12; Nalty, *Tactics and Techniques of Electronic Warfare*, 31.

9. Whitten, *Silent Heroes*, 66; Price, *The History of US Electronic Warfare*, 3:36.
10. Steven J. Zaloga, *Red SAM: The SA-2 Guideline Anti-Aircraft Missile* (Oxford: Osprey, 2007), 17; Van Staaveren, *Gradual Failure*, 163, 165; Nalty, *Tactics and Techniques of Electronic Warfare*, 31; Dan Hampton, *The Hunter Killers* (New York: William Morrow, 2015), 101; Burch, "Tactical Electronic Warfare Operations in SEA, 1962–1968," 28.
11. Van Staaveren, *Gradual Failure*, 166–67.
12. Cagle, "Task Force 77 in Action off Vietnam," 76; "Francis Dee Roberge," Veterans Day Commemoration, accessed January 29, 2023, https://veteransday.utah.edu/honorees/francis-dee-roberge/.
13. Van Staaveren, *Gradual Failure*, 167–69; Cagle, "Task Force 77 in Action off Vietnam," 76; Nalty, *Tactics and Techniques of Electronic Warfare*, 32.
14. Hannah, *Striving for Air Superiority*, 78 note 22, 122; Van Staaveren, *Gradual Failure*, 169, 195; Barker, "The SA-2 and Wild Weasel," 62; Price, *The History of US Electronic Warfare*, 3:43; Larry Davis, *Wild Weasel: The SAM Suppression Story* (Carrollton, TX: Squadron/Signal Publications, 1986), 8.
15. Barker, "The SA-2 and Wild Weasel," 66.
16. "Julian S. Lake, Rear Admiral, USN (Ret.)," summary of service, Early and Pioneer Naval Aviators Association, accessed November 14, 2021, http://epnaao.com/BIOS_files/EMERITUS/Lake-%20Julian%20S.pdf; Price, *The History of US Electronic Warfare*, 3:30.
17. Price, *The History of US Electronic Warfare*, 3:30; Peter E. Davies, *A-4 Skyhawk vs North Vietnamese AAA: North Vietnam 1964–72* (Oxford: Osprey, 2020), 60.
18. Price, *The History of US Electronic Warfare*, 3:30; "Julian S. Lake, Rear Admiral, USN (Ret.)"; Davies, *A-4 Skyhawk vs North Vietnamese AAA*, 60; Julian Lake, quoted in Price, *The History of US Electronic Warfare*, 3:30.
19. Price, *The History of US Electronic Warfare*, 3:49.
20. Price, *The History of US Electronic Warfare*, 3:49; Davies, *A-4 Skyhawk vs North Vietnamese AAA*, 60.
21. U.S. Navy, CinCPacFlt, "Analysis Staff Study 8–68: An Analysis of SA-2 Missile Activity in North Vietnam from July 1965 through March 1968," October 31, 1968 (FPO San Francisco: Commander in Chief United States Pacific Fleet, 1968), Figure X, "Installation Periods of ECM Equipment, in U.S. Navy and U.S. Air Force Aircraft," 39.
22. Price, *The History of US Electronic Warfare*, 3:60.
23. Pribbenow, "The 'Ology War," 179.

CHAPTER 5. ECM PODS, RHAWs, AND CHAFF DISPENSERS

1. Price, *The History of US Electronic Warfare*, 3:79–80.
2. Price, *The History of US Electronic Warfare*, 3:25.
3. Price, *The History of US Electronic Warfare*, 3:80.
4. Price, *The History of US Electronic Warfare*, 3:81; Joe Telford, "Story by Joe W. Telford, WW # 256," in *First In, Last Out: Stories by the Wild Weasels*, ed. Edward T. Rock (Bloomington, IN: AuthorHouse, 2005), 435.
5. Price, *The History of US Electronic Warfare*, 3:81; Nalty, *Tactics and Techniques of Electronic Warfare*, 54.
6. Nalty, *Tactics and Techniques of Electronic Warfare*, 54.
7. Nalty, *Tactics and Techniques of Electronic Warfare*, 54; Price, *The History of US Electronic Warfare*, 3:81–82.
8. Price, *The History of US Electronic Warfare*, 3:83. Note: The Seventh Air Force was activated on March 28, 1966.
9. Price, *The History of US Electronic Warfare*, 3:83. Nalty, *Tactics and Techniques of Electronic Warfare*, 54–55; Burch, "Tactical Electronic Warfare Operations in SEA, 1962–1968," 36.
10. Nalty, *Tactics and Techniques of Electronic Warfare*, 54.
11. Nalty, *Tactics and Techniques of Electronic Warfare*, 56; Marshall L. Michel III, *Clashes: Air Combat over North Vietnam* (Annapolis, MD: Naval Institute Press, 1997), 59–60.
12. Price, *The History of US Electronic Warfare*, 3:85.
13. Nalty, *Tactics and Techniques of Electronic Warfare*, 58; Price, *The History of US Electronic Warfare*, 3:85–86.
14. Maj. Gen. John A. Corder, quoted in Price, *The History of US Electronic Warfare*, 3:87–88. General Price flew 100 missions over North Vietnam in F-4D Phantoms while assigned to the 497th Tactical Fighter Squadron. By January 1966, the Seventh Air Force had enough ALQ-71s to begin equipping F-4 Phantoms as well as the F-105Ds. See Nalty, *Tactics and Techniques of Electronic Warfare*, 59.
15. Price, *The History of US Electronic Warfare*, 3:88; Jeffrey D. Glasser, *The Secret Vietnam War: The United States Air Force in Thailand, 1961–1975* (Jefferson, NC: McFarland, 1995), 86.
16. Col. William Chairsell, in Price, *The History of US Electronic Warfare*, 3:88.
17. Pribbenow, "The 'Ology War," 185, 195.
18. Price, *The History of US Electronic Warfare*, 3:16.
19. Price, *The History of US Electronic Warfare*, 3:19; Nalty, *Tactics and Techniques of Electronic Warfare*, 63.

20. Nalty, *Tactics and Techniques of Electronic Warfare*, 63.
21. Nalty, *Tactics and Techniques of Electronic Warfare*, 63–64.
22. Peter Davies, *US Navy F-4 Phantom II Units of the Vietnam War 1964–68* (Oxford: Osprey 2016), 59.
23. "Capt. Trent Richard Powers," Military Hall of Honor, accessed March 3, 2023, https://militaryhallofhonor.com/honoree-record.php?id=271335; Price, *The History of US Electronic Warfare*, 3:38, Davies, *A-4 Skyhawk vs North Vietnamese AAA*, 28; U.S. Navy, CinCPacFlt, "Analysis Staff Study 8-68," figure X, "Installation Periods of ECM Equipment in the U.S. Navy and U.S. Air Force Aircraft," 339.
24. Michael France and Craig Kaston. "AN/APR-24 C/S/X-Band Radar Homing and Warning System (RHAW)," Phantom Phacts, May 2, 2014, https://phantomphacts.blogspot.com/p/usn-ew-equipment.html; John D. Sherwood, *Afterburner: Naval Aviators and the Vietnam War* (New York: New York University Press, 2004), 107.
25. France and Kaston, "AN/APR-24."
26. France and Kaston, "AN/APR-24"; "ATI AN/APR-25 S/C/X-Band Radar Homing & Warning System (Mod 1)," Phantom Phacts, March 26, 2014, https://phantomphacts.blogspot.com/p/anapr-25n-scx-band-radar-homing-warning.html; "Udorn—ECM." International F-104 Society, accessed March 5, 2023, https://www.i-f-s.nl/udorn-ecm/.
27. Sherwood, *Afterburner*, 108.
28. "CMR-312," Phantom Phacts, July 3, 2018, https://phantomphacts.blogspot.com/p/an.html; "Miniature, Portable Microwave Receiver," *Electronics* 39, no. 1 (1966): 212; Davies, *US Navy F-4 Phantom II Units of the Vietnam War 1964–68*, 54–55; "Capt. John (Smash) Nash, USN Flight Log ID: 2438," Naval Aviation Museum Foundation, accessed February 7, 2023, https://navalaviationfoundation.org/ways-to-give/national-flight-log/national-flight-log-entry/?id=2438; Roy A. Grossnick, *United States Naval Aviation 1910–1995* (Washington, DC: Naval Historical Center, 1997), 798; John S. Attinclio, "Air-to-Air Encounters in Southeast Asia (U), Vol. I: Account of F-4 and F-8 Events Prior to 1 March 1967 (U)," Weapons Systems Evaluation Group Report 116, October 20, 1967, DTIC No. ADC003627 (Washington, DC: Office of the Director of Defense Research and Engineering, 1967), 53–54.
29. U.S. Navy, CinCPacFlt, "Analysis Staff Study 8-68," figure X, "Installation Periods of ECM Equipment in the U.S. Navy and U.S. Air Force Aircraft," 339.

30. "Tracor AN/ALE-29 Countermeasures Dispensing Set," Phantom Phacts, 2014. Accessed January 31, 2023, https://phantomphacts.blogspot.com/p/anale-29.html; Davies, *US Navy F-4 Phantom II Units of the Vietnam War 1964–68*, 61.

CHAPTER 6. SUPPRESSING SAM SITES

1. Cagle, "Task Force 77 in Action off Vietnam," 79; Davies, *A-4 Skyhawk vs North Vietnamese AAA*, 59–60.
2. Thomas Stovall, "Surface-to-Air Missiles (SAMs)," Global Security, accessed February 21, 2023, https://www.globalsecurity.org/military/world/vietnam/nva-ad-sam.htm; Cagle, "Task Force 77 in Action off Vietnam," 79.
3. Central Intelligence Agency, Directorate of Science and Technology, Office of Scientific Intelligence, "The Soviet SA-2 Surface-to-Surface-to-Air Missile System," [Excerpt], *Scientific Intelligence Digest*, June 1965, National Security Archive, accessed February 26, 2023, https://nsarchive.gwu.edu/document/16270-document-18-sa-2-surface-air-missile-excerpt, 21–22; Cahill, "Strategic Air Command SIGINT Support to the Vietnam War," 33.
4. Cahill, "Strategic Air Command SIGINT Support to the Vietnam War," 33; David Axe, "In 1966, U.S. Air Force Drones Tricked North Vietnamese Missileers into Giving Up Their Secrets," *The National Interest*, April 27, 2020, https://nationalinterest.org/blog/buzz/1966-us-air-force-drones-tricked-north-vietnamese-missileers-giving-their-secrets-148426; Price, *The History of US Electronic Warfare*, 3:58; "Ryan Model 147," Wikipedia, accessed February 25, 2023, https://en.wikipedia.org/wiki/Ryan_Model_147#References.
5. Cahill, "Strategic Air Command SIGINT Support to the Vietnam War," 33; "Ryan Model 147"; David Hambling, "The 200 Millisecond Mission: Inside the Secret CIA Plan to Steal Soviet Missile Data," *Popular Mechanics*, October 28, 2020, https://www.popularmechanics.com/military/aviation/a34386117/suicide-drone-cia-sa-2/.
6. Cahill, "Strategic Air Command SIGINT Support to the Vietnam War," 33; Carl O. Schuster, "Lightning Bug War over North Vietnam," Historynet, July 13, 2017, https://www.historynet.com/lightning-bug-war-north-vietnam/.
7. R. Cargill Hall, "Reconnaissance Drones: Their First Use in the Cold War," *Air Power History* 61, no. 3 (Fall 2014): 24.
8. Van Staaveren, *Gradual Failure*, 167–74.

9. Van Staaveren, *Gradual Failure*, 174; Cahill, "Strategic Air Command SIGINT Support to the Vietnam War," 33. Note: Van Staaveren's account appears more reliable based on the primary source listed in his endnote no. 35.
10. Cahill, "Strategic Air Command SIGINT Support to the Vietnam War," 33; David Axe, *Drone War Vietnam* (Yorkshire, UK: Pen & Sword, 2021), 80.
11. Davies, *US Navy F-4 Phantom II Units of the Vietnam War 1964–68*, 59.
12. Cagle, "Task Force 77 in Action off Vietnam," 76; Van Staaveren, *Gradual Failure*, 145; Mark Morgan and Rick Morgan, *Intruder: The Operational History of Grumman's A-6* (Atglen, PA: Schiffer Military History, 2004), 27–28; John D. Sherwood, *Nixon's Trident: Naval Power in Southeast Asia, 1968–1972* (Washington, DC: Naval History and Heritage Command, 2009), 10; Rick Morgan, *A-6 Intruder Units of the Vietnam War* (Oxford: Osprey, 2012), 7, 16.
13. Cagle, "Task Force 77 in Action off Vietnam," 76.
14. Cahill, "Strategic Air Command SIGINT Support to the Vietnam War," 33; Axe, "In 1966, U.S. Air Force Drones Tricked North Vietnamese Missileers into Giving Up Their Secrets"; Chief Plans for Field Activities, OSA, CIA [author redacted]. "Memorandum for the Record." Subject: Long Arm Drone Modification, August 13, 1965, National Security Archive, accessed February 26, 2023, https://nsarchive.gwu.edu/document/28370-document-12-chief-plans-field-activities-osa-cia-memorandum-record-subject-long-arm.
15. Van Staaveren, *Gradual Failure*, 192; Price, *The History of US Electronic Warfare*, 3:38; "Capt. Trent Richard Powers," Military Hall of Honor, accessed February 5, 2023. https://militaryhallofhonor.com/honoree-record.php?id=271335.
16. Van Staaveren, *Gradual Failure*, 192; "Capt. Trent Richard Powers"; Price, *The History of US Electronic Warfare*, 3:38; Davies, *A-4 Skyhawk vs North Vietnamese AAA*, 28.
17. George Hearing, "CINCPACFLT Analysis Staff Study 12–65: Attacks on SA-2 Sites 27 July–8 November 1965," November 15, 1965, 3.
18. Howard W. Plunkett, "Coming To Grips with SAM Sites July – Dec 1965," 7th Triennial Vietnam Symposium, March 11, 2011, The Vietnam Center & Sam Johnson Vietnam Archives, Texas Tech University, accessed March 3, 2023, https://www.vietnam.ttu.edu/events/presentations/4c-Plunkett.pdf.82; Davies, *A-4 Skyhawk vs North Vietnamese AAA*, 61.
19. Van Staaveren, *Gradual Failure*, 195–96; Davies, *B/EB-66 Destroyer Units in Combat*, 44, 48, 57.
20. Chris Pocock, "Commentary by British U-2 Historian Chris Pocock on the CIA History," National Security Archives, accessed January 20, 2023, https://

nsarchive2.gwu.edu/NSAEBB/NSAEBB434/docs/W-P%20History%20Notes%20by%20CP%20on%20redactions%20lifted%20Aug13.pdf; Central Intelligence Agency, "Electronic Equipment—U-2 Program, 1955–1966," Top Secret-Byeman, April 1, 1969, National Security Archive, accessed March 6, 2023, https://nsarchive.gwu.edu/document/27524-document-06-cia-electronic-equipment-u-2-program-1955–1966-top-secret-byeman-1-april, 8–9.

21. Central Intelligence Agency, "The Soviet SA-2 Surface-to-Surface-to-Air Missile System," 21–22; Chief Plans for Field Activities, OSA, CIA [author redacted], "Memorandum for the Record"; James A. Cunningham, "Memorandum for Deputy Director for Science and Technology," Subject: Significant OSA [Office of Special Activities] Activities, August 25, 196[4], Central Intelligence Agency, accessed February 26, 2023, https://www.cia.gov/readingroom/docs/CIA-RDP33–02415A000800290025–4.pdf.

22. Chief Plans for Field Activities, OSA, CIA [author redacted], "Memorandum for the Record."

23. Schuster, "Lightning Bug War over North Vietnam"; Cahill, "Strategic Air Command SIGINT Support to the Vietnam War," 33; Axe, "In 1966, U.S. Air Force Drones Tricked North Vietnamese Missileers into Giving Up Their Secrets."

CHAPTER 7. BLACK BOXES AND WILD WEASELS

1. "New Air Power Concepts Tested in U.S. Exercise," *New York Times*, November 11, 1964, 13; Davis, *Wild Weasel*, 8; Robert Sparkman, "Exercise Gold Fire I," *Air University Review* 16, no. 3 (March–April 1965): 39.
2. Price, *The History of US Electronic Warfare*, 3:45; Davis, *Wild Weasel*, 8.
3. Bill Sweetman, "First In . . . Still Here," *Journal of Electronic Defense*, November 1, 1999, accessed March 9, 2023, https://www.thefreelibrary.com/_/print/PrintArticle.aspx?id=58064582; Frederick Scaruffi, "Frederick Terman," *A History of Silicon Valley*, accessed March 9, 2023, https://www.scaruffi.com/svhistory/silicon/terman.html; Price, *The History of US Electronic Warfare*, 3:43–44.
4. Price, *The History of US Electronic Warfare*, 3:44.
5. Ed Chapman, quoted in Price, *The History of US Electronic Warfare*, 3:45; the meeting's date was provided by John L. Grisby, "The Wild Weasel Story: A Contractor's View" (Arlington, VA: Association of Old Crows, 2015), 4.
6. Price, *The History of US Electronic Warfare*, 3:45; Grisby, "The Wild Weasel Story," 4.

7. Grisby, "The Wild Weasel Story," 4; Hewitt, "Planting the Seeds of SEAD," 13; Barker, "The SA-2 and Wild Weasel," 89; Robert P. Breault, interview at his home in Tucson, AZ, July 20, 2024.
8. Grisby, "The Wild Weasel Story," 5.
9. Alfred W. Price, *War in the Fourth Dimension: US Electronic Warfare, from the Vietnam War to the Present* (London: Greenway, 2001), 42; Davis, *Wild Weasel*, 10; Harold E. Johnson, "Of Bears, Weasels, Ferrets, and Eagles," *Air University Review* 33, no. 2 (January–February 1982): 87; Hampton, *The Hunter Killers*, 62.
10. Davis, *Wild Weasel*, 8; Grisby, "The Wild Weasel Story," 5; Hannah, *Striving for Air Superiority*, 80.
11. Grisby, "The Wild Weasel Story," 5–6; Hampton, *The Hunter Killers*, 61; Davis, *Wild Weasel*, 8, 10.
12. Grisby, "The Wild Weasel Story," 6; Ed Chapman, as quoted by Price, *The History of US Electronic Warfare*, 3:48–45.
13. Grisby, "The Wild Weasel Story," 6–7; Marcelle S. Knaack, *Post–World War II Fighters 1945–1973* (Washington, DC: Office of Air Force History, 1986), 131; Hewitt, "Planting the Seeds of SEAD," 13.
14. Davis, *Wild Weasel*, 10; Peter E. Davies and David W. Menard, *F-100 Super Sabre Units of the Vietnam War* (Oxford, UK: Osprey, 2012), 73, 75.
15. Davis, *Wild Weasel*, 10; Hampton, *The Hunter Killers*, 82; Nalty, *Tactics and Techniques of Electronic Warfare*, 34–35; Price, *The History of US Electronic Warfare*, 3:46–47, 65–66; William Nance, "Quality ELINT," *Studies in Intelligence* 12, no. 2 (1968), Central Intelligence Agency, https://www.cia.gov/static/c54483c4718fbb7f4b3a31756717ead9/quality-elint.pdf; Telford, "Story by Joe W. Telford, WW # 258," in *First In, Last Out: Stories by the Wild Weasels*, ed. Edward T. Rock (Bloomington, IN: AuthorHouse 2005), 120–33.
16. Nalty, *Tactics and Techniques of Electronic Warfare*, 34–35; Hampton, *The Hunter Killers*, 62.
17. Hampton, *The Hunter Killers*, 65; Price, *The History of US Electronic Warfare*, 3:46.
18. Davis, *Wild Weasel*, 11; Nalty, *Tactics and Techniques of Electronic Warfare*, 35–36; Van Staaveren, *Gradual Failure*, 196; Hewitt, "Planting the Seeds of SEAD," 14–15.
19. Davis, *Wild Weasel*, 11; Van Staaveren, *Gradual Failure*, 196; Hewitt, "Planting the Seeds of SEAD," 14; Edward White, "Story by Ed White, WW Charter # 15," in *First In, Last Out: Stories by the Wild Weasels*, ed. Edward T. Rock (Bloomington, IN: AuthorHouse, 2005), 142.

20. Nalty, *Tactics and Techniques of Electronic Warfare*, 35–36; Davies and Menard, *F-100 Super Sabre Units of the Vietnam War*, 9, 13 (Kindle edition); Keith Ferris, "Story by Keith Ferris, WW # 2021," in *First In, Last Out: Stories by the Wild Weasels*, ed. Edward T. Rock (Bloomington, IN: AuthorHouse, 2005), 143.
21. John Schlight, "Rules of Engagement (U)—1 January 1966—1 November 1969," Contemporary Historical Evaluation of Combat Operations (CHECO) Report, August 31, 1969, HQ PACAF, Directorate, Tactical Evaluation, CHECO Division, accessed July 24, 2024, https://apps.dtic.mil/sti/tr/pdf/ADA602256.pdf.
22. Allen T. Lamb, quoted by Davies and Menard, *F-100 Super Sabre Units of the Vietnam War*, 76.
23. Nalty, *Tactics and Techniques of Electronic Warfare*, 37; Van Staaveren, *Gradual Failure*, 197.
24. Davies and Menard, *F-100 Super Sabre Units of the Vietnam War*, 76; Porter, "Air Tactics against NVN Air/Ground Defenses," 20; Van Staaveren, *Gradual Failure*, 197.
25. Hewitt, "Planting the Seeds of SEAD," 19–20.
26. Nalty, *Tactics and Techniques of Electronic Warfare*, 40–41; Van Staaveren, *Gradual Failure*, 269; Hannah, *Striving for Air Superiority*, 81–82; James L. Young Jr., "United States Air Force Defense Suppression Doctrine, 1968–1972" (Masters thesis, Kansas State University, 2008), 48; Breault, interview.

CHAPTER 8. AGM-45 SHRIKE

1. Van Staaveren, *Gradual Failure*, 176.
2. Elizabeth Babcock, *Magnificent Mavericks: Transition of the Naval Ordnance Test Station from Rocket Station to Research, Development, Test, and Evaluation Center, 1948–58* (Washington, DC: Naval Historical Center, 2008), 445; Cliff Lawson, *The Station Comes of Age: Satellites, Submarines, and Special Operations in the Final Years of the Naval Ordnance Test Station, 1959–1967* (China Lake, CA: Naval Air Warfare Center Western Division, 2017), 180, 183–84.
3. Babcock, *Magnificent Mavericks*, 446.
4. Lawson, *The Station Comes of Age*, 186.
5. Babcock, *Magnificent Mavericks*, 446; U.S. House of Representatives, *Department of Defense Appropriations for 1962*, Hearings before the Subcommittee of the Committee on Appropriations House of Representatives, 87th Cong., 1st Sess., Part 3 (Washington, DC: Government Printing Office, 1962), 323; Lawson, *The Station Comes of*

Age, 187; Duane J. "Jack" Russell, interview, ARM History Project, circa 1984, 4, as cited in Lawson, *The Station Comes of Age*, 187.
6. Lawson, *The Station Comes of Age*, 184, 189; Andreas Parsch, "AGM-45 *Shrike*," Directory of U.S. Military Rockets and Missiles, May 21, 2002, https://www.designation-systems.net/dusrm/m-45.html; "AGM-45 Shrike," Global Security, January 7, 2012, https://www.globalsecurity.org/military/systems/munitions/agm-45.htm. Note: The loggerhead shrike is a highly intelligent predator that is known in mythology and folklore for its tenacious hunting skills. Where the idea about pecking out the eyes of its prey came from was not specifically identified by Babcock and remains a mystery.
7. Lawson, *The Station Comes of Age*, 196, 201, 204.
8. Roy A. Grossnick, *Dictionary of American Naval Aviation Squadrons*, vol. 1 (Washington, DC: Naval Historical Center, 1995), 33; Van Staaveren, *Gradual Failure*, 176.
9. Van Staaveren, *Gradual Failure*, 201.
10. Franklin H. Knemeyer, interview, China Lake, California, November 20 and December 13, 1991, No. S-200, Oral History Collection, Naval Weapons Center China Lake, CA, 92–93; Hewitt, "Planting the Seeds of SEAD," 23–24.
11. George M. "Bud" Biery II, quoted in Lawson, *The Station Comes of Age*, 204; Glenn E. Bugos, *Engineering the F-4B Phantom II: Parts into Systems* (Annapolis, MD: Naval Institute Press, 1996), 71; Ernest Mares, "Shooting Stars," China Lake Alumni, accessed March 19, 2023, http://www.chinalakealumni.org/Downloads/1966%20SHRIKE%20SIDS.pdf; Parsch, "AGM-45 *Shrike*"; National Museum of the United States Air Force, "AGM-45 Shrike Anti-Radar Missile," nationalmuseum.af, accessed July 26, 2024, https://www.nationalmuseum.af.mil/Visit/Museum-Exhibits/Fact-Sheets/Display/Article/196035/agm-45-shrike-anti-radar-missile/; Breault, interview.
12. Van Staaveren, *Gradual Failure*, 176, 212; Parsch, "AGM-45 *Shrike*"; "AGM-45 Shrike." WeaponSystemnet, accessed March 21, 2023, https://weaponsystems.net/system/1066-HH08%20-%20AGM-45%20Shrike.
13. Lawson, *The Station Comes of Age*, 212–13; Denny Sapp, quoted in Davies, *A-4 Skyhawk vs North Vietnamese AAA*, 28–29.
14. Van Staaveren, *Gradual Failure*, 179, 239; U.S. Senate, *Fiscal Year 1974 Authorization for Military Procurement, Research and Development, Construction Authorization for the Safeguard ABM and Active Duty and Selected Reserve Strengths, Hearings before the Committee on Armed Services United States Senate*, 93rd Cong., 1st Sess. on S 1263 Part 6 Tactical Air Power (Washington, DC: Government Printing Office,

1973), Shrike AGM 45A Master Schedule, 4291; Parsch, "AGM-45 *Shrike*"; Michael A. McMaster, "AGM-45-7A Shrike: Final Test Report" (MS thesis, California State University Northridge, January 1977), 39–40.
15. See statement of Pierre Levy as quoted in Price, *The History of US Electronic Warfare*, 3:70–71; France and Kaston, "AN/APR-24 C/S/X-Band Radar Homing and Warning System (RHAW)"; "US Navy F-4 EW Development— Revisited," Phantom Phacts, May 2, 2014, https://phantomphacts.blogspot.com/p/usn-ew-equipment.html.
16. Porter, "Air Tactics against NVN Air/Ground Defenses, 32–41; Henry E. Stephenson, "The Pioneers: Wild Weasel and the F-100F," National Museum of the United States Air Force, accessed May 5, 2023, https://www.nationalmuseum.af.mil/Visit/Museum-Exhibits/Fact-Sheets/Display/Article/197491/the-pioneers-wild-weasel-and-the-f-100f/.
17. U.S. Senate, *Fiscal Year 1974 Authorization for Military Procurement, Research and Development, Construction Authorization for the Safeguard ABM and Active Duty and Selected Reserve Strengths*, Shrike AGM 45A Master Schedule, 4291; Parsch, "AGM-45 *Shrike*."

CHAPTER 9. WILD WEASELS II AND III

1. Nalty, *Tactics and Techniques of Electronic Warfare*, 40; Grisby, "The Wild Weasel Story," 8; Robert Beechy, "Ferrets, Ravens & Weasels Radar Countermeasures and SAM Suppression," ProHosting, accessed April 28, 2023, http://hud607.fire.prohosting.com/uncommon/reference/usa/sead.html; Joe W. Telford, "Story by Joe W. Telford, WW # 258," in *First In, Last Out: Stories by the Wild Weasels*, ed. Edward T. Rock (Bloomington, IN: AuthorHouse, 2005), 133.
2. Telford, "Story # 258," 133.
3. Telford, "Story # 258," 134; Grisby, "The Wild Weasel Story," 11–12.
4. Nalty, *Tactics and Techniques of Electronic Warfare*, 40.
5. Davis, *Wild Weasel*, 14; Nalty, *Tactics and Techniques of Electronic Warfare*, 40.
6. Grisby, "The Wild Weasel Story," 13–14.
7. Edward T. Rock, "Wild Weasel III-2 Memories," in *First In, Last Out: Stories by the Wild Weasels*, ed. Edward T. Rock (Bloomington, IN: AuthorHouse, 2005), 228–30.
8. Rock, "Wild Weasel III-2 Memories," 230.
9. Davis, *Wild Weasel*, 14.
10. Peter E. Davies, *F-105 Thunderchief Units of the Vietnam War* (Oxford: Osprey, 2012), Google Books, accessed May 10, 2023, https://www.google.com/

books/edition/F_105_Thunderchief_Units_of_the_Vietnam/HsTvCwAA QBAJ?hl=en&gbpv=1&dq=wild+weasels+korat&printsec=frontcover; Nalty, *Tactics and Techniques of Electronic Warfare*, 41; Henry E. Stephenson, "Route Packs in Vietnam," Air Force History and Museums Program, accessed May 10, 2023, https://www.afhistoryandmuseums.af.mil/About-Us/Fact-Sheets/Article/639570/route-packs-in-vietnam/; Walter Boyne, "Route Package 6," *Air and Space Forces Magazine*, November 1, 1999, https://www.airandspaceforces.com/article/1199pack/.

11. Davis, *Wild Weasel*, 14; Grisby, "The Wild Weasel Story," 15; Hannah, *Striving for Air Superiority*, 82.
12. Nalty, *Tactics and Techniques of Electronic Warfare*, 41; Rock, "Wild Weasel III-2 Memories," 233–37.
13. Rock, "Wild Weasel III-2 Memories," 239–41.
14. Nalty, *Tactics and Techniques of Electronic Warfare*, 43.
15. Young, "United States Air Force Defense Suppression Doctrine, 1968–1972," 49.
16. Porter, "Air Tactics against NVN Air/Ground Defenses," 34–35.
17. Nalty, *Tactics and Techniques of Electronic Warfare*, 43.
18. Nalty, *Tactics and Techniques of Electronic Warfare*, 44.
19. John C. Pratt, "Air Tactics against NVN Air Ground Defenses December 1966—1 November 1968 (U)," Contemporary Historical Evaluation of Combat Operations (CHECO) Report, August 30, 1969, HQ PACAF, Directorate, Tactical Evaluation, CHECO Division, 44–45.
20. Nalty, *Tactics and Techniques of Electronic Warfare*, 45; Porter, "Air Tactics against NVN Air/Ground Defenses," 29–30.
21. Porter, "Air Tactics against NVN Air/Ground Defenses," 32. See also 52 note 49.
22. Pribbenow, "The 'Ology War," 178. See also note 10.
23. James Pierson, "Electronic Warfare in SEA 1964–1968" (USAF Security Service, 1973), reproduced in Price, *The History of US Electronic Warfare*, 3:149.
24. Hannah, *Striving for Air Superiority*, 83; Davis, *Wild Weasel*, 41.
25. Davis, *Wild Weasel*, 31; Streetly, *Airborne Electronic Warfare*, 86.
26. Hannah, *Striving for Air Superiority*, 83.

CHAPTER 10. AGM-78 STANDARD ANTIRADIATION MISSILE

1. Davis, *Wild Weasel*, 23; *Testimony of Dr. R. L Frosch, Assistant Secretary of the Navy for R&D, April 4, 1967, U.S. House of Representatives, Department of Defense*

Appropriations for 1968, Hearings before a Subcommittee of the Committee on Appropriations House of Representatives, 90th Cong. 1st Sess. Part 1, Military Personnel (Washington, DC: Government Printing Office, 1967), 417; Andreas Parsch, "AGM-78," Directory of U.S. Missiles and Rockets and Missiles, accessed May 18, 2023, https://www.designation-systems.net/dusrm/m-78.html; Andreas Parsch, "RIM-66," Directory of U.S. Military Rockets and Missiles, accessed May 18, 2023, https://www.designation-systems.net/dusrm/m-66.html; Streetly, *Airborne Electronic Warfare*, 88; *Testimony of Rear Adm. R. L. Townsend, Commander, Naval Air Systems Command, July 13, 1967, U.S. Senate. Department of Defense Appropriations for 1968, Hearings before a Subcommittee of the Committee on Appropriations United States Senate*, 90th Cong. 1st Sess. on H.R. 10738, Part 2 (Washington, DC: Government Printing Office, 1968), 500.

2. Morgan and Morgan, *Intruder*, 68, 207.
3. Morgan and Morgan, *Intruder*, 68; "USS Kitty Hawk CV 63," US Carriers, accessed May 21, 2023, http://www.uscarriers.net/cv63history.htm; J. E. Davis, Commanding Officer, USS Kitty Hawk (CVA 63), to CNO, Subj. USS Kitty Hawk (CVA 63) Command History, 1968, Naval History and Heritage Command, accessed May 20, 2023, https://www.history.navy.mil/content/dam/nhhc/research/archives/command-operation-reports/vietnam/KITTY%20HAWK%201968.pdf.
4. Davis, *Wild Weasel*, 23, 30, 41; National Museum of the U.S. Air Force, "'Something Better Than the Shrike': The First USAF AGM-78," National Museum of the U.S. Air Force, accessed May 19, 2023, https://www.nationalmuseum.af.mil/Visit/Museum-Exhibits/Fact-Sheets/Display/Article/196908/something-better-than-the-shrike-the-first-usaf-agm-78-standard-mission/.
5. Warren J. Kerzon, *Throw a Nickel on the Grass: A Fighter Pilot's Life Narrative* (Raleigh, NC: Lulu, 2016), 101–2.
6. Kerzon, *Throw a Nickel on the Grass*, 101; National Museum of the U.S. Air Force, "Something Better Than the Shrike."
7. Melvin F. Porter, "Second Generation Weaponry in SEA," September 10 1970, HQ PACAF, Directorate, Tactical Evaluation CHECO Division, 63; "31 March 1968: President Lyndon B. Johnson Announces Bombing Halt in Vietnam," Vietnam The Art of War, accessed May 21, 2023, https://vietnamtheartofwar.com/1968/03/31/31-march-1968-president-lyndon-b-johnson-announces-bombing-halt-in-vietnam/; "The President's Address to the Nation upon Announcing His Decision to Halt the Bombing of North Vietnam, October 31, 1968," The American Presidency Project, accessed

May 21, 2023, https://www.presidency.ucsb.edu/documents/the-presidents-address-the-nation-upon-announcing-his-decision-halt-the-bombing-north.
8. Warren E. Thompson, "Wild Duel: Weasel vs SAMs over Dong Hoi," historynet, March 19, 2013, https://www.historynet.com/wild-duel-weasels-vs-sams-over-dong-hoi/.
9. Thompson, "Wild Duel: Weasel vs SAMs over Dong Hoi."
10. Davis, *Wild Weasel*, 44; Streetly, *Airborne Electronic Warfare*, 88.
11. Streetly, *Airborne Electronic Warfare*, 88; Norman Friedman, *U.S. Naval Weapons: Every Gun, Missile, Mine and Torpedo Used by the U.S. Navy from 1883 to the Present Day* (Annapolis, MD: Naval Institute Press, 1988), 210–11; Parsch, "AGM-78"; Porter, "Second Generation Weaponry in SEA," 65–66.
12. Hewitt, "Planting the Seeds of SEAD," 26; Porter, "Second Generation Weaponry in SEA," 63.
13. USAF Chief of Staff John D. Ryan to Tactical Air Command, October 3, 1969, reproduced in Hewitt, "Planting the Seeds of SEAD," 42 note 87.

CHAPTER 11. DEALING WITH THE MIG THREAT

1. Burch, "Tactical Electronic Warfare Operations in SEA, 1962–1968," 15; Carl O. Schuster, "The Rise of North Vietnam's Air Defense," Historynet, 2017. Accessed June 3, 2023, https://www.historynet.com/rise-north-vietnams-air-defense/.
2. Burch, "Tactical Electronic Warfare Operations in SEA, 1962–1968," 15.
3. Thomas R. Johnson, *American Cryptology during the Cold War, 1945–1989. Book II: Centralization Wins 1960–1972*, Center for Cryptologic History, National Security Agency, 1995, National Security Agency, accessed May 28, 2023, https://www.nsa.gov/portals/75/documents/news-features/declassified-documents/cryptologic-histories/cold_war_ii.pdf; William Cahill, "The Short but Interesting Life of a Plane Called Rivet Top," *Air Power History* 54, no. 3 (Fall 2007): 25.
4. Martin Bowman, *The Men Who Flew the F-4 Phantom* (Barnsley, South Yorkshire, UK: Pen & Sword Aviation), Google Books, accessed May 27, 2023, https://www.google.com/books/edition/The_Men_Who_Flew_the_F_4_Phantom/iLPNDwAAQBAJ?hl=en&gbpv=1&dq=c-130+queen+bee&pg=PT84&printsec=frontcover.
5. National Security Agency, "IRONHORSE: A Tactical SIGINT System," *Cryptolog* II, no. 10 (October 1975) [NSA], https://nsarchive.gwu.edu/sites/default/files/documents/5301807/National-Security-Agency-Cryptolog-Vol-2-No-10.pdf, 24; Robert J. Hanyok, *(U) Spartans in Darkness: American*

SIGINT and the Indochina War, 1945–1978, Center for Cryptologic History, National Security Agency, 2002, 239, 258.

6. Hanyok, *(U) Spartans in Darkness*, 251.
7. Hanyok, *(U) Spartans in Darkness*, 250; van Geffen, "The Air War against North Vietnam, 10.
8. Chris Hobson and David Lovelady, "North Vietnamese MIGs (Apr 1965)," Vietnam Air Losses, accessed May 26, 2023, https://www.vietnamairlosses.com/index.php/sidelines/1965/apr65; John T. Correll, "Against the MiGs in Vietnam," *Air & Space Forces Magazine*, October 1, 2019, https://www.airandspaceforces.com/article/against-the-migs-in-vietnam/; "Mikoyan-Gurevich MiG-17 (Fresco)," Military Factory, accessed May 31, 2023, https://www.militaryfactory.com/aircraft/detail.php?aircraft_id=31; Michel, *Clashes*, 76.
9. Correll, "Against the MiGs in Vietnam"; Hobson and Lovelady, "North Vietnamese MIGs (Apr 1965)"; Hallion, *Rolling Thunder 1965–68*, 26.
10. John S. Attinclio, "Air-to-Air Encounters in Southeast Asia (U): Volume 1, Account of F-4 and F-8 Events to 1 March 1967 (U)," WSEG Report 116, October 1967, Institute for Defense Analyses, Systems Evaluation Division, Arlington, Virginia, 33; Hallion, *Rolling Thunder 1965–68*, 48.
11. Hallion, *Rolling Thunder 1965–68*, 48; van Geffen, "The Air War against North Vietnam," 10–11; Vern Kulla as quoted in van Geffen, "The Air War against North Vietnam," 11.
12. Burch, "Tactical Electronic Warfare Operations in SEA, 1962–1968," 19; Van Nederveen, "Sparks over Vietnam," 35–36; U.S. Army, Center of Military History, *History of Strategic Air and Ballistic Missile Defense Volume II: 1956–1972* (Washington, DC: Center of Military History, 2009), 276; "Radar (Spoon Rest C [P-12])—SA-2," Command: Modern Operations/Modern Air Naval Operations, accessed June 2, 2023, http://cmano-db.com/pdf/facility/19/.
13. U.S. Army, Center of Military History, *History of Strategic Air and Ballistic Missile Defense Volume II*, 276; John A. Schell, "The SA-2 and U-2: The Rest of the Story," *Journal of the Air Force Historical Foundation* 70, no. 2 (Summer 2023): 39; Hanyok, *(U) Spartans in Darkness*, 236–37; Schuster, "The Rise of North Vietnam's Air Defense."
14. Carl W. Reddel, "College Eye," November 1, 1968, Project CHECO HQ PACAF, Directorate, Tactical Evaluation CHECO Division, 1.
15. Reddel, "College Eye," 2–3; Nalty, *Tactics and Techniques of Electronic Warfare*, 116–17; Jerold R. Mack and Richard M. Williams, "552d Airborne Early Warning and Control Wing in Southeast Asia: A Case Study in Airborne Command and Control," *Air University Review* 25, no. 1 (November–December

1973): 72. Note: Nalty lists the task force as departing with five aircraft. I have chosen to use the number provided by Reddel, which I believe is more of an original source.

16. Mack and Williams, "552d Airborne Early Warning and Control Wing in Southeast Asia," 71, Cahill, "The Short but Interesting Life of a Plane Called Rivet Top," 25; Nalty, *Tactics and Techniques of Electronic Warfare*, 118; Reddel, "College Eye," 11–12.
17. Robert E. Morrison, "How VQ-1 Supported Military Actions with SIGINT during Vietnam (Parts 1 of 7) Guest Post," Station HYPO, accessed February 12, 2023, https://stationhypo.com/2018/04/22/how-vq-1-supported-military-actions-with-sigint-during-vietnam-part-1-of-7-guest-post/#more-9343; Mack and Williams, "552d Airborne Early Warning and Control Wing in Southeast Asia," 72.
18. Mack and Williams, "552d Airborne Early Warning and Control Wing in Southeast Asia," 72; Nalty, *Tactics and Techniques of Electronic Warfare*, 120–22; Don E. Born, quoted in Gabriele Barison, "That Time a USAF EC-121 Warning Star Almost Landed on a U.S. Navy Aircraft Carrier," The Aviation Geek Club, July 18, 2018, https://theaviationgeekclub.com/that-time-a-usaf-ec-121-warning-star-almost-landed-on-a-u-s-navy-aircraft-carrier/.
19. Nalty, *Tactics and Techniques of Electronic Warfare*, 1201; Kenneth P. Werrell, *Chasing the Silver Bullet: U.S. Air Force Weapons Development from Vietnam to Desert Storm* (Washington, DC: Smithsonian Books, 2003), 193.
20. Nalty, *Tactics and Techniques of Electronic Warfare*, 124; Hallion, *Rolling Thunder 1965–68*, 50.
21. Nalty, *Tactics and Techniques of Electronic Warfare*, 124; Michel, *Clashes*, 50.
22. Mack and Williams, "552d Airborne Early Warning and Control Wing in Southeast Asia," 72; Nalty, *Tactics and Techniques of Electronic Warfare*, 120–22.
23. Charles K. Hopkins, *SAC Tanker Operations of the Southeast Asia War* (Omaha, NE: Office of the Historian, Headquarters Strategic Air Command, 1987), 98, 100.
24. Hopkins, *SAC Tanker Operations of the Southeast Asia War*, 100; Reddel, "College Eye," 1. The name change took effect on March 13, 1967.
25. Nalty, *Tactics and Techniques of Electronic Warfare*, 122; Michel, *Clashes*, 48; George F. Schreader, *Hognose Silent Warrior: The USAF's Airborne Intelligence War in the Final Air Campaigns of Vietnam* (Parker, CO: Outskirts Press, 2018, Kindle edition), 3559.
26. Nalty, *Tactics and Techniques of Electronic Warfare*, 126.
27. Hanyok, *(U) Spartans in Darkness*, 250; Johnson, *American Cryptology during the Cold War, 1945–1989*.

28. Hanyok, *(U) Spartans in Darkness*, 250–51.
29. Johnson, *American Cryptology during the Cold War*, 1945–1989: Nalty, *Tactics and Techniques of Electronic Warfare*, 127–28; Hanyok, *(U) Spartans in Darkness*, 250; Michael R. Patterson, "Robert Gordon Owens, Jr.—Major General, United States Marine Corps," Arlington National Cemetery, accessed June 11, 2023, https://www.arlingtoncemetery.net/rgowens.htm.
30. Nalty, *Tactics and Techniques of Electronic Warfare*, 129; Hanyok, *(U) Spartans in Darkness*, 251.

CHAPTER 12. BIG SAFARI, COLLEGE EYE, AND COMBAT APPLE

1. Reddel, "College Eye," 18.
2. Bill Grimes, *The History of Big Safari* (Bloomington, IN: Archway Publishing, 2014), 1–3, 162; Axe, *Drone War Vietnam*, 30; Cahill, "The Short but Interesting Life of a Plane Called Rivet Top," 24.
3. Reddel, "College Eye," 18–19; Hanyok, *(U) Spartans in Darkness*, 255; Ruben Urribarres and Mike Little, "The Cuban MiGs," The Latin American Aviation Historical Society, April 15, 2018, https://www.laahs.com/the-cuban-migs/.
4. Reddel, "College Eye," 5, 18–19; Hanyok, *(U) Spartans in Darkness*, 255; Price, *The History of US Electronic Warfare*, 3:104.
5. Reddel, "College Eye," 19–21, part 2, 62; Cahill, "The Short but Interesting Life of a Plane Called Rivet Top," 24–25; Hanyok, *(U) Spartans in Darkness*, 255; Grimes, *The History of Big Safari*, 162; Price, *The History of US Electronic Warfare*, 3:104; Thomas W. Morris, "CORONA HARVEST Interview No. 92," by Capt. Richard Clement, February 4, 1969, United States Air Force Oral History Program, K230.0512-092, Air Force Historical Research Agency, Maxwell Air Force Base, AL, 17.
6. Cahill, "The Short but Interesting Life of a Plane Called Rivet Top," 24; Hank Maifeld, "MacDill AFB, Florida: Detachment 2 Tactical Air Warfare Center," Geocities, June 14, 2023, https://www.geocities.ws/hmaifeld/macdill.html; Hank Maifeld, "Osan Air Base," Geocities, accessed June 15, 2023, http://www.maifeldfamily.net/hank/osan.html.
7. Reddel, "College Eye," 8, 19; Hanyok, *(U) Spartans in Darkness*, 255; Price, *The History of US Electronic Warfare*, 3:104; Nalty, *Tactics and Techniques of Electronic Warfare*, 136.
8. Reddel, "College Eye," 20, 63 note 26.

9. Reddel, "College Eye," 9; Morris, "CORONA HARVEST Interview No. 92," 2.
10. Nalty, *Tactics and Techniques of Electronic Warfare*, 135; Hanyok, *(U) Spartans in Darkness*, 255–56.
11. Mack and Williams, "552d Airborne Early Warning and Control Wing in Southeast Asia," 75–76; Litton Industries, "AN/GPA-122," *News & Opinion* 12, no. 2 (February 19, 1973): 3; Reddel, "College Eye," 16, 25.
12. Nalty, *Tactics and Techniques of Electronic Warfare*, 135; Reddel, "College Eye," 26; Graham M. Simons, *Lockheed Constellation: A History* (Yorkshire: Pen & Sword, 2021), Google Books, accessed June 20, 2023, https://www.google.com/books/edition/Lockheed_Constellation/Myo1EAAAQBAJ?hl=en&gbpv=1&dq=rivet+gym&pg=PT483&printsec=frontcover. Note the discrepancy in the date between the Reddel and Nalty accounts.
13. Nalty, *Tactics and Techniques of Electronic Warfare*, 139–40.
14. Nalty, *Tactics and Techniques of Electronic Warfare*, 141; Mack and Williams, "552d Airborne Early Warning and Control Wing in Southeast Asia," 76.
15. Charles K. Hopkins, *SAC Tanker Operations in the Southeast Asia War* (Omaha, NE: Office of the Historian, Headquarters Strategic Air Command, 1987), 97; Cahill, "Strategic Air Command SIGINT Support to the Vietnam War," 34; Bob Archer, *Super Snooper: The Evolution and Service Career of the Specialist Boeing C-135 Series with the 55th Wing and Associated Units* (Stroud, UK: Fonthill Media, 2020), 97.
16. Schreader, *Hognose Silent Warrior*, 1241–42, 1310, 1347, 2441.
17. Schreader, *Hognose Silent Warrior*, 2137, 2494.
18. Michel, *Clashes*, 7; Schreader, *Hognose Silent Warrior*, 2473.
19. "Radio Communication Station Hanoi/Bac Mai Airfield," April 19, 1968, CIA, National Photographic Interpretation Center, accessed July 11, 2023, https://www.cia.gov/readingroom/docs/CIA-RDP83-01074R000100030002-5.pdf; Schreader, *Hognose Silent Warrior*, 2484.

CHAPTER 13. COMBAT MARTIN AND AN AGM-45 FRIENDLY FIRE INCIDENT

1. Michel, *Clashes*, 134–37; Price, *The History of US Electronic Warfare*, 3:143.
2. *U.S. House of Representatives. Hearings on Military Posture and H.R. 3818 and H.R. 8687 . . . Before the Committee on Armed Services House of Representatives*, 92nd Cong., 1st Sess., Part 2 (Washington, DC: Government Printing Office, 1971), 4759; Marcelle S. Knaack, *Encyclopedia of U.S Air Force Aircraft and Missile*

Systems, Volume I: *Post-World War II Fighters 1945–1973* (Washington, DC: Officer of Air Force History, United States Air Force, 1978), 202; Nalty, *Tactics and Techniques of Electronic Warfare*, 64; Thomas Wildenberg, *Fighting in the Electromagnetic Spectrum: U.S. Navy and Marine Corps Electronic Warfare Aircraft, Operations, and Equipment* (Annapolis, MD: Naval Institute Press, 2023), 164.

3. Price, *The History of US Electronic Warfare*, 3:143; Andreas Parsch, "ORC - Equipment Listing," Designations of U.S. Military Electronic and Communications Equipment, accessed March 7, 2023, http://www.designation-systems.net/usmilav/jetds/qrc.html; Nalty, *Tactics and Techniques of Electronic Warfare*, 64–65; Greg Goebel, "Republic F-105 Thunderchief," AirVectors, March 1, 2023, https://www.airvectors.net/avf105.html.

4. Marty Selmanovitz, quoted by Price, in *The History of US Electronic Warfare*, 3:143. Price never asked, and I was not able to contact Selmanovitz, as he passed away in 2018.

5. Howard W. Plunkett, *F-105 Thunderchiefs: A 29-Year Illustrated Operational History* (Jefferson, NC: McFarland and Company, 2001), 240, 249; Price, *The History of US Electronic Warfare*, 3:143.

6. Marshal L. Michel III, *Operation Linebacker I 1972: The First High-Tech Air War* (Oxford: Osprey, 2019), 35; U.S. Air Force, Historical Report Division, "1972 Operations Linebacker I," accessed August 16, 2023, https://www.afhistory.af.mil/FAQs/Fact-Sheets/Article/458990/1972-operation-linebacker-i/; Michel, *Clashes*, 201; Kenneth V. Jack, *Eyes of the Fleet over Vietnam: RF-8 Crusader Combat Photo-Reconnaissance Missions* (Philadelphia: Casemate, 2021), 151.

7. George Guy Thomas, email to the author, April 21, 2023.

8. George Guy Thomas, *A Silent Warrior Steps Out of the Shadows* (Self-published, Alpha Book Publisher, 2021), Kindle Edition, 61.

9. George B. Schick Jr. [commanding officer], "Command History of USS WORDEN 01 January 1972 to 31 December 1972," OpNav Report 5750-11, Naval History and Heritage Command, accessed August 15, 2023, https://www.history.navy.mil/content/dam/nhhc/research/archives/command-operation-reports/vietnam/Worden%201972.pdf, 1–3; USS *Worden* to NWL [Naval Weapons Laboratory], Dahlgren, VA [Naval Message], November 12, 1972, Naval History and Heritage Command, accessed August 15, 2023, https://www.history.navy.mil/content/dam/nhhc/research/archives/command-operation-reports/vietnam/Worden%201972.pdf; Thomas, *A Silent Warrior Steps Out of the Shadows*, 66.

CHAPTER 14. AUTOMATING THE TACTICAL AIR CONTROL SYSTEM

1. U.S. National Security Agency, "IRONHORSE: A Tactical SIGINT System," *Cryptolog* II, no. 10 (October 1975): 224–26; Hanyok, *(U) Spartans in Darkness*, 258.
2. U.S. National Security Agency, "IRONHORSE: A Tactical SIGINT System," 25; Hanyok, *(U) Spartans in Darkness*, 259.
3. Hanyok, *(U) Spartans in Darkness*, 259; K. Sams, J. Schlight, R. F. Knot, M. J. Medelson, and P. D. Caine, "The Air War in Vietnam 1968–1969 (U)," April 1, 1970, Project CHECO HQ PACAF, Directorate, Tactical Evaluation CHECO Division, 64; Frank M. Machovec, "Southeast Asia Tactical Systems Interface (U)," January 1, 1975, CHECO/Corona Harvest Division, Operations Analysis Office, Headquarters Pacific Air Forces, 5.
4. U.S. Army, "'Seek Dawn' Aircraft Control System," *Military Review* XLVIII, no. 10 (October 1968): 97; John J. Lane Jr., *Command and Control and Communications Structures in Southeast Asia* (Maxwell Air Force Base, AL: Airpower Research Institute, 1981), 77; Walter Boyne, "The Teaball Tactic," *Air Force Magazine* 91, no. 7 (July 2008): 69.
5. Sherwood, *Afterburner*, 82; Edward J. Marolda, "Forged in Battle," *Naval History* 28, no. 4 (July 2014), https://www.usni.org/magazines/naval-history-magazine/2014/july/forged-battle; Gilles K. Van Nederveen, "Wizardry for Air Campaigns Signals Intelligence Support to the Cockpit," Research Paper 2002-03 (Maxwell Air Force Base, AL: Airpower Research Institute, 2001), 23.
6. Garette E. Lockee, "PIRAZ," U.S Naval Institute *Proceedings* 95, no. 4 (April 1969): 143; John Gargas, "The Greatest Naval Deception of the Vietnam War, Part Two," *Naval History Magazine* 36, no. 3 (June 2022), https://www.usni.org/magazines/naval-history-magazine/2022/june/greatest-naval-deception-vietnam-war; Robert Morrison, "PIRAZ in the Gulf of Tonkin—NGS's Role." Station HYPO, accessed January 18, 2023, https://stationhypo.com/2021/04/17/piraz-in-the-gulf-of-tonkin-nsgs-role-guest-post/.
7. David L. Boslaugh, *When Computers Went to Sea: The Digitization of the United States Navy* (Los Alamitos, CA: IEEE Computer Society, 1999), 297.
8. William Bryant and Heath I. Hermaine, "History of Naval Fighter Direction," *CIC Magazine*, April, May, and June, as cited by David L. Boslaugh, *First-Hand: No Damned Computer is Going to Tell Me What to DO—The Story of the Naval Tactical Data System, NTDS*, May 12, 2021, https://ethw.org/Firstand:No_Damned_Computer_is_Going_to_Tell_Me_What_to_DO_The_Story_of_the_Naval_Tactical_Data_System,_NTDS, chapter 1 (no page numbers).

9. Paul W. Cherington, "Systems-Acquisition and the Utilization of Scientific and Engineering Manpower (Requirements and Program-Determination Contracts and Grants)," in Committee on Utilization of Scientific and Engineering Manpower, National Academy of Sciences, *Toward Better Utilization of Scientific and Engineering Talent* (Washington, DC: National Academy of Sciences, 1964), 123; Boslaugh, *First-Hand*, chap. 3.
10. Lockee, "PIRAZ," 144.
11. Lockee, "PIRAZ," 145–46; Boslaugh, *First-Hand*, chap. 7.
12. Malcolm Muir Jr., *Black Shoes and Blue Water: Surface Warfare in the United States Navy, 1945–1975* (Washington, DC: Naval Historical Center, 1996), 157–58; Boslaugh, *First-Hand*, chap. 7.
13. Copalman, "Glory Days// EA-6A in Vietnam"; Kristina Panos, "Retrotechtacular: Radar Jamming," *Hackaday*, April 28, 2015, https://hackaday.com/tag/analq-55/.
14. Tom Carter, quoted in Copalman, "Glory Days// EA-6A in Vietnam."
15. Muir, *Black Shoes and Blue Water*, 157–58; Boslaugh, *First-Hand*, chap. 7.
16. Copalman, "Glory Days// EA-6A in Vietnam"; Nalty, *Tactics and Techniques of Electronic Warfare*, 65.
17. Nalty, *Tactics and Techniques of Electronic Warfare*, 65.
18. E. Hartsook, *The Air Force in Southeast Asia: Shield for Vietnamization and Withdrawal 1971* (Washington, DC: Office of Air Force History, 1976), 33–34; Hanyok, *(U) Spartans in Darkness*, 260.

CHAPTER 15. PROUD DEEP ALPHA, RGM-8H TALOS, AND OPERATION LINEBACKER I

1. Melvin F. Porter. "Linebacker: Overview of the First 120 Days," September 27, 1973. HQ PACAF, Directorate of Operation Analysis, CHECO/CORONA HARVEST Division, 7.
2. Hanyok, *(U) Spartans in Darkness*, 269; Correll, "Take It Down! The Wild Weasels in Vietnam," 69; Michel, *Clashes*, 188.
3. Hartsook, *The Air Force in Southeast Asia*, 31.
4. Hartsook, *The Air Force in Southeast Asia*, 31, 34.
5. Hartsook, *The Air Force in Southeast Asia*, 30–32; Theo van Geffen, "U.S. Mini-Air War against the North Vietnam: Protective Reaction Strikes, 1968–1972," *Air Power History* 66, no. 2 (Summer 2019): 34.
6. Philip Hays, "Details of the First RGM-8H Anti-Radiation Missile Combat Firing," Okieboat, June 30, 2020, https://www.okieboat.com/Talos%20

antiradiation%20shot.html; Earl H. Tilford Jr., *Setup: What the Air Force Did in Vietnam and Why* (Maxwell Air Force Base, AL: Air University Press, 1991), 86; Van Staaveren, *Gradual Failure*, 239.
7. Hays, "Details of the First RGM-8H Anti-Radiation Missile Combat Firing"; Andreas Parsch, "Bendix SAM-N-6/IM-70/RIM-8 *Talos*," Directory of U.S. Military Rockets and Missiles, accessed September 22, 2023, https://www.designation-systems.net/dusrm/m-8.html; William Garden Jr. and Frank A. Dean, "Evolution of the Talos Missile," *Johns Hopkins APL Technical Digest* 3, no. 2 (1982): 121.
8. Hays, "Details of the First RGM-8H Anti-Radiation Missile Combat Firing"; Parsch, "Bendix SAM-N-6/IM-70/RIM-8 *Talos*."
9. Hays, "Details of the First RGM-8H Anti-Radiation Missile Combat Firing."
10. Porter, "Linebacker: Overview of the First 120 Days," 12–13.
11. U.S. Navy, Naval History and Heritage Command, "H-Gram 070: The Easter Offensive-Vietnam 1971(1)," April 27, 2022, https://www.history.navy.mil/about-us/leadership/director/directors-corner/h-grams/h-gram-070.html; "Operation Freedom Train, Operation Linebacker I." Global Security, accessed May 19, 2023, https://www.globalsecurity.org/military/ops/linebacker-1.htm; U.S. Department of Defense, Vietnam War Commemoration, "North Vietnamese Easter Offensive, March 30, 1972," The United States of America Vietnam War Commemoration, accessed July 14, 2023, https://www.vietnamwar50th.com/1972–1974_negotiations_and_passing_the_torch/North-Vietnamese-Easter-Offensive/; Warren L. Harris, "The Linebacker Campaigns: An Analysis" (Maxwell Air Force Base, AL: Air University Press, 1987), 7.
12. Porter, "Linebacker: Overview of the First 120 Days," 14–15; Harris, "The Linebacker Campaigns: An Analysis," 7–8; Richard M. Nixon, *The Memoirs of Richard Nixon*, vol. 2 (New York: Warner Books, 1978), 64.
13. W. Hays Parks, "Linebacker and the Law of War," *Air University Review* 34, no. 2 (January–February 1983): 5; Bernard C. Nalty, "1972—Operation Linebacker I," Air Force Historical Support Division, accessed July 15, 2023, https://www.afhistory.af.mil/FAQs/Fact-Sheets/Article/458990/1972-operation-linebacker-i/; Michel, *Clashes*, 201; Michel, *Operation Linebacker I 1972*, 34, 56; Yancy Mailes, "B-52 Played Major Role in Operations Freedom Train and Linebacker I," Air Force Global Strike Command, accessed May 21, 2023, https://www.afgsc.af.mil/News/Article-Display/Article/454706/b-52-played-major-role-in-operations-freedom-train-and-linebacker-i/; Jeffery N. Meyer, "Andersen

AFB's Legacy: Operation Linebacker II," Anderson Air Force Base, accessed July 16, 2023, https://www.andersen.af.mil/News/Commentaries/Display/Article/416815/andersen-afbs-legacy-operation-linebacker-ii/.
14. Nalty, *Tactics and Techniques of Electronic Warfare*, 14; Michel, *Operation Linebacker I 1972*, 73; Steven A. Fino, *Tiger Check: Automating the US Air Force Fighter Pilot in Air-to-Air Combat, 1950–1980* (Baltimore: Johns Hopkins University Press, 2017), 152.
15. Zaloga, *Red SAM*, 19–20.
16. Zaloga, *Red SAM*, 20.
17. Max Steel, "Vietnam War: Soviet Air Defense Systems," Russian Defense Forum, February 8, 2016, https://www.russiadefence.net/t2474-vietnam-war-soviet-air-defence-systems; Zaloga, *Red SAM*, 21; Mandeep Singh, *Anti-Aircraft Artillery in Combat 1950–1972: Air Defense in the Jet Age* (Philadelphia, PA: Pen & Sword, 2020), Google Books, accessed August 18, 2023, https://www.google.com/books/edition/Anti_Aircraft_Artillery_in_Combat_1950_1/sxnhDwAAQBAJ?hl=en&gbpv=1&dq=sa-2+optical+tracking&pg=PT102&printsec=frontcover.
18. Johnson, *American Cryptology during the Cold War, 1945–1989*, 579–80.

CHAPTER 16. TEABALL, NTDS, AND RED CROWN

1. Doyle Larson, "Direct Intelligence Combat Support in Vietnam Project Teaball," *American Intelligence Journal* 15, no. 1 (Spring/Summer 1994): 56; Johnson, *American Cryptology during the Cold War*, 580; Michel, *Operation Linebacker I 1972*, 77.
2. Porter, "Linebacker: Overview of the First 120 Days," 70–71; Johnson, *American Cryptology during the Cold War*, 580; Sharon Maneki, "Delmar C. Lang: A SIGINT Innovator," Cryptographic Almanac 50th Anniversary, accessed July 17, 2023, https://media.defense.gov/2021/Jul/01/2002754126/-1/-1/0/DELMAR-C-LANG.PDF; National Security Agency, "TEABALL: Some Personal Observations of SIGINT at War," *Cryptology Quarterly*, Winter 1991, accessed September 20, 2023, https://www.nsa.gov/.portals/75/documents/news-features/declassified-documents/cryptologic-quarterly/teaball.pdf, 92.
3. National Security Agency, "TEABALL," 93.
4. Van Nederveen, "Wizardry for Air Campaigns, 25.
5. Larson, "Direct Intelligence Combat Support in Vietnam Project Teaball," 56–57; Van Nederveen, "Wizardry for Air Campaigns," 25.
6. Larson, "Direct Intelligence Combat Support in Vietnam Project Teaball," 56–57; Van Nederveen, "Wizardry for Air Campaigns," 25.

7. Van Nederveen, "Wizardry for Air Campaigns," 24, 28; Larson, "Direct Intelligence Combat Support in Vietnam Project Teaball," fig. 3, page 57; National Security Agency, "TEABALL," 93.
8. Michel, *Operation Linebacker I 1972*, 82–83.
9. Christian Wolff, "AN/SPS-48E," Radar Tutorial, accessed July 20, 2023, https://www.radartutorial.eu/19.kartei/07.naval/karte010.en.html; Michel, *Clashes*, 46; Michel, *Operation Linebacker I 1972*, 73.
10. Robert E. Morrison, "PIRAZ in the Gulf of Tonkin—NGS's Role," Station HYPO, accessed January 18, 2023, https://stationhypo.com/2021/04/17/piraz-in-the-gulf-of-tonkin-nsgs-role-guest-post/; Mario Valcano, "Out of the Shadows—The Art of Time-Sensitive Reporting (Part 6 of 7)," Station HYPO, accessed July 4, 2023, https://stationhypo.com/2021/04/02/out-of-the-shadows-the-art-of-time-sensitive-reporting-part-6-of-7/.
11. Michel, *Clashes*, 226; U.S. Navy, Naval History and Heritage Command. "*Chicago* III (CA-136)." Naval History and Heritage Command, accessed July 22, 2023, https://www.history.navy.mil/research/histories/ship-histories/danfs/c/chicago-iii.html. According to Guy Thomas, the *Chicago* departed Subic Bay on May 2, 1972. Thomas, email to author, July 24, 1923.
12. Larry Nowell, "Sea Control 315—Fighter Control over Vietnam with Master Chief Larry Nowell (ret.), Pt. 1," CIMSEC, February, 3, 2022, https://cimsec.org/sea-control-315-fighter-control-over-vietnam-with-master-chief-larry-nowell-ret/; Marc Whetstone, "Dedication & Extensive Training Lead to Distinguished Service Medal," *All Hands* no. 672 (January 1973), 10; Henry E. Stephenson, ed., "FAAWTRACEN San Diego Trains Air Intercept Controllers for Pacific Fleet," *Navy Training Bulletin*, Winter 1969–70, 30.
13. Larry Nowell, "Navy Red Crown Air Intercept Controller Seeks F-4 Pilots," Flying the F-4 Phantom in Combat, accessed July 7, 2023, https://www.f-4phantom.com/red-crown/; Larry Nowell as quoted in "Radarman Awarded Distinguished Service Medal," Orders & Medal Society of America, accessed July 7, 2023, http://www.omsa.org/files/jomsa_arch/Splits/1973/22543_JOMSA_Vol24_5_22.pdf; Michel, *Operation Linebacker I 1972*, 73; unidentified pilot quoted in "Radarman Awarded Distinguished Service Medal"; Valcano, "Out of the Shadows—The Art of Time-Sensitive Reporting; Mario Valcano, "Out of the Shadows—First 'ACE' of the Naval Security Group (Part 7 of 7)," Station HYPO, accessed July 4, 2023, https://stationhypo.com/2021/04/03/out-of-the-shadows-first-ace-of-the-naval-security-group-part-7-of-7/. See Nowell's comment about communicating

his needs in the chiefs' mess given during his interview with CIMSET (Nowell, "Sea Control 315—Fighter Control over Vietnam).
14. Guy Thomas, quoted by Valcano, "Out of the Shadows—The Art of Time-Sensitive Reporting"; Guy Thomas, email to the author, July 24, 2023.
15. Nowell, "Navy Red Crown Air Intercept Controller Seeks F-4 Pilots"; "Radarman Awarded Distinguished Service Medal"; Valcano, "Out of the Shadows—First 'ACE.'"
16. "Radarman Awarded Distinguished Service Medal."

CHAPTER 17. COMBAT TREE PHANTOMS, PROWLERS, AND WILD WEASEL IVs

1. Michel, *Operation Linebacker I 1972*, 60; Michel, *Clashes*, 181, 194; Kevin O'Rourke and Joe Peters, *Taking Fire Saving Captain Aikman: A Story of the Vietnam War* (Havertown, PA: Casemate, 2013), Google Books, accessed June 9, 2023, https://www.google.com/books/edition/Taking_Fire/5JfUAgAAQBAJ?hl=en&gbpv=1&dq=combat+tree&pg=PT38&printsec=frontcover; John Stillion, *Trends in Air-to-Air Combat: Implications for Future Air Superiority* (Washington, DC: Center for Strategic and Budgetary Assessment, 2015), 18; Peter E. Davies, *USAF F-4 Phantom II Mig Killers 1972–73* (Oxford: Osprey, 2005), 16.
2. Dan D'Costa, "How Combat Tree Made the F-4 Phantom II the Deadliest Fighter over Vietnam in the 1970s," Tacairnet, January 2, 2017, https://tacairnet.com/2017/01/02/how-combat-tree-made-the-f-4-phantom-ii-the-deadliest-fighter-over-vietnam-in-the-1970s/; Peter Davies, *USAF McDonnell Douglas F-4 Phantom II* (Oxford: Osprey, 2013), 32.
3. Michel, *Clashes*, 194.
4. "QRC-248 and 'Combat Tree' AN/APX-81," *Secret Projects* (blog), unidentified post, November 19, 2006, https://www.secretprojects.co.uk/threads/qrc-248-and-combat-tree-an-apx-81.1028/.
5. Michel, *Operation Linebacker I 1972*, 77–78.
6. Davis, *Wild Weasel*, 45; Marcelle S. Knaack, *Encyclopedia of U.S Air Force Aircraft and Missile Systems*, vol. II (Washington, DC: Office of Air Force History, United States Air Force, 1978), 202; Parsch, "ORC - Equipment Listing"; Theo van Geffen and Gerald Arruda, *Republic F-105 Thunderchief Peacetime Operations* (Haverton, PA: Key Publishing, 2021), chap. 6, [ebook] Google Books, accessed August 7, 2023, https://www.worldcat.org/search?q=ti%3ARepublic+F-105+Thunderchief+Peacetime+Operations; Price, *The History of US Electronic Warfare*, 3:118; Streetly, *Airborne Electronic Warfare*, 88.

7. Davis, *Wild Weasel*, 45; Michel, *Operation Linebacker I 1972*, 20.
8. Michel, *Operation Linebacker I 1972*, 81–82; "GBU-52," Global Security, July 7, 2011, https://www.globalsecurity.org/military/systems/munitions/cbu-52.htm.
9. U.S. Navy, Naval Air Force, U.S. Pacific Fleet, "Squadron History," Electronic-Attack-Squadron-VAQ-132, accessed July 25, 2023, https://www.airpac.navy.mil/Organization/Electronic-Attack-Squadron-VAQ-132/About-Us/History/; Greg Goebel, "The Grumman A-6 Intruder and EA-6B Prowler," AirVectors, December 1, 2022, https://www.airvectors.net/ava6.html.
10. Streetly, *Airborne Electronic Warfare*, 88; John Pike, "EA-6B Prowler," Federation of American Scientists, 2000. Accessed December 6, 2021, https://irp.fas.org/program/collect/ea-6b_prowler.htm.
11. U.S. Navy, Naval Air Force, U.S. Pacific Fleet, Electronic-Attack-Squadron-VAQ-132, "Squadron History"; *DOD Appropriations for 1974*, 4272.
12. Testimony of Capt. Albert A. Gallotta Jr., USN, EA-6B Program Coordinator, March 15, 1973, *DOD Appropriations for 1974*, 4273; statement of Maj. Gen. John J. Burns, USAF, March 16, 1973, *DOD Appropriations for 1974*, 4606; statement of Vice Adm. William D. Houser, USN, deputy CNO for Air Warfare, *DOD Appropriations for 1974*, 4269, 4301.
13. Carl O. Schuster, "The EA-6B Prowler: Outwitting Hanoi's Air Defenses," Historynet, 2018. Accessed November 29, 2022, https://www.historynet.com/ea-6b-prowler-outwitting-hanois-air-defenses.htm; statement of Cdr. D. H. Westbrock, USN, EA-6B Project Manager, *DOD Appropriations for 1974*, 4281–82; Michael F. Hake, "Stealth, the End of Dedicated Electronic Attack Aircraft," Monograph, School of Advanced Studies United States Army Command and General Staff College, Fort Leavenworth, Kansas, 1999, 13; "A Tribute to the Raven EF-111AA," *Journal of Electronic Defense*, May 1, 1998, https://www.thefreelibrary.com/A+tribute+to+the+Raven+EF-111AA.-a020791918.
14. Don Logan, *The 388th Tactical Fighter Wing at Korat Royal Thai Air Base 1972* (Atglen, PA: Schiffer, 1995), 33; Davis, *Wild Weasel*, 46, 50.
15. Davis, *Wild Weasel*, 50–51.

CHAPTER 18. LINEBACKER II: THE PRELIMINARIES

1. Philip S. Michael, "The Strategic Significance of Linebacker II: Political, Military, and Beyond," U.S. Army War College Strategy Research Project, U.S. Army War College, Carlisle Barracks, Pennsylvania, April 7, 2003, 7–8;

William P. Head, "War from Above the Clouds: B-52 Operations during the Second Indochina War and the Effects of the Air on Theory and Doctrine" [Fairchild Paper] (Maxwell Air Force Base, AL: Air University Press, 2002), 75–76.
2. Head, "War from Above the Clouds," 76–78; Peter Grier, "The Nightmare before Christmas," *Air Force Magazine* 98, no. 10 (October 2015): 58.
3. Head, "War from Above the Clouds," 78; Grier, "The Nightmare before Christmas," 58; James R. McCarthy and George B. Allison, *Linebacker II: A View from the Rock* (Barksdale Air Force Base, LA: History and Museums Program Air Force Global Strike Command, 2018), 4, 31.
4. Knaack, *Encyclopedia of U.S. Air Force Aircraft and Missile Systems Volume II*, 356.
5. Knaack, *Encyclopedia of U.S. Air Force Aircraft and Missile Systems Volume II*, 271; Joe Baugher, "Boeing B-52-G Stratofortress," Joebaugher.com, April 12, 2021, https://www.joebaugher.com/usaf_bombers/b52_15.html; Michel, *The 11 Days of Christmas*, 32–33.
6. Knaack, *Encyclopedia of U.S. Air Force Aircraft and Missile Systems Volume II*, 279; Price, *The History of US Electronic Warfare*, 3:114; Baugher, "Boeing B-52-G Stratofortress."
7. Michel, *The 11 Days of Christmas*, 25; Dana Drenkowski and Lester W. Grau, "Patterns and Predictability: The Soviet Evaluation of Operation Linebacker II," Admiralty Trilogy, accessed August 29, 2023, http://www.admiraltytrilogy.com/read/Soviet_view_of_Linebacker_II.pdf, 6.
8. Price, *The History of US Electronic Warfare*, 3:197–98; Price, *War in the Fourth Dimension*, 197–98; Michel, *The 11 Days of Christmas*, 43–44. Note: Price's timing for the Giant Stride tests, which was based on an oral history interview years after the fact, is incorrect based on the listing I found in the Technical Abstract Bulletin for 1968, the only other source on Giant Stride I could locate.
9. Michel, *The 11 Days of Christmas*, 67.
10. Michel, *The 11 Days of Christmas*, 36, 38.
11. McCarthy and Allison, *Linebacker II*, 29–30; Michel, *The 11 Days of Christmas*, 45–46, 56.
12. Calvin R. Johnson, "Linebacker Operations September–December 1972 (U)," December 31, 1978. Project CHECO, Office of History, Headquarter Pacific Air Force, 32–33; Michel, *The 11 Days of Christmas*, 46; McCarthy and Allison, *Linebacker II*, 33–34; statement of Maj. Adam Rech, radar navigator, in Johnson, "Linebacker Operations," 33; Kenneth S. Katta, "Terror Night over Vinh." *Air Power History* 66, no. 2 (Summer 2019): 48–49.
13. Johnson, "Linebacker Operations," 32.

14. Johnson, "Linebacker Operations," 33.
15. McCarthy and Allison, *Linebacker II*, 39.
16. Michel, *The 11 Days of Christmas*, 60–61.
17. Steven Brown, quoted in Michel, *The 11 Days of Christmas*, 73; Dwight Moore, quoted in Michel, *The 11 Days of Christmas*, 81.

CHAPTER 19. LINEBACKER II: INTO THE DRAGON'S TEETH

1. Head, "War from Above the Clouds," 73, 79; Johnson, "Linebacker Operations," 96, 102.
2. Price, *The History of US Electronic Warfare*, 3:208; Michel, *The 11 Days of Christmas*, 91; Van Nederveen, *Sparks over Vietnam*, 78–79.
3. Price, *The History of US Electronic Warfare*, 3:208; Johnson, "Linebacker Operations," 65; Michel, *The 11 Days of Christmas*, 91, 102; Rudy Smart, quoted in Price, *The History of US Electronic Warfare*, 3:208. Note: Price gives the altitude at 45,000 feet, Michel 35,000. The latter is per interview and is more consistent with the operational characteristics of the F-4.
4. Michel, *The 11 Days of Christmas*, 37, 92; Price, *The History of US Electronic Warfare*, 3:207, 209; McCarthy and Allison, *Linebacker II*, 44–46.
5. Michel, *The 11 Days of Christmas*, 86–87; Lady Borton, "Linebacker II: Dien Bien Phu in the Air," *The VA Veteran Online*, January/February 2014, accessed September 1, 2023, https://vvaveteran.org/34–1/34_1_borton.html.
6. Marshall L. Michel III, *Operation Linebacker II 1972: The B-52s Are Sent to Hanoi* (Oxford: Osprey, 2018), Kindle version, 43.
7. Michel, *The 11 Days of Christmas*, 88–89.
8. Michel, *The 11 Days of Christmas*, 94, 98.
9. Michel, *The 11 Days of Christmas*, 99.
10. Price, *The History of US Electronic Warfare*, 3:209; Michel, *The 11 Days of Christmas*, 96–97.
11. Andy Vittoria, quoted in Price, *The History of US Electronic Warfare*, 3:209.
12. Ed Wildeboor, quoted in Michel, *The 11 Days of Christmas*, 159.
13. Bill Gilbert, *Air Power: Heroes and Heroism in American Missions 1916 to Today* (New York: Citadel Press, 2003), 233; Michel, *The 11 Days of Christmas*, 101–2, 104–5.
14. Michel, *The 11 Days of Christmas*, 105; Price, *The History of US Electronic Warfare*, 3:207, 210.
15. Michel, *The 11 Days of Christmas*, 106.
16. Price, *The History of US Electronic Warfare*, 3:210; Michel, *The 11 Days of Christmas*, 183.

17. Price, *The History of US Electronic Warfare*, 3:210; Michel, *The 11 Days of Christmas*, 106–7; Gilbert, *Air Power*, 233; Johnson, "Linebacker Operations," 95.
18. Michel, *The 11 Days of Christmas*, 117–18.
19. Nguyen Thang, quoted in Michel, *The 11 Days of Christmas*, 105.
20. Michel, *The 11 Days of Christmas*, 119.
21. Robert O. Harder, "Operation Linebacker II: The 11-Day War," Historynet, December 29, 2020, https://www.historynet.com/the-11-day-war/; Johnson, "Linebacker Operations," 96; Michel, *The 11 Days of Christmas*, 119–20; Bob Steffen, quoted in Michel, *The 11 Days of Christmas*, 120.
22. Michel, *The 11 Days of Christmas*, 124–25; Price, *The History of US Electronic Warfare*, 3:211.
23. Michel, *The 11 Days of Christmas*, 126.
24. Price, *The History of US Electronic Warfare*, 3:211; Michel, *The 11 Days of Christmas*, 127–28.
25. McCarthy and Allison, *Linebacker II*, 65.
26. McCarthy and Allison, *Linebacker II*, 64–65.

CHAPTER 20. LINEBACKER II: LOSSES LEAD TO NEW TACTICS

1. Johnson, "Linebacker Operations," 96; Michel, *The 11 Days of Christmas*, 137.
2. Lien-Hang T. Nguyen, *Hanoi's War: An International History of the War for Peace in Vietnam* (Chapel Hill: University of North Carolina Press, 2012), 271; Michel, *The 11 Days of Christmas*, 43, 60, 148; Johnson, "Linebacker Operations," 96. Three F-111s also targeted the radio station that night: one each on three separate occasions.
3. Michel, *The 11 Days of Christmas*, 139.
4. Michel, *The 11 Days of Christmas*, 140.
5. Michel, *The 11 Days of Christmas*, 148, 152; Wayne Thompson, *To Hanoi and Back: The United States Air Force and North Vietnam, 1966–1973* (Washington, DC: Air Force History and Museums Program, U.S. Air Force, 2000), 306; Michel, *Operation Linebacker II 1972*, 57.
6. Michel, *The 11 Days of Christmas*, 144–45.
7. Price, *The History of US Electronic Warfare*, 3:216; Johnson, "Linebacker Operations," 96; Michel, *The 11 Days of Christmas*, 148.
8. Michel, *The 11 Days of Christmas*, 148–51; Price, *The History of US Electronic Warfare*, 3:216–17; McCarthy and Allison, *Linebacker II*, 79; Johnson, "Linebacker Operations," 96. Note: Numerous discrepancies exist in the first three sources.

When in doubt, I have used the information provided in Johnson's tables detailing the specifics of both the aircraft lost and the aircraft involved in each target.

9. Michel, *The 11 Days of Christmas*, 154; Price, *The History of US Electronic Warfare*, 3:217.
10. Price, *The History of US Electronic Warfare*, 3:217; Michel, *The 11 Days of Christmas*, 156; McCarthy and Allison, *Linebacker II*, 81.
11. Michel, *The 11 Days of Christmas*, 157.
12. Price, *The History of US Electronic Warfare*, 3:217; Michel, *The 11 Days of Christmas*, 159–60.
13. Michel, *The 11 Days of Christmas*, 158, 160–61; Johnson, "Linebacker Operations," 95–96.
14. Johnson, "Linebacker Operations," 95; Price, *The History of US Electronic Warfare*, 3:217–18; Michel, *The 11 Days of Christmas*, 161, 185.
15. Price, *The History of US Electronic Warfare*, 3:218; Central Intelligence Agency, Directorate of Science and Technology, Office of Scientific Intelligence, "The Soviet SA-3 Missile System," OSI-SR/TCS/65–10, July 8 1965, https://www.cia.gov/readingroom/docs/CIA-RDP78T05439A000500210055-7.pdf; "S-125 SA-3 GOA," Federation of American Scientists, July 3, 1998, https://nuke.fas.org/guide/russia/airdef/s-125.htm; U.S. Army, "*S-125 Neva/Pechora (SA-3 Goa) Russian 6x6 Surface-to-Air Missile System*," ODIN, accessed September 11, 2023, https://odin.tradoc.army.mil/WEG/Asset/S-125_Neva:Pechora_(SA-3_Goa)_Russian_6x6_Surface-to-Air_Missile_System; Michel, *The 11 Days of Christmas*, 177; Adam R. Grissom, Caitlin Lee, and Karl P. Mueller, *Innovation in the United States Air Force: Evidence from Six Cases* (Santa Monica, CA: Rand, 2016), 232.
16. Michel, *The 11 Days of Christmas*, 166; Price, *The History of US Electronic Warfare*, 3:224.
17. Price, *The History of US Electronic Warfare*, 3:220; Michel, *The 11 Days of Christmas*, 162, 188; Grier, "The Nightmare before Christmas," 61.
18. Michel, *The 11 Days of Christmas*, 188; Price, *The History of US Electronic Warfare*, 3:220.
19. Michel, *The 11 Days of Christmas*, 170–73.
20. Michel, *The 11 Days of Christmas*, 173–74, 178; Leonard D. G. Teixeira, "Linebacker II: A Strategic and Tactical Case Study," Air War College, Air University, Maxwell Air Force Base, AL, April 1990, 18; Johnson, "Linebacker Operations," 98–99; Price, *The History of US Electronic Warfare*, 3:223.
21. Michel, *The 11 Days of Christmas*, 179–80.

22. Johnson, "Linebacker Operations," 99; Michel, *The 11 Days of Christmas*, 181–82; Price, *The History of US Electronic Warfare*, 3:223; Teixeira, "Linebacker II," 19.
23. Johnson, "Linebacker Operations," 99; Michel, *The 11 Days of Christmas*, 183; Price, *The History of US Electronic Warfare*, 3:223.
24. Michel, *The 11 Days of Christmas*, 185; David Sjolund, quoted in Price, *The History of US Electronic Warfare*, 3:224.
25. Michel, *The 11 Days of Christmas*, 185; Zaloga, *Red SAM*, 21.
26. Teixeira, "Linebacker II," 20; Walter J. Boyne, "Linebacker II," *Air Force Magazine* 80, no. 11 (November 1997): 56; Carl O. Schuster, "The Rise of North Vietnam's Air Defense"; Michel, *The 11 Days of Christmas*, 189.
27. Michel, *The 11 Days of Christmas*, 188; Johnson, "Linebacker Operations," 100; Price, *The History of US Electronic Warfare*, 3:224–25.
28. Stanley J. Dougherty, "Defense Suppression: Building Some Operation Concepts," Air University Press, Maxwell Air Force Base, Alabama, May 1992, 13; Earl H. Tilford Jr., *Crosswinds: The Air Force's Setup in Vietnam* (College Station, TX: Texas A&M University Press, 2009), 168; Parks, "Linebacker and the Law," 20; Michel, *The 11 Days of Christmas*, 210, 212.
29. Parks, "Linebacker and the Law," 20; Michel, *The 11 Days of Christmas*, 212. Note: For a detailed analysis of LORAN and its use over Vietnam, see Michel, *The 11 Days of Christmas*, 210–12.
30. Price, *The History of US Electronic Warfare*, 3:226; Michel, *The 11 Days of Christmas*, 201.
31. Schuster, "The Rise of North Vietnam's Air Defense"; Mark Clodfelter, *The Limits of Airpower: The American Bombing of North Vietnam* (New York: The Free Press, 1989), 188; Michel, *The 11 Days of Christmas*, 204.
32. Price, *The History of US Electronic Warfare*, 3:226; Clodfelter, *The Limits of Airpower*, 189; Michel, *The 11 Days of Christmas*, 208.
33. Clodfelter, *The Limits of Airpower*, 189, 198.

CHAPTER 21. LOOKING BACK

1. Schuster, "The Rise of North Vietnam's Air Defense"; Boyne, "Linebacker II," 56; Johnson, "Linebacker Operations," 55; Michel, *The 11 Days of Christmas*, 248. Note: I have been unable to locate a copy of the "Comfy Coat Evaluation," which Michel failed to list in his sources.
2. John M. Curatola, "'Black Thursday' October 14, 1943: The Second Schweinfurt Bombing Raid," The National WWII Museum, August

17, 2022, https://www.nationalww2museum.org/war/articles/schweinfurt-regensburg-raid-august-17-1943.

3. Price, *The History of US Electronic Warfare*, 3:230.

4. Gary D. Joiner and Ashley E. Dean, "Operation Linebacker II: A Retrospective Part 6: Linebacker II," report of the LSU Shreveport Unit for the SAC Symposium, December 2, 2017, Shreveport, LA: Strategy Alternatives Center Louisiana State University, 2017, 41–48; Johnson, "Linebacker Operations," 95; McCarthy and Allison, *Linebacker II*, 175; Momyer, *Airpower in Three Wars*, 243; U. S. Grant Sharp, *Strategy for Defeat: Vietnam in Retrospect* (San Rafael, CA: Presidio Press, 1978), 259–60. Note: The figures provided by Joiner and Dean include one KIA lost in an OV-10 over South Vietnam and thirteen KIA lost in an A-130 over Laos, neither of which are included in Johnson's tally of the losses over North Vietnam.

5. Sharp, *Strategy for Defeat*, 261; "Fall of Saigon," Britannica, October 12, 2023, https://www.britannica.com/event/Fall-of-Saigon.

6. Wildenberg, *Fighting in the Electromagnetic Spectrum*, 152–53.

7. Michael Herman, *Intelligence Power in Peace and War* (Cambridge: Royal Institute of International Affairs, 1996), 45, as cited by Christopher Ford and Davis Rosenberg, *The Admirals' Advantage: U.S. Navy Operational Intelligence in World War II and the Cold War* (Annapolis, MD: Naval Institute Press, 2005), 14; John Winton, *Ultra at Sea* (New York: William Morrow, 1988), 47, as cited by Ford and Rosenberg, *The Admirals' Advantage*, 14; Ford and Rosenberg, *The Admirals, Advantage*, 54.

8. McCarthy and Allison, *Linebacker II*, 175.

BIBLIOGRAPHY

BOOKS

Archer, Bob. *Super Snooper: The Evolution and Service Career of the Specialist Boeing C-135 Series with the 55th Wing and Associated Units.* Stroud, UK: Fonthill Media, 2020.

Axe, David. *Drone War Vietnam.* Yorkshire, UK: Pen & Sword, 2021.

Babcock, Elizabeth. *Magnificent Mavericks: Transition of the Naval Ordnance Test Station from Rocket Station to Research, Development, Test, and Evaluation Center, 1948–58.* Washington, DC: Naval Historical Center, 2008.

Borgiaz, William S. *The Strategic Air Command: Evolution and Consolidation of Nuclear Forces, 1945–1955.* Westport, CT: Greenwood Publishers, 1996.

Boslaugh, David L. *First-Hand: No Damned Computer Is Going to Tell Me What to DO - The Story of the Naval Tactical Data System, NTDS*, May 12, 2021. Engineering and Technology History Wiki. https://ethw.org/Firstand:No_Damned_Computer_is_Going_to_Tell_Me_What_to_DO_The_Story_of_the_Naval_Tactical_Data_System,_NTDS.

———. *When Computers Went to Sea: The Digitization of the United States Navy.* Los Alamitos, CA: IEEE Computer Society, 1999.

Bowman, Martin. *The Men Who Flew the F-4 Phantom.* Barnsley, South Yorkshire, UK: Pen & Sword Aviation. Google Books. Accessed May 27, 2023. https://www.google.com/books/edition/The_Men_Who_Flew_the_F_4_Phantom/iLPNDwAAQBAJ?hl=en&gbpv=1&dq=c-130+queen+bee&pg=PT84&printsec=frontcover.

Boyle, Michael J. *The Drone Age: How Drone Technology Will Change War and Peace.* New York: Oxford University Press, 2020.

Broughton, Jack. *Thud Ridge: F-105 Thunderchief Missions over Vietnam.* Manchester, UK: Crecy Publishing, 2006.

Bugos, Glenn E. *Engineering the F-4B Phantom II: Parts into Systems*. Annapolis, MD: Naval Institute Press, 1996.
Burgess, Richard R., and Rosario M. Rausa. *U.S. Navy A-1 Skyraider Units of the Vietnam War*. London: Osprey, 2013.
Clodfelter, Mark. *The Limits of Airpower: The American Bombing of North Vietnam*. New York: The Free Press, 1989.
Clodfelter, Mark, and Berry Craig. *River Rats: Red River Valley Fighter Pilots of Vietnam*. Paducah, KY: Turner Publishing, 1989.
Committee on Utilization of Scientific and Engineering Manpower, National Academy of Sciences. *Toward Better Utilization of Scientific and Engineering Talent*. Washington, DC: National Academy of Sciences, 1964.
Cooling, Benjamin F., ed. *Case Studies in Achievement of Air Superiority*. Washington, DC: Center for Air Force History, 1991.
Davies, Peter E. *A-4 Skyhawk vs North Vietnamese AAA: North Vietnam 1964–72*. Oxford: Osprey, 2020.
———. *B/EB-66 Destroyer Units in Combat*. Oxford: Osprey, 2021.
———. *F-4 Phantom II Wild Weasel Units in Combat*. Oxford: Osprey, 2023.
———. *F-105 Thunderchief Units of the Vietnam War*. Oxford: Osprey, 2012.
———. *F-105 Wild Weasel vs SA-2 "Guideline" SAM Vietnam 1965–73*. Oxford: Osprey, 2011.
———. *USAF McDonnell Douglas F-4 Phantom II*. Oxford, Osprey, 2013.
———. *USAF F-4 Phantom II MiG Killers, 1965–68*. Oxford: Osprey, 2014.
———. *USAF F-4 Phantom II MiG Killers, 1972–73*. Oxford: Osprey, 2005.
———. *US Navy F-4 Phantom II Units of the Vietnam War 1964–68*. Oxford: Osprey 2016.
Davies, Peter E., and David W. Menard. *F-100 Super Sabre Units of the Vietnam War*. Oxford: Osprey, 2012.
Davies, Peter E., with Jim Laurier and Gareth Hector. *RF-101 Voodoo Units in Combat*. London: Bloomsbury, 2019.
Davis, John, and D. Willison. *Wings over Vietnam*. Independently published (June 3, 2019).
Davis, Larry. *Wild Weasel: The SAM Suppression Story*. Carrollton, TX: Squadron/Signal Publications, 1986.
Eisel, Braxton, and Jim Schreiner. *Magnum! The Wild Weasels in Desert Storm*. Barnsley, South Yorkshire, UK: Pen & Sword Aviation, 2009. Google Books. Accessed March 3, 2023. https://www.google.com/books/edition/Magnum_The_Wild_Weasels_in_Desert_Storm/97fNDwAAQBAJ?hl=en&gbpv=1&dq=an/apr-23&pg=PT219&printsec=frontcover.

Fey, Peter. *Bloody Sixteen: The USS Oriskany and Air Wing 16 during the Vietnam War*. Lincoln: University of Nebraska Press, 2018. Google Books. https://www.google.com/books/edition/Bloody_Sixteen/WLtSDwAAQBAJ?hl=en&gbpv=1&dq=project+shoehorn&pg=PT82&printsec=frontcover.

Fino, Steven A. *Tiger Check: Automating the US Air Force Fighter Pilot in Air-to-Air Combat, 1950–1980*. Baltimore, MD: Johns Hopkins University Press, 2017.

Ford, Christopher, and Davis Rosenberg. *The Admirals' Advantage: U.S. Navy Operational Intelligence in World War II and the Cold War*. Annapolis, MD: Naval Institute Press, 2005.

Fox, Roger P. *Air Base Defense in the Republic of Vietnam 1961–1973*. Washington, DC: Office of Air Force History, 1979.

Frankum, Ronald B. Jr. *Like Rolling Thunder: The Air War in Vietnam, 1964–1975*. Lanham, MD: Rowman & Littlefield, 2005.

Friedman, Norman. *U.S. Naval Weapons: Every Gun, Missile, Mine and Torpedo Used by the U.S. Navy from 1883 to the Present Day*. Annapolis, MD: Naval Institute Press, 1988.

Futrell, Robert F. *The United States Air Force in Southeast Asia: The Advisory Years to 1965*. Washington, DC: Office of Air Force History, United States Air Force, 1981.

Gilbert, Bill. *Air Power: Heroes and Heroism in American Missions, 1916 to Today*. New York: Citadel Press, 2003.

Glasser, Jeffrey D. *The Secret Vietnam War: The United States Air Force in Thailand, 1961–1975*. Jefferson, NC: McFarland and Company, 1995.

Goulden, Joseph C. *Truth Is the First Casualty*. New York: James B. Adler Inc., 1969.

Grimes, Bill. *The History of Big Safari*. Bloomington, IN: Archway Publishing, 2014.

Grissom, Adam R., Caitlin Lee, Karl P. Mueller. *Innovation in the United States Air Force: Evidence from Six Cases*. Santa Monica, CA: Rand, 2016.

Grossnick, Roy A. *Dictionary of American Naval Aviation Squadrons*, vol. 1. Washington, DC: Naval Historical Center, 1995.

Hallion, Richard P. *Rolling Thunder 1965–68: Johnson's Air War over Vietnam*. Oxford: Osprey, 2018.

Hampton, Dan. *The Hunter Killers*. New York: William Morrow, 2015.

Hannah, Craig C. *Striving for Air Superiority: The Tactical Air Command in Vietnam*. College Station: Texas A&M University Press, 2002.

Hanyok, Robert J. *Spartans in Darkness: American SIGINT and the Indochina War, 1945–1975*. Fort George G. Meade, MD: Center for Cryptologic History, National Security Agency, 2002.

Hartsook, E. *The Air Force in Southeast Asia: Shield for Vietnamization and Withdrawal 1971*. Washington, DC: Office of Air Force History, 1976.

Herman, Michael. *Intelligence Power in Peace and War*. Cambridge: Royal Institute of International Affairs, 1996.

Hopkins, Charles K. *SAC Tanker Operations in the Southeast Asia War*. Omaha, NE: Office of the Historian, Headquarters, Strategic Air Command, 1987.

Jack, Kenneth V. *Eyes of the Fleet over Vietnam: RF-8 Crusader Combat Photo-Reconnaissance Missions*. Philadelphia: Casemate, 2021.

Johnson, Thomas R. *American Cryptology during the Cold War, 1945–1989. Book II: Centralization Wins 1960–1972*. Center for Cryptologic History, National Security Agency, 1995. NSA. Accessed May 28, 2023. https://www.nsa.gov/portals/75/documents/news-features/declassified-documents/cryptologic-histories/cold_war_ii.pdf.

Kerzon, Warren J. *Post-World War II Fighters 1945–1973*. Washington, DC: Office of Air Force History, 1986.

———. *Throw a Nickel on the Grass: A Fighter Pilot's Life Narrative*. Raleigh, NC: Lulu, 2016.

Knaack, Marcelle S. *Encyclopedia of U.S Air Force Aircraft and Missile Systems Volume I: Post–World War II Fighters 1945–1973*. Washington, DC: Office of Air Force History, United States Air Force, 1978.

———. *Encyclopedia of U.S. Air Force Aircraft and Missile Systems Volume II: Post–World War II Bombers 1945–1973*. Washington, DC: Office of Air Force History, 1978.

———. *Post–World War II Fighters 1945–1973*. Washington, DC: Office of Air Force History, 1986.

Krepinevich, Andrew F. Jr. *The Origins of Victory: How Disruptive Innovation Determines the Fates of Great Powers*. New Haven, CT: Yale University Press, 2023.

Lane, John J. Jr. *Command and Control and Communications Structures in Southeast Asia*. Maxwell Air Force Base, AL: Airpower Research Institute, 1981.

Lawson, Cliff. *The Station Comes of Age: Satellites, Submarines, and Special Operations in the Final Years of the Naval Ordnance Test Station, 1959–1967*. China Lake, CA: Naval Air Warfare Center Western Division, 2017.

Logan, Don. *The 388th Tactical Fighter Wing at Korat Royal Thai Air Base 1972*. Atglen, PA: Schiffer, 1995.

Longacre, Edward G. *Strategic Air Command: The Formative Years (1944–1949)*. Offutt Air Force Base, NE: Office of the Historian, Headquarters Strategic Air Command, 1990.

Love, Robert W. Jr. *History of the U.S. Navy: Volume Two 1942–1991*. Lanham, MD: Stockpole Books, 1992.

Marolda, Edward J. *The Approaching Storm: Conflict in Asia, 1945–1965*. Washington, DC: Naval History and Heritage Command, 2009.
McCarthy, James R., and George B. Allison. *Linebacker II: A View from the Rock*. Barksdale Air Force Base, LA: History and Museums Program Air Force Global Strike Command, 2018.
Mersky, Peter B. *Whitey: The Story of Rear Admiral E. L. Feightner, A Navy ACE*. Annapolis, MD: Naval Institute Press, 2014.
Michel, Marshal L. III. *The 11 Days of Christmas: America's Last Vietnam Battle*. San Francisco, CA: Encounter Books, 2002.
———. *Clashes: Air Combat over North Vietnam*. Annapolis, MD: Naval Institute Press, 1997.
———. *Operation Linebacker I 1972: The First High-Tech Air War*. Oxford: Osprey, 2019.
———. *Operation Linebacker II 1972: The B-52s Are Sent to Hanoi*. Oxford: Osprey, 2018. Kindle.
Mobley, Richard A., and Edward J. Marolda. *Knowing the Enemy*. Washington, DC: Naval History and Heritage Command, 2015.
Momyer, William W. *Airpower in Three Wars*. Washington, DC: Government Printing Office, 1978.
Morgan, Mark, and Rick Morgan. *Intruder: The Operational History of Grumman's A-6*. Atglen, PA: Schiffer Military History, 2004.
Morgan, Rick. *A-3 Skywarrior Units of the Vietnam War*. Oxford: Osprey, 2015.
———. *A-6 Intruder Units of the Vietnam War*. Botley, Oxford: Osprey, 2012.
Morrison, Robert E. *Naval Communications Station Philippines Fleet Support Detachment Da Nang, Republic of Vietnam (Det Bravo) Command History*. Published by the author, 2017. VQ Association. Accessed February 13, 2023. https://vqassociation.org/wp-content/uploads/2020/12/038-History-of-Det-Bravo-Da-Nang.pdf.
Morton, Tyler. *From Kites to Cold War: The Evolution of Manned Airborne Reconnaissance*. Annapolis, MD: Naval Institute Press, 2019.
Muir, Malcolm Jr., *Black Shoes and Blue Water: Surface Warfare in the United States Navy, 1945–1975*. Washington, DC: Naval Historical Center, 1996.
Nalty, Bernard C. *Tactics and Techniques of Electronic Warfare: Electronic Countermeasures in the Air War against North Vietnam 1965–1973*. Washington, DC: Office of Air Force History, 1977.
National Security Agency. *National Security Agency during the Cold War*, Chapter 13. National Security Archive. Accessed January 18, 2023. https://nsarchive2.gwu.edu/NSAEBB/NSAEBB441/docs/doc%203%202008-021%20Burr%20Release%20Document%201-%20Part%20C.pdf.

Nguyen, Lien-Hang T. *Hanoi's War: An International History of the War for Peace in Vietnam*. Chapel Hill: University of North Carolina Press, 2012.

Nichols, John B., and Barrett Tillman. *On Yankee Station: The Naval Air War over Vietnam*. Annapolis, MD: Naval Institute Press, 1987.

Nixon, Richard M. *The Memoirs of Richard Nixon* vol. 2. New York: Warner Books, 1978.

O'Brien, J. T. *Top Secret: An Informal History of Electronic & Photographic Reconnaissance in Marine Corps Aviation 1940–2000*. Anaheim, CA: Equidata Publishing, 2004.

O'Rourke, Kevin, and Joe Peters. *Taking Fire: Saving Captain Aikman: A Story of the Vietnam War*. Havertown, PA: Casemate, 2013. Google Books. Accessed June 9, 2023. https://www.google.com/books/edition/Taking_Fire/5JfUAgAAQBAJ?hl=en&gbpv=1&dq=combat+tree&pg=PT38&printsec=frontcover.

Pedlow, Gregory W., and Donald E. Welzenbach. *The Central Intelligence Agency and Overhead Reconnaissance: The U-2 and OXCART Programs, 1954–1974*. Washington, DC: Central Intelligence Agency, 1992.

Plunkett, W. Howard. *F-105 Thunderchiefs: A 29-Year Illustrated Operational History*. Jefferson, NC: McFarland and Company, 2001.

Price, Alfred W. *The History of US Electronic Warfare*, vol. II, *The Renaissance Years, 1946 to 1964*. Arlington, VA: Association of Old Crows, 1989.

———. *The History of US Electronic Warfare*, vol. III, *Rolling Thunder through Allied Force, 1964 to 2000*. Arlington, VA: Association of Old Crows, 2000.

———. *War in the Fourth Dimension: US Electronic Warfare, from the Vietnam War to the Present*. London: Greenway, 2001.

Rock, Edward T., ed. *First In, Last Out: Stories by the Wild Weasels*. Bloomington, IN: AuthorHouse, 2005.

Roland, Alex. *War and Technology: A Very Short Introduction*. New York: Oxford University Press, 2016.

Romano, Angelo, and John D. Herndon. *From Bats to Rangers: A Pictorial History of Electronic Countermeasures Squadron Two (ECMRON2-2) Fleet Air Reconnaissance Squadron Two (VQ-2)*. Simi Valley, CA: Ginter Books, 2017.

St. John, Philip M. *50 Years USAF: A Look at the Air Force, Air Force Association and Commemorative Las Vegas Reunion*. Paducah, KY: Turner, 1998.

Samuel, Wolfgang W. E. *Glory Days: The Untold Story of the Men Who Flew the B-66 Destroyer into the Face of Fear*. Atglen, PA: Schiffer Publishing, 2008.

Schlight, John. *The War in South Vietnam: The Years of the Offensive 1964–1968*. Washington, DC: Air Force History and Museums Program, 1999.

Schreader, George F. *Hognose Silent Warrior: The USAF's Airborne Intelligence War in the Final Air Campaigns of Vietnam*. Parker, CO: Outskirts Press, 2018. Kindle edition.

Sharp, U. S. Grant. *Strategy for Defeat: Vietnam in Retrospect*. San Rafael, CA: Presidio Press, 1978.
Sherwood, John D. *Afterburner: Naval Aviators and the Vietnam War*. New York: New York University Press, 2004.
———. *Nixon's Trident: Naval Power in Southeast Asia, 1968–1972*. Washington, DC: Naval History and Heritage Command, 2009.
Shulimson, Jack, and Charles M. Johnson. *U.S. Marines in Vietnam: The Landing and the Buildup 1965*. Washington, DC: History and Museums Division, Headquarters, U.S. Marine Corps, 1978.
Simons, Graham M. *Lockheed Constellation: A History*. Yorkshire, UK: Pen & Sword, 2021. Google Books. Accessed June 20, 2023. https://www.google.com/books/edition/Lockheed_Constellation/Myo1EAAAQBAJ?hl=en&gbpv=1&dq=rivet+gym&pg=PT483&printsec=frontcover.
Singh, Mandeep. *Anti-Aircraft Artillery in Combat 1950–1972: Air Defense in the Jet Age*. Philadelphia: Pen & Sword, 2020. Google Books. Accessed August 18, 2023. https://www.google.com/books/edition/Anti_Aircraft_Artillery_in_Combat_1950_1/sxnhDwAAQBAJ?hl=en&gbpv=1&dq=sa-2+optical+tracking&pg=PT102&printsec=frontcover.
Sloggett, Dave. *Drone Warfare: The Development of Unmanned Aerial Conflict*. South Yorkshire, UK: Pen & Sword, 2014.
Starostina, Natalia, ed. *Between Memory and Mythology: The Construction of Memory of Modern Wars*. Newcastle upon Tyne, UK: Cambridge Scholars, 2015.
Stillion, John. *Trends in Air-to-Air Combat: Implications for Future Air Superiority*. Washington, DC: Center for Strategic and Budgetary Assessment, 2015.
Streetly, Martin. *Airborne Electronic Warfare: History, Techniques and Tactics*. London: Jane's, 1988.
Swanborough, Gordon, and Peter M. Bowers. *United States Naval Aircraft since 1911*. London: Putnam Aeronautical Books, 1990.
Thomas, George Guy. *A Silent Warrior Steps Out of the Shadows*. Self-published, Alpha Book Publisher, 2021, Kindle Edition.
Thompson, Wayne. *To Hanoi and Back: The United States Air Force and North Vietnam, 1966–1973*. Washington, DC: Air Force History and Museums Program, U.S. Air Force, 2000.
Tilford, Earl H. Jr. *Crosswinds: The Air Force's Setup in Vietnam*. College Station: Texas A&M University Press, 2009.
———. *Setup: What the Air Force Did in Vietnam and Why*. Maxwell Air Force Base, AL: Air University Press, 1991.

Tovy, Tal. *The Gulf of Tonkin: The United States and the Escalation in the Vietnam War.* New York: Routledge, 2021. Google Books. Accessed August 7, 2023. https://www.google.com/books/edition/The_Gulf_of_Tonkin/je8eEAAAQBAJ?hl=en&gbpv=1&dq=%22operation+pierce+arrow%22&pg=PT99&printsec=frontcover.

Van Creveld, Martin. *Technology and War from 2000 B.C. to the Present.* New York: The Free Press, 1991.

Van Geffen, Theo, and Gerald Arruda. *Republic F-105 Thunderchief Peacetime Operations.* Haverton, PA: Key Publishing, 2021. [ebook] Google Books. Accessed August 7, 2023. https://www.worldcat.org/search?q=ti%3ARepublic+F-105+Thunderchief+Peacetime+Operations.

Van Staaveren, Jacob. *Gradual Failure: The Air War over North Vietnam 1965–66.* Washington, DC: Air Force History and Museums Program, 2002.

Werrell, Kenneth P. *Chasing the Silver Bullet: U.S. Air Force Weapons Development from Vietnam to Desert Storm.* Washington, DC: Smithsonian Books, 2003.

Whitten, H. Wayne. *Silent Heroes: U.S. Marines and Airborne Electronic Warfare 1950–2012.* Lutz, FL: Colonel H. Wayne Whitten and Associates, 2011.

Wildenberg, Thomas. *Fighting in the Electromagnetic Spectrum: U.S. Navy and Marine Corps Electronic Warfare Aircraft, Missions, and Equipment.* Annapolis, MD: Naval Institute Press, 2023.

Winton, John. *Ultra at Sea.* New York: William Morrow, 1988.

Withington, Thomas. *Wild Weasel Fighter Attack: The History of the Suppression of Enemy Air Defense Mission.* Barnsley, UK: Pen & Sword Aviation, 2008.

Worden, Mike. *Rise of the Fighter Generals: The Problem of Air Force Leadership 1945–1982.* Maxwell Air Force Base, AL: Air University Press, 1998.

Zaloga, Steven J. *Red SAM: The SA-2 Guideline Anti-Aircraft Missile.* Oxford: Osprey, 2007.

ARTICLES

"31 March 1968: President Lyndon B. Johnson Announces Bombing Halt in Vietnam." Vietnam The Art of War, September 10, 1970. Accessed May 21, 2023. https://vietnamtheartofwar.com/1968/03/31/31-march-1968-president-lyndon-b-johnson-announces-bombing-halt-in-vietnam/.

"AGM-45 Shrike." Global Security. January 7, 2012. https://www.globalsecurity.org/military/systems/munitions/agm-45.htm.

"AGM-45 Shrike." WeaponSystemnet. Accessed March 21, 2023. https://weaponsystems.net/system/1066-HH08%20-%20AGM-45%20Shrike.

BIBLIOGRAPHY

"AN/ALE-38." Military Periscope. Accessed August 8, 2023. https://www.military periscope.com/weapons/sensorselectronics/electronic-support-measures electronic-warfare/anale-38/overview/.

"AN/ALE-38 Chaff/Flare Pod." Command: Modern Operations/Modern Air Naval Operations. Accessed August 8, 2023. http://cmano-db.com/pdf/weapon/605/.

"AN/APR-24 C/S/X-Band Radar Homing and Warning System (RHAW)." Phantom Phacts, 2014. Accessed April 23, 2023. https://phantomphacts.blogspot.com/p/usn-ew-equipment.html.

"Analog Goes Electronic." Computer History Museum. Accessed March 28, 2023. https://www.computerhistory.org/revolution/analog-computers/3/150.

"ATI AN/APR-25 S/C/X-Band Radar Homing & Warning System (Mod 1)." Phantom Phacts, March 26, 2014. https://phantomphacts.blogspot.com/p/anapr-25n-scx-band-radar-homing-warning.html.

Axe, David. "In 1966, U.S. Air Force Drones Tricked North Vietnamese Missileers into Giving Up Their Secrets." *The National Interest*, April 27, 2020. https://nationalinterest.org/blog/buzz/1966-us-air-force-drones-tricked-north-vietnamese-missileers-giving-their-secrets-148426.

"B-47 Stratojet: Historical Snapshot." Boeing Corporation. Accessed February 16, 2023. https://www.boeing.com/history/products/b-47-stratojet.page.

Bailey, Bruce M. "55th Strategic Reconnaissance Wing." In *50 Years USAF: A Look at the Air Force, Air Force Association and Commemorative Las Vegas Reunion*, edited by Philip A. St. John. Paducah, KY: Turner, 1998.

Barison, Gabriele. "That Time a USAF EC-121 Warning Star Almost Landed on a U.S. Navy Aircraft Carrier." The Aviation Geek Club, July 18, 2018. https://theaviationgeekclub.com/that-time-a-usaf-ec-121-warning-star-almost-landed-on-a-u-s-navy-aircraft-carrier/.

Bass, Willard B. "USS Wainright Has Lots of PIRAZ." *All Hands* no. 612 (January 1968): 16–17.

Baucom, Donald R. "Editorial Note." *Air University Review* 33, no. 2 (January–February 1982): 33. https://babel.hathitrust.org/cgi/pt?id=mdp.39015081905674&view=1up&seq=195&size=125&q1=face%20defenses%20consisting.

Baugher, Joe. "Boeing B-52-D Stratofortress." Joebaugher, June 30, 2000. http://www.joebaugher.com/usaf_bombers/b52_9.html.

———. "Boeing B-52-G Stratofortress." Joebaugher, April 12, 2021. https://www.joebaugher.com/usaf_bombers/b52_15.html.

———. "McDonnell EF-4C Phantom II." Joebauger, December 28, 1999. https://www.joebaugher.com/usaf_fighters/f4_8.html.

———. "Republic F-105G Thunderchief." Joebauger, January 5, 2003. http://www.joebaugher.com/usaf_fighters/f105_9.html.

Beechy, Robert. "Ferrets, Ravens & Weasels: Radar Countermeasures and SAM Suppression." ProHosting. Accessed April 28, 2023. http://hud607.fire.prohosting.com/uncommon/reference/usa/sead.html.

"Boris Vasilievich Bunkin." Global Security. Accessed January 17, 2023. https://www.globalsecurity.org/wmd/world/russia/bunkin.htm.

Borton, Lady. "Linebacker II: Dien Bien Phu in the Air." *The VA Veteran Online*, January/February 2014. Accessed September 1, 2023. https://vvaveteran.org/34-1/34-1_borton.html.

Boyne, Walter J. "Linebacker II." *Air Force Magazine* 80, no. 11 (November 1997): 50–57.

———. "Momyer: The No-Nonsense General Made Airpower Work in Some of the Toughest Environments." *Air Force Magazine* 96, no. 8 (August 2011): 62–68.

———. "Route Pack 6." *Air and Space Forces Magazine*, November 1, 1999. https://www.airandspaceforces.com/article/1199pack/.

———. "The Teaball Tactic." *Air Force Magazine* 91, no. 7 (July 2008): 67–70.

Brosh, Larry D. "EC-121 Layout," VQ Association. Accessed February 15, 2023. https://vqassociation.org/history/.

Brown, Steven. "Lt. Colonel Steve Brown, Vietnam Veteran, Pilot." St. Charles County Veteran's Museum. Accessed August 25, 2023. https://stcharlescountyveteransmuseum.org/wp-content/uploads/2021/02/Steve-Brown.pdf.

"Burroughs BUIC – AN/GSA-51 SAGE Backup." Burroughs Corporation. Southwest Museum of Engineering, Communications, and Computation. Accessed June 21, 2023. https://www.smecc.org/burroughs_buic_-__an_gsa-51__sage_backup.htm.

Cagle, Malcolm W. "Task Force 77 in Action off Vietnam." U.S. Naval Institute *Proceedings* 98, no. 5 (May 1972): 66–109.

Cahill, William. "The Short but Interesting Life of a Plane Called Rivet Top." *Air Power History* 54, no. 3 (Fall 2007): 22–29.

———. "Strategic Air Command SIGINT Support to the Vietnam War." *Air Power History* 66, no. 4 (Winter 2019): 29–42.

"Capt. John (Smash) Nash, USN Flight Log ID: 2438." Naval Aviation Museum Foundation. Accessed February 7, 2023. https://navalaviationfoundation.org/ways-to-give/national-flight-log/national-flight-log-entry/?id=2438.

"Capt. Trent Richard Powers." Military Hall of Honor. Accessed March 3, 2023. https://militaryhallofhonor.com/honoree-record.php?id=271335.

Cherington, Paul W. "Systems-Acquisition and the Utilization of Scientific and Engineering Manpower (Requirements and Program-Determination Contracts and Grants.)" In *Toward Better Utilization of Scientific and Engineering Talent: A Program for Action*. Committee on Utilization of Scientific and Engineering Manpower, National Academy of Science. Washington, DC: National Academy of Science, 1964, 112–34.

"CMR-312." Phantom Phacts, July 3, 2018. https://phantomphacts.blogspot.com/p/an.html.

Coles, Joe. "Enter the Skynight: Hornet Pilot Shares the Dark History of the Douglas F3D 'Night Killer.'" *Hush-Kit: The Alternative Aviation Magazine*, December 29, 2020. https://hushkit.net/2020/12/29/enter-the-skyknight-hornet-pilot-shares-the-dark-history-of-the-douglas-f3d-night-killer/.

Copalman, Joe. "Glory Days// EA-6A in Vietnam." *Combat Aircraft Journal*. Accessed June 6, 2019. https://www.keymilitary.com/article/electric-intruder-war.

Correll, John T. "Against the MiGs in Vietnam." *Air & Space Forces Magazine*, October 1, 2019. https://www.airandspaceforces.com/article/against-the-migs-in-vietnam/.

———. "Take It Down! The Wild Weasels in Vietnam." *Air Force Magazine*, July 1, 2010, 66–69.

———. "With Waveform and Wits." *Air and Space Forces Magazine*, June 1, 1987. https://www.airandspaceforces.com/article/0687electronic/.

Curatola, John M. "'Black Thursday' October 14, 1943: The Second Schweinfurt Bombing Raid." The National WWII Museum, August 17, 2022. https://www.nationalww2museum.org/war/articles/schweinfurt-regensburg-raid-august-17-1943.

D'Costa, Dan. "How Combat Tree Made the F-4 Phantom II the Deadliest Fighter over Vietnam in the 1970s." Tacairnet, January 2, 2017. https://tacairnet.com/2017/01/02/how-combat-tree-made-the-f-4-phantom-ii-the-deadliest-fighter-over-vietnam-in-the-1970s/.

Drenkowski, Dana, and Lester W. Grau. "Patterns and Predictability: The Soviet Evaluation of Operation Linebacker II." Admiralty Trilogy. Accessed August 29, 2023. http://www.admiraltytrilogy.com/read/Soviet_view_of_Linebacker_II.pdf.

Duncan, Scott. "The Combat History of the F-105." *Aerospace Historian* 22, no. 3 (1975): 121–28. http://www.jstor.org/stable/44523414.

"EC-121 Layout." VQ Association. Accessed February 13, 2023. https://vqassociation.org/history/.

"F105F/G Thunderchief." Aviation – Airports, Aircraft, Helicopters. Accessed April 27, 2023. http://ourairports.biz/?p=6174.

"Fall of Saigon." Britannica, October 12, 2023. https://www.britannica.com/event/Fall-of-Saigon.

Ferris, Keith. "Story by Keith Ferris, WW # 2021." In *First In, Last Out: Stories by the Wild Weasels*, edited by Edward T. Rock, 136–39. Bloomington, IN: AuthorHouse, 2005.

France, Michael, and Craig Kaston. "AN/APR-24 C/S/X-Band Radar Homing and Warning System (RHAW)." Phantom Phacts, May 2, 2014. https://phantomphacts.blogspot.com/p/usn-ew-equipment.html.

Garden, William Jr., and Frank A. Dean. "Evolution of the Talos Missile." *Johns Hopkins APL Technical Digest* 3, no. 2 (1982): 117–22.

Gargas, John. "The Greatest Naval Deception of the Vietnam War, Part Two." *Naval History Magazine* 36, no. 3 (June 2022). https://www.usni.org/magazines/naval-history-magazine/2022/june/greatest-naval-deception-vietnam-war.

"GBU-52." Global Security, July 7, 2011. https://www.globalsecurity.org/military/systems/munitions/cbu-52.htm.

"General Joseph J. Nazzaro." Air Force. Accessed January 27, 2023. https://www.af.mil/About-Us/Biographies/Display/Article/106057/general-joseph-j-nazzaro/.

Goebel, Greg. "Douglas A3D Skywarrior & B-66 Destroyer." AirVectors. Accessed January 9, 2023. http://www.airvectors.net/avskywar.html.

———. "The Grumman A-6 Intruder and EA-6B Prowler," AirVectors, December 1, 2022. https://www.airvectors.net/ava6.html.

———. "Republic F-105 Thunderchief." AirVectors, March 1, 2023. https://www.airvectors.net/avf105.html.

Graham, David E. "In Memoriam W. Hays Parks: A Law of Armed Conflict Icon." *Army Lawyer* no. 4 (2021): 15–17.

Grier, Peter. "The Nightmare before Christmas." *Air Force Magazine* 98, no. 10 (October 2015): 56–61.

"Gulf of Tonkin Incident (Part 4) and Operation Pierce Arrow August 5, 1964." Vietnam War Commemoration. Accessed December 16, 2022. https://www.vietnamwar50th.com/history_and_legacy/.

Hall, R. Cargill. "Reconnaissance Drones: Their First Use in the Cold War." *Air Power History* 61, no. 3 (Fall 2014): 20–27.

Hambling, David. "The 200 Millisecond Mission: Inside the Secret CIA Plan to Steal Soviet Missile Data." *Popular Mechanics*, October 28, 2020. https://

www.popularmechanics.com/military/aviation/a34386117/suicide-drone-cia-sa-2/.

Hankins, Michael. "The Teaball Solution: The Evolution of Air Combat Technology in Vietnam, 1968–1972." *Air Power History* 63, no. 3 (Fall 2016): 7–24.

Hanson, Robert. "Counterpunch." *Air and Space Magazine*, September 1998. https://www.smithsonianmag.com/air-space-magazine/counterpunch-206258/.

Hanyok, Robert J. "Skunks, Bogies, Silent Hounds, and Flying Fish: The Gulf of Tonkin Mystery, 2–4 August 1964." Naval History and Heritage Command. Accessed December 22, 2022. https://www.history.navy.mil/research/library/online-reading-room/title-list-alphabetically/s/skunks-bogies-silent-hounds-flying-fish.html.

Harder, Robert O. "Operation Linebacker II: The 11-Day War." Historynet, December 29, 2020. https://www.historynet.com/the-11-day-war/.

Hawkens, Kristen. "Loggerhead Spiritual Meaning, Symbolism and Totem." Spirit & Symbolism, March 9, 2023. https://spiritandsymbolism.com/.

Hays, Philip. "Details of the First RGM-8H Anti-Radiation Missile Combat Firing." Okieboat, June 30, 2020. https://www.okieboat.com/Talos%20antiradiation%20shot.html.

Hehs, Eric. "Carmine Vito: U-2 Pilot." *Code One* 17 no. 1 (2002). Accessed October 27, 2021. https://www.codeonemagazine.com/article.html?item_id=167.

Hobson, Chris, and David Lovelady. "The B-66 Variants in Southeast Asia (Oct 1965)." Vietnam Air Losses. Accessed October 2, 2023. https://www.vietnamairlosses.com/index.php/sidelines/1965/oct65.

———. "North Vietnamese MIGs (Apr 1965)." Vietnam Air Losses. Accessed May 26, 2023. https://www.vietnamairlosses.com/index.php/sidelines/1965/apr65.

Hone, Thomas. "Southeast Asia." In *Case Studies in Achievement of Air Superiority*, edited by Benjamin F. Cooling, 505–60. Washington, DC: Center for Air Force History, 1991.

"John A. Corder." Veteran Tributes. Accessed February 3, 2023. http://veterantributes.org/TributeDetail.php?recordID=439.

Johnson, Harold E. "Of Bears, Weasels, Ferrets, and Eagles." *Air University Review* 33, no. 2 (January–February 1982): 87–88.

"Julian S. Lake, Rear Admiral, USN (Ret.)" [summary of service]. Early and Pioneer Naval Aviators Association. Accessed November 14, 2021. http://epnaao.com/BIOS_files/EMERITUS/Lake-%20Julian%20S.pdf.

Katta, Kenneth S. "Terror Night over Vinh." *Air Power History* 66, no. 2 (Summer 2019): 45–49.

King, Tom. "People, Let Me Tell You 'Bout My Best Friend.'" *Air Reservist* 36, no. 3 (March/April 1984): 6, 30.

Kopp, Carlo. "Almaz S-75 Diva/Desna/Volkov Air Defense System/HQ-2A/B / CSA-1/SA-2 Guideline." Air Power Australia. Accessed October 27, 2021. http://www.ausairpower.net/APA-S-75-Volkhov.html.

———. "Engagement and Fire Control Radars." Air Power Australia, Accessed October 27, 2021. http://www.ausairpower.net/APA-Engagement-Fire-Control.html#mozTocId517282.

———. "SNR-75M3 Fan Song E Engagement Radar." Air Power Australia. Accessed January 25, 2023. https://www.ausairpower.net/APA-SNR-75-Fan-Song.html.

"Kosygin Hailed in Hanoi: He Praises the Vietcong." *New York Times*, February 7, 1965, 1, 6.

Kristan, Kevin. "Colonel Aton Durham Brees USAF-RET." Accessed January 16, 2023. https://www.kristanfuneralhome.com/obituary/colonel-anton-durham-brees-usaf-ret.

Lamb, Allen. "The First Wild Weasel and the First Wild Weasel SAM Kill." In *First In, Last Out: Stories by the Wild Weasels*, edited by Edward T. Rock, 87–100. Bloomington, IN: AuthorHouse, 2005.

Lang, Katie. "Highlighting History: How 'Tet' Began the End of Vietnam." DOD News, February 7, 2023. https://www.defense.gov/News/Feature-Stories/story/Article/3291950/highlighting-history-how-tet-began-the-end-of-vietnam/.

Larson, Doyle. "Direct Intelligence Combat Support in Vietnam Project Teaball." *American Intelligence Journal* 15, no. 1 (Spring/Summer 1994): 56–57.

"Late Years of the War (1964–1975)." Studienet.dk. Accessed October 27, 2022. https://www.studienet.dk/the-vietnam-war/late-years-1964-1975.

Leone, Dario. "A Boss Weapon: It's the AGM-78 Standard Anti-radiation Missile." *The National Interest,* July 29, 2019. https://nationalinterest.org/blog/buzz/boss-weapon-its-agm-78-standard-anti-radiation-missile-70056.

———. "F-105 Pilot Explains How SAC Bomber Philosophy Imposed to Thud Crews F-105s High Loss Rate over Vietnam." The Aviation Geek Club. July 11, 2023. https://theaviationgeekclub.com/f-105-pilot-explains-how-the-sac-bomber-philosophy-imposed-to-thud-crews-contributed-to-the-f-105s-high-loss-rate-over-vietnam/.

———. "RB-47H Down: When a Soviet MiG-19 Downed a USAF Elint Plane over the Barents Sea." The Aviation Geek Club. August 19, 2018. https://theaviationgeekclub.com/rb-47h-shot-down-when-a-soviet-mig-19-downed-a-usaf-elint-plane-over-the-barents-sea/.

Litton Industries. "AN/GPA-122." *News & Opinion* 12, no. 2 (February 19, 1973).

Lockee, Garette E. "PIRAZ." U.S Naval Institute *Proceedings* 95, no. 4 (April 1969): 143–46.

"Lockheed DC-130." Military Hall of Honor. Accessed February 20, 2023. https://military-history.fandom.com/wiki/Lockheed_DC-130.

"Lockheed EC-121 Warning Star." Airborne Early Warning Association. Accessed June 4, 2023. https://www.aewa.org/Library/ec121-info.html.

Mack, Jerold R., and Richard M. Williams. "552d Airborne Early Warning and Control Wing in Southeast Asia: A Case Study in Airborne Command and Control." *Air University Review* 25, no. 1 (November-December 1973): 70–78.

Maifeld, Hank. "MacDill A F B, Florida: Detachment 2 Tactical Air Warfare Center." Geocities. June 14, 2023. https://www.geocities.ws/hmaifeld/macdill.html.

———. "Osan Air Base," Geocities. Accessed June 15, 2023. http://www.maifeldfamily.net/hank/osan.html.

March, Lee. "Going to War in Korea and Vietnam: The Decisions of Harry Truman and Lyndon Johnson." Chapter 4 in *Between Memory and Mythology: The Construction of Memory of Modern Wars*, edited by Natalia Starostina, 53–63. Newcastle upon Tyne, UK: Cambridge Scholars, 2015.

Mares, Ernest. "Shooting Stars." China Lake Alumni. Accessed March 19, 2023. http://www.chinalakealumni.org/Downloads/1966%20SHRIKE%20SIDS.pdf.

Marion, Forrest L. "A Hot Day in a Cold War: An RB-47 vs. MiG-17s, April 28, 1965." *Air Power History* 53, no. 3 (Fall 2006): 26–33.

Marolda, Edward J. "Forged in Battle." *Naval History* 28, no. 4 (July 2014). https://www.usni.org/magazines/naval-history-magazine/2014/july/forged-battle.

Merlin, Peter W., and Tony Moore. "Wild Weasel." The X-Hunters. Accessed May 15, 2023. https://www.thexhunters.com/xpeditions/f-105g_aircraft.html.

"Mikoyan-Gurevich MiG-17 (Fresco)." Military Factory. Accessed May 31, 2023. https://www.militaryfactory.com/aircraft/detail.php?aircraft_id=31.

"Miniature, Portable Microwave Receiver." *Electronics* 39, no. 1 (1966): 212.

Mitsotakis, Spyridon. "Israel Saved Many American Lives in Vietnam." *Washington Examiner*, February 20, 2020. https://www.washingtonexaminer.com/opinion/op-eds/israel-saved-many-american-lives-in-vietnam.

Morrison, Robert E. "March 2, 1965 VQ-1 Supports Operation Rolling Thunder." Station HYPO. Accessed February 12, 2023. https://stationhypo.com/2019/03/02/march-2-1965-vq-1-supports-operation-rolling-thunder/.

———. "How VQ-1 Supported Military Actions with SIGINT during Vietnam (Parts 1 of 7) Guest Post." Station HYPO. Accessed February 12, 2023. https://stationhypo.com/2018/04/22/how-vq-1-supported-military-actions-with-sigint-during-vietnam-part-1-of-7-guest-post/#more-9343.

———. "How VQ-1 Supported Military Actions with SIGINT during Vietnam Parts 2 of 7." Station HYPO. Accessed January 7, 2023. https://stationhypo.com/2022/02/07/early-sigint-efforts-in-vietnam-part-2-of-7/.

———. "PIRAZ in the Gulf of Tonkin – NGS's Role." Station HYPO. Accessed January 18, 2023. https://stationhypo.com/2021/04/17/piraz-in-the-gulf-of-tonkin-nsgs-role-guest-post/.

———. "September 3, 1965 – VQ-1 First Deployed EC-121M to Da Nang Full Time." Station HYPO. Accessed February 12, 2023. https://stationhypo.com/2022/09/03/september-3-1965-vq-1-first-deployed-ec-121m-to-da-nang-full-time-2/#more-20256.

Nance, William. "Quality ELINT." *Studies in Intelligence* 12, no. 2 (1968). Central Intelligence Agency. https://www.cia.gov/static/c54483c4718fbb7f4b3a31756717ead9/quality-elint.pdf.

"New Air Power Concepts Tested in U.S. Exercise." *New York Times*, November 11, 1964, 13.

"Nowadays, the Indispensable Jamming Pod for Fighters, at the Beginning of Its Birth, Why Did the U.S. Air Force Not Like It?" INEWS, September 30, 2023. https://inf.news/en/military/da8d44cfb3d64a5860b2de35641f0437.html.

Nowell, Larry. "Navy Red Crown Air Intercept Controller Seeks F-4 Pilots." Flying the F-4 Phantom in Combat. Accessed July 7, 2023. https://www.f-4phantom.com/red-crown/.

"Obituary: Dr. John L. Grisby Passes." eCrow, April 8, 2015. https://www.ecrow.org/articles/index-v7.asp?aid=315900&issueID=37759.

"Operation Freedom Train, Operation Linebacker I." Global Security. Accessed May 19, 2023. https://www.globalsecurity.org/military/ops/linebacker-1.htm.

Osborne, Arthur M. "Air Defense for the Mining of Haiphong." U.S. Naval Institute *Proceedings* 100, no. 4 (September 1974): 113–15.

Panos, Kristina. "Retrotechtacular: Radar Jamming." *Hackaday*, April 28, 2015. https://hackaday.com/tag/analq-55/.

Parks, W. Hays. "Linebacker and the Law of War." *Air University Review* 34, no. 2 (January–February 1983): 2–30.

Parsch, Andreas. "AGM-45 *Shrike*." Directory of U.S. Military Rockets and Missiles. Accessed May 21, 2002. https://www.designation-systems.net/dusrm/m-45.html.

———. "AGM-78." Directory of U.S. Military Rockets and Missiles. Accessed May 23, 2023. https://www.designation-systems.net/dusrm/m-78.html.

———. "Bendix SAM-N-6/IM-70/RIM-8 Talos." Directory of U.S. Military Rockets and Missiles. Accessed September 22, 2023. https://www.designation-systems.net/dusrm/m-8.html

———. "ORC - Equipment Listing." Designations of U.S. Military Electronic and Communications Equipment. Accessed March 7, 2023. http://www.designation-systems.net/usmilav/jetds/qrc.html.

———. "RIM-66," Directory of U.S. Military Rockets and Missiles. Accessed May 18, 2023. https://www.designation-systems.net/dusrm/m-66.html

Paterson, Pat. "The Truth about Tonkin." *Naval History Magazine* 22, no. 1 (February 2008). https://www.usni.org/magazines/naval-history-magazine/2008/february/truth-about-tonkin.

Patterson, Michael R. "Robert Gordon Owens, Jr. — Major General, United States Marine Corps." Arlington National Cemetery. Accessed June 11, 2023. https://www.arlingtoncemetery.net/rgowens.htm.

"Phúc Yên Air Base." Military-History Fandom. Accessed May 31, 2023. https://military-history.fandom.com/wiki/Ph%C3%BAc_Y%C3%AAn_Air_Base.

Pike, John. "EA-6B Prowler." Federation of American Scientists, 2000. Accessed December 6, 2021. https://irp.fas.org/program/collect/ea-6b_prowler.htm.

———. "RB-66 Destroyer." Federation of American Scientists. Accessed January 8, 2023. https://irp.fas.org/program/collect/rb-66.htm.

———. "S-60 Anti-Aircraft Artillery." Federation of American Scientists. Accessed January 7, 2023. https://man.fas.org/dod-101/sys/land/row/s-60.htm.

"The Pioneers: Wild Weasel I and the F-100F." Misawa Air Base. Accessed March 13, 2023. https://www.misawa.af.mil/News/Photos/igphoto/2000558681/.

Piowaty, John P. "Reflections of a Thud Driver." *Air University Review* 33 no. 2 (January-February 1982): 52–53. https://www.google.com/books/edition/Air_University_Review/vFTD2gFd99QC?hl=en&gbpv=1&dq=thunderstick+II&pg=PA57&printsec=frontcover.

Plunkett, Howard W. "Coming to Grips with SAM Sites July – Dec 1965." 7th Triennial Vietnam Symposium, March 11, 2011. The Vietnam Center &

Sam Johnson Vietnam Archives, Texas Tech University. Accessed March 3, 2023. https://www.vietnam.ttu.edu/events/presentations/4c-Plunkett.pdf.

Pocock, Chris. "Commentary by British U-2 Historian Chris Pocock on the CIA History." National Security Archive. Accessed January 20, 2023. https://nsarchive2.gwu.edu/NSAEBB/NSAEBB434/docs/W-P%20History%20Notes%20by%20CP%20on%20redactions%20lifted%20Aug13.pdf.

Polska, Wersala. "Douglas EKA-3B 'Skywarrior/Whale.'" Virtual Museum of the Vietnam War. Accessed May 16, 2023. https://vietnam.net.pl/EKA3Ben.htm.

"The President's Address to the Nation upon Announcing His Decision to Halt the Bombing of North Vietnam, October 31, 1968." The American Presidency Project. Accessed May 21, 2023. https://www.presidency.ucsb.edu/documents/the-presidents-address-the-nation-upon-announcing-his-decision-halt-the-bombing-north.

Pribbenow, Merle L. "The 'Ology War: Technology and Ideology in the Vietnamese Defense of Hanoi." *Journal of Military History* 67, no. 1 (January 2003): 175–200.

"Radar (Spoon Rest C [P-12]) - SA-2." Command: Modern Operations/Modern Air Naval Operations. Accessed June 2, 2023. http://cmano-db.com/pdf/facility/19/.

"Radarman Awarded Distinguished Service Medal." Orders & Medal Society of America. Accessed July 7, 2023. http://www.omsa.org/files/jomsa_arch/Splits/1973/22543_JOMSA_Vol24_5_22.pdf.

"The Radar Warning Story." Northrop-Grumman. Arboga Elektronikhistoriska Förening AEF. Accessed January 28, 2023. https://www.aef.se/Avionik/Artiklar/Motmedel/Nya_hotbilder/RadarWarnStory.pdf.

Rock, Edward T. "Wild Weasel III-2 Memories." In *First In, Last Out: Stories by the Wild Weasels*, edited by Edward T. Rock, 228–55. Bloomington, IN: AuthorHouse, 2005.

Rose, Scott D., ed. "Republic F-105 Thunderchief Design & Development." Warbirds Resource Group. Accessed August 12, 2023. http://vietnam.warbirdsresourcegroup.org/f105-design.html.

"S-125 SA-3 GOA." Federation of American Scientists. July 3, 1998. https://nuke.fas.org/guide/russia/airdef/s-125.htm.

Scaruffi, Frederick. "Frederick Terman." A History of Silicon Valley. Accessed March 9, 2023. https://www.scaruffi.com/svhistory/silicon/terman.html.

Schuster, Carl O. "The EA-6B Prowler: Outwitting Hanoi's Air Defenses." Historynet, 2018. Accessed November 29, 2022. https://www.historynet.com/ea-6b-prowler-outwitting-hanois-air-defenses.htm.

———. "The EB-66 Destroyer: A Life-Saving Intelligence Aircraft." Historynet, November 29, 2022. https://www.historynet.com/eb-66-destroyer/.

———. "Lightning Bug War over North Vietnam," Historynet. July 13, 2017. https://www.historynet.com/lightning-bug-war-north-vietnam/.

———. "The Rise of North Vietnam's Air Defense." Historynet, 2017. Accessed June 3, 2023. https://www.historynet.com/rise-north-vietnams-air-defense/.

Schell, John A. "The SA-2 and U-2: The Rest of the Story." *Journal of the Air Force Historical Foundation* 70, no. 2 (Summer 2023): 37–46.

Sears, David. "A 'Piercing Arrow' Strikes North Vietnam." *VFW Magazine*, August 2014. http://digitaledition.qwinc.com/publication/?i=215523&article_id=1749279&view=articleBrowser.

Seefluth, August. "The Other Jammer." *Air Force Magazine* 75, no. 3 (March 1992): 74–77.

Smetek, Ronald T. "Tactical Intelligence: Green Door to Battlefield." *Journal of Electronic Defense* (January 1964): 44–58.

"Soviet Aid to North Vietnam." Global Security. Accessed December 28, 2022. https://www.globalsecurity.org/military/world/vietnam/hist-2nd-indochina-ussr.htm.

Sparkman, Robert. "Exercise Gold Fire I." *Air University Review* 16, no. 3 (March–April 1965): 22–44.

"Special Electronic Search Project (forerunner of VQ-1) Established October 15, 1951." Station HYPO. Accessed October 9, 2021. https://stationhypo.com/2019/10/15/vq-1-fleet-air-reconnaissance-squadron-one-established-october-15-1951/.

Steel, Max. "Vietnam War: Soviet Air Defense Systems." Russian Defense Forum. February 8, 2016. https://www.russiadefence.net/t2474-vietnam-war-soviet-air-defence-systems.

Sterner, Doug. "John A. Corder." Hall of Valor: The Military Medals Database. Accessed February 3, 2023. https://valor.militarytimes.com/hero/3534.

Stovall, Thomas. "My Time in the Gulf of Tonkin (Guest Post)." Station HYPO. Accessed July 4, 2023. https://stationhypo.com/2018/12/05/my-time-in-the-gulf-of-tonkin-guest-post/.

———. "Surface-to-Air Missiles (SAMs)," Global Security. Accessed February 21, 2023. https://www.globalsecurity.org/military/world/vietnam/nva-ad-sam.htm.

Sweetman, Bill. "First In . . . Still Here." *Journal of Electronic Defense*, November 1, 1999. Accessed March 9, 2023. https://www.thefreelibrary.com/_/print/PrintArticle.aspx?id=58064582.

Swopes, Bryan. "22 November 1972." This Day in Aviation. Accessed August 24, 2023. https://www.thisdayinaviation.com/22-november-1972/.

Telford, Joseph W. "Story by Joe W. Telford, WW # 258." In *First In, Last Out: Stories by the Wild Weasels*, edited by Edward T. Rock, 101–39. Bloomington, IN: AuthorHouse, 2005.

———. "Story by Joe W. Telford, WW # 256 [sic]." In *First In, Last Out: Stories by the Wild Weasels*, edited by Edward T. Rock, 418–45. Bloomington, IN: AuthorHouse, 2005.

Thompson, Warren E. "Wild Duel: Weasel vs SAMs over Dong Hoi." Historynet, March 19, 2013. https://www.historynet.com/wild-duel-weasels-vs-sams-over-dong-hoi/.

Tilford, Earl Jr. "Linebacker II: The Christmas Bombing." The VA Veteran Online, January–February 2014. https://vvaveteran.org/34-1/34-1_tilford.html.

Tillman, Barrett. "Sam SA-2: The Aviator's Real Enemy." *Flight Journal* (December 2014): 56–57. https://www.flightjournal.com/wp-content/uploads/2016/06/SAM-SA-2.pdf.

Tiwary, Deepak, Preeti Rathor, Pradeep Vishwakarma, Mh. Firoz Warsi. "Implementation and Application of Radar Signal Simulator Based on FPGA." *International Journal of Engineering and Innovative Technology* 1, no. 6 (June 2012): 40–42.

"Tracor AN/ALE-29 Countermeasures Dispensing Set." Phantom Phacts, 2014. Accessed January 31, 2023, https://phantomphacts.blogspot.com/p/anale-29.html.

"A Tribute to the Raven EF-111AA." *Journal of Electronic Defense*, May 1, 1998. https://www.thefreelibrary.com/A+tribute+to+the+Raven+EF-111AA.-a020791918.

Turner, Theodore, J. "Seventh Air Force History." Seventh Air Force. Accessed January 28, 2021. https://www.7af.pacaf.af.mil/About-Us/Fact-Sheets/Display/Article/408382/seventh-air-force-history/.

"Udorn – ECM." International F-104 Society. Accessed March 5, 2023. https://www.i-f-s.nl/udorn-ecm/.

Urribarres, Ruben, and Mike Little. "The Cuban MiGs." Latin American Aviation Historical Society. April 15, 2018. https://www.laahs.com/the-cuban-migs/.

"US Navy F-4 EW Development—Revisited." Phantom Phacts. Accessed April 23, 2023. https://phantomphacts.blogspot.com/2014/03/us-navy-f-4-ew-development-revisited.html.

"USS Kitty Hawk CV 63." US Carriers. Accessed May 21, 2023. http://www.uscarriers.net/cv63history.htm.

"V-75 SA-2 Guideline." Federation of American Scientists. Accessed January 25, 2023. https://nuke.fas.org/guide/russia/airdef/v-75.htm.

Valcano, Mario. "Out of the Shadows—The Art of Time-Sensitive Reporting (Part 6 of 7)." Station HYPO. Accessed July 4, 2023. https://stationhypo.com/2021/04/02/out-of-the-shadows-the-art-of-time-sensitive-reporting-part-6-of-7/.

———. "Out of the Shadows – First 'ACE' of the Naval Security Group (Part 7 of 7)." Station HYPO. Accessed July 4, 2023. https://stationhypo.com/2021/04/03/out-of-the-shadows-first-ace-of-the-naval-security-group-part-7-of-7/.

Van Geffen, Theodore [Jr?]. "The Air War against North Vietnam: The Thanh Hoa Railroad and Highway Bridge, Part 1." *Air Power History* 65, no. 2 (Summer 2018): 7–13.

———. "U.S. Mini-Air War against North Vietnam: Protective Reaction Strikes." *Air Power History* 66, no. 2 (Summer 2019): 31–44.

Van Geffen, Theodore Jr., and Gerald C. Arruda. "Thunderchief." *Air University Review* 33 no. 2 (January–February 1982): 46–51. https://babel.hathitrust.org/cgi/pt?id=mdp.39015081905674&view=1up&seq=189&size=125&q1=face%20defenses%20consisting.

White, Edward. "Story by Ed White, WW Charter # 15." In *First In, Last Out: Stories by the Wild Weasels*, edited by Edward T. Rock, 140–48. Bloomington, IN: AuthorHouse, 2005.

Whetstone, Marc. "Dedication & Extensive Training Led to Distinguished Service Medal." *All Hands* no. 672 (January 1973): 10–12.

Whitten, H. Wayne. "Marine Composite Reconnaissance Squadron One (VMCJ-1) History." Marine Corps Aviation Reconnaissance Association. Accessed June 6, 2019. https://www.mcara.us/VMCJ-1.html.

———. "MCARA Aircraft > F3D-2Q/EF-10B Skyknight- History." Marine Corps Aviation Reconnaissance Association. Accessed July 28, 2023. https://www.mcara.us/EF-10B_history.html.

———. "MCARA Aircraft > Grumman EA-6A Intruder – History." Marine Corps Aviation Reconnaissance Association. Accessed July 8, 2023. https://www.mcara.us/EA-6A_history.html.

Wolff, Christian. "AN/SPS-48E." Radar Tutorial. Accessed July 20, 2023. https://www.radartutorial.eu/19.kartei/07.naval/karte010.en.html.

———. "Backward-Wave Oscillator." Radar Tutorial. Accessed September 30, 2023. https://www.radartutorial.eu/08.transmitters/tx20.en.html.

Zaloga, Steven J. "Defending the Kremlin: The First Generation of Soviet Strategic Air Defense Systems 1950–60." New York Military Affairs Symposium. Accessed January 23, 2023. https://nymas.org/defendingthekremlin.htm.

RESEARCH REPORTS, UNPUBLISHED DOCUMENTS, DISSERTATIONS, AND ORAL HISTORIES

Attinclio, John S. "Air-to-Air Encounters in Southeast Asia (U), Vol. I: Account of F-4 and F-8 Events Prior to 1 March 1967 (U)." Weapons Systems Evaluation Group Report 116, October 20, 1967 DTIC No. ADC003627. Washington, DC: Office of the Director of Defense Research and Engineering, 1967.

Barker, Patrick K. "The SA-2 and Wild Weasel: The Nature of Technological Change in Military Systems." Thesis, Lehigh University, 1994.

Bittencourt, Raul Pereira. "Electronic Warfare Technology." Thesis, Naval Postgraduate School, 1976.

Dougherty, Stanley J. "Defense Suppression: Building Some Operation Concepts." Maxwell Air Force Base, AL: Air University, May 1992.

Drew, Dennis W. "Rolling Thunder 1965: Anatomy of a Failure." Report No. AU-ARI-CF-86-3. Maxwell Air Force Base, AL: Air University, 1986.

Grisby, John L. "The Wild Weasel Story: A Contractor's View." Arlington, VA: Association of Old Crows, 2015.

Hake, Michael F. "Stealth, the End of Dedicated Electronic Attack Aircraft." Monograph, School of Advanced Studies, United States Army Command and General Staff College, Fort Leavenworth, Kansas, 1999.

Harris, Warren L. "The Linebacker Campaigns: An Analysis." Maxwell Air Force Base, AL: Air University, 1987.

Head, William P. "War From above the Clouds: B-52 Operations during the Second Indochina War and the Effects of the Air on Theory and Doctrine." [Fairchild Paper] Maxwell Air Force Base, AL: Air University, 2002.

Hewitt, William A. "Planting the Seeds of SEAD: The Wild Weasel in Vietnam." Thesis, School of Advanced Air Power Studies, Maxwell Air Force Base, AL: Air University, 1992.

Johnson, Calvin R. "Linebacker Operations September - December 1972 (U)." December 31, 1978. Project CHECO, Office of History, Headquarters Pacific Air Force.

Joiner, Gary D., and Ashley E. Dean. "Operation Linebacker II: A Retrospective Part 6: Linebacker II." Report of the LSU Shreveport Unit for the SAC Symposium, December 2, 2017. Shreveport, LA: Strategy Alternatives Center, Louisiana State University, 2017.

Knemeyer, Franklin H. "Interview of." China Lake, California, November 20 and December 13, 1991, No. S-200, Oral History Collection, Naval Weapons Center, China Lake, CA.

"List of U.S. Department of Defense and Partner Code Names." Wikipedia. Accessed February 20, 2023. https://en.wikipedia.org/wiki/List_of_U.S._Department_of_Defense_and_partner_code_names#O.

Machovec, Frank M. "Southeast Asia Tactical Systems Interface (U)." January 1, 1975, CHECO/Corona Harvest Division, Operations Analysis Office, Headquarters Pacific Air Forces.

McMaster, Michael A. "AGM-45-7A Shrike: Final Test Report." MS thesis, California State University Northridge, 1977.

Michael, Philip S. "The Strategic Significance of Linebacker II: Political, Military, and Beyond." U.S. Army War College Strategy Research Project, U.S. Army War College, Carlisle Barracks, Pennsylvania, April 7, 2003.

Moore, Courtland C. "EB-66C Out-Country Electronic Reconnaissance, 1965–67, A Case Study." Report No. 3655. Maxwell Air Force Base, AL, 1968.

Morris, Thomas W., Col. USAF. "CORONA HARVEST Interview No. 92," by Capt. Richard Clement, February 4, 1969. United States Air Force Oral History Program, K230.0512-092, Air Force Historical Research Agency, Maxwell Air Force Base, AL.

Porter, Melvin F. "Air Tactics against NVN Air/Ground Defenses." February 27, 1967, Project CHECO 7th AF, Directorate of Operations Analysis.

———. "Linebacker: Overview of the First 120 Days." September 27, 1973. HQ PACAF, Directorate of Operation Analysis, CHECO/CORONA HARVEST Division.

———. "Second Generation Weaponry in SEA." September 10 1970. HQ PACAF, Directorate, Tactical Evaluation CHECO Division.

"Radio Communication Station Hanoi/Bac Mai Airfield." April 19, 1968. National Photographic Interpretation Center. Central Intelligence Agency. Accessed July 11, 2023. https://www.cia.gov/readingroom/docs/CIA-RDP83-01074R000100030002-5.pdf.

Reddel, Carl W. "College Eye." November 1, 1968. Project CHECO HQ PACAF, Directorate, Tactical Evaluation CHECO Division.

"Ryan Model 147." Wikipedia. Accessed February 25, 2023. https://en.wikipedia.org/wiki/Ryan_Model_147#References.

Sams, K., J. Schlight, R. F. Knot, M. J. Medelson, and P. D. Caine. "The Air War in Vietnam 1968–1969 (U)." April 1, 1970. Project CHECO HQ PACAF, Directorate, Tactical Evaluation CHECO Division.

Teixeira, Leonard D. G. "Linebacker II: A Strategic and Tactical Case Study." Air War College, Maxwell Air Force Base, AL: Air University, April 1990.

Van Nederveen, Gilles K. "Sparks over Vietnam: The EB-66 and the Early Struggle of Tactical Electronic Warfare." ARI Paper 2000-03, College of Aerospace Doctrine, Research and Education, Airpower Research Institute, Maxwell Air Force Base, AL: Air University, 2000.

———. "Wizardry for Air Campaigns Signals Intelligence Support to the Cockpit." Research Paper 2002-03. Maxwell Air Force Base, AL: Airpower Research Institute, 2001.

Williams, Gary H. "Operation Linebacker II: An Analysis in Operational Design." Naval War College, Newport, RI, June 13, 1997.

Young, James L. Jr. "United States Air Force Defense Suppression Doctrine, 1968–1972." Masters thesis, Kansas State University, 2008.

GOVERNMENT DOCUMENTS AND PUBLICATIONS

Bailey, Carl E. "17 Weapons Squadron (ACC)." Air Force Historical Research Agency. September 10, 2021. https://www.afhra.af.mil/About-Us/Fact-Sheets/Display/Article/433945/17-weapons-squadron-acc/.

Burch, Robert M. "Tactical Electronic Warfare Operations in SEA, 1962–1968." Contemporary Historical Evaluation of Combat Operations (CHECO) Report, February 10, 1969, HQ PACAF, Directorate, Tactical Evaluation, CHECO Division.

Central Intelligence Agency. "Electronic Equipment – U-2 Program, 1955–1966." Top Secret-Byeman, April 1, 1969, National Security Archive. Accessed March 6, 2023. https://nsarchive.gwu.edu/document/27524-document-06-cia-electronic-equipment-u-2-program-1955-1966-top-secret-byeman-1-april.

———. "Le Duan and the Post-Ho Chi Minh Leadership [Intelligence Report]." April 1974. Internet Archive. Accessed September 15, 2023. https://ia600607.us.archive.org/28/items/CIA-RDP85T00353R000100040011-0/CIA-RDP85T00353R000100040011-0_text.pdf.

Central Intelligence Agency, Directorate of Science and Technology, Office of Scientific Intelligence. "The Soviet SA-2 Surface-to-Surface-to-Air Missile System." Scientific Intelligence Digest, June 1965 [Excerpt]. National Security Archive. Accessed February 26, 2023. https://nsarchive.gwu.edu/document/16270-document-18-sa-2-surface-air-missile-excerpt, pp. 20-25.

———. "The Soviet SA-3 Missile System." OSI-SR/TCS/65-10, July 8 1965. https://www.cia.gov/readingroom/docs/CIA-RDP78T05439A000500210055-7.pdf.

Central Intelligence Agency, National Photographic Interpretation Center. "Radio Communication Station Hanoi/Bac Mai Airfield." April 19, 1968. Central Intelligence Agency. Accessed July 11, 2023. https://www.cia.gov/reading room/docs/CIA-RDP83-01074R000100030002-5.pdf.

Chief Plans for Field Activities, OSA, CIA [author redacted]. "Memorandum for the Record." Subject: Long Arm Drone Modification, August 13, 1965, National Security Archive. Accessed February 26, 2023. https://nsar chive.gwu.edu/document/28370-document-12-chief-plans-field-activities -osa-cia-memorandum-record-subject-long-arm.

Cunningham, James A. "Memorandum for Deputy Director for Science and Technology." Subject: Significant OSA [Office of Special Activities] Activities, August 25, 196[4], Central Intelligence Agency. Accessed February 26, 2023. https://www.cia.gov/readingroom/docs/CIA-RDP33-02415A0008 00290025-4.pdf.

Davis, J. E., Commanding Officer, USS *Kitty Hawk* (CVA 63), to CNO, Subj. USS *Kitty Hawk* (CVA 63) Command History, 1968, Naval History and Heritage Command. Accessed May 20, 2023. https://www.history.navy.mil/con tent/dam/nhhc/research/archives/command-operation-reports/vietnam/ KITTY%20HAWK%201968.pdf.

Department of Defense, Defense Supply Agency, Defense Documentation Center. *Technical Abstract Bulletin Personal Author Annual Index for 1968*.

"General Hunter Harris." U.S. Air Force. Accessed June 13, 2023. https://www.af.mil/ About-Us/Biographies/Display/Article/106828/general-hunter-harris/.

Hearing, George. "CINCPACFLT Analysis Staff Study 12-65: Attacks on SA-2 Sites 27 July–8 November 1965." November 15, 1965.

Intelligence Advisory Committee. "Soviet Capabilities and Probable Programs in the Guided Missile Field." *National Intelligence Estimate 11-5-57*, TOP SECRET, 12 March 1957, 9, 14–15. National Security Archive. Accessed January 23, 2023. https://nsarchive.gwu.edu/document/19840 -national-security-archive-doc-14-intelligence.

Lang, Delmar. "Memorandum for the Record," October 14, 1964. Subject: Chronology of Events of 2–5 August 1964 in the Gulf of Tonkin. U.S. Department of Defense. Accessed September 27, 2023. https://media.defense. gov/2021/Jul/14/2002762861/-1/-1/0/REL1_LANG.PDF.

Mailes, Yancy. "B-52 Played Major Role in Operations Freedom Train and Linebacker I." Air Force Global Strike Command. Accessed May 21, 2023. https://www.afgsc.af.mil/News/Article-Display/Article/454706/b -52-played-major-role-in-operations-freedom-train-and-linebacker-i/.

Maneki, Sharon. "Delmar C. Lang: A SIGINT Innovator." Cryptographic Almanac 50th Anniversary. Accessed July 17, 2023. https://media.defense.gov/2021/Jul/01/2002754126/-1/-1/0/DELMAR-C-LANG.PDF.

Meyer, Jeffery N. "Andersen AFB's Legacy: Operation Linebacker II." Anderson Air Force Base. Accessed July 16, 2023. https://www.andersen.af.mil/News/Commentaries/Display/Article/416815/andersen-afbs-legacy-operation-linebacker-ii/.

Miller, George C. "Memorandum for Assistant Director for Special Activities; Subject: System XVII Study." Internet Archive. Accessed March 6, 2023. https://archive.org/details/CIA-RDP67R00587A000100030047-2/page/n1/mode/2up.

———. "Memorandum for Deputy Director for Science and Technology; Subject: System XVII Contract." Internet Archive. Accessed March 6, 2023. https://archive.org/details/CIA-RDP71B00185A000100060217-7/mode/1up.

Nalty, Bernard C. "1972 – Operation Linebacker I." Air Force Historical Support Division. Accessed July 15, 2023. https://www.afhistory.af.mil/FAQs/Fact-Sheets/Article/458990/1972-operation-linebacker-i/.

National Archives and Records Administration. "Tonkin Gulf Resolution." Public Law 88-408, 88th Congress, August 7, 1964. General Records of the United States Government. Record Group 11. https://usapoliticaldatabase.weebly.com/tonkin-gulf-resolution-1964.htm.

———. "Tonkin Gulf Resolution (1964)." Accessed July 28, 2023. https://www.archives.gov/milestone-documents/tonkin-gulf-resolution.

National Museum of the United States Air Force. "AGM-45 Shrike Anti-Radar Missile." Accessed July 26, 2024. https://www.nationalmuseum.af.mil/Visit/Museum-Exhibits/Fact-Sheets/Display/Article/196035/agm-45-shrike-anti-radar-missile/.

———. "The Pioneers: Wild Weasel and the F-100F." Accessed May 5, 2023. https://www.nationalmuseum.af.mil/Visit/Museum-Exhibits/Fact-Sheets/Display/Article/197491/the-pioneers-wild-weasel-and-the-f-100f/.

———. "'Something Better Than the Shrike': The First USAF AGM-78." Accessed May 19, 2023. https://www.nationalmuseum.af.mil/Visit/Museum-Exhibits/Fact-Sheets/Display/Article/196908/something-better-than-the-shrike-the-first-usaf-agm-78-standard-mission/.

National Security Agency. "Gulf of Tonkin Incidents Additional SEAHAG Inputs (U)." Accessed September 27, 2023. https://www.nsa.gov/portals/75/documents/news-features/declassified-documents/gulf-of-tonkin/history-of-southeast-asia/release-2/rel2_gulf_tonkin_incidents.pdf.

———. "IRONHORSE: A Tactical SIGINT System." *Cryptolog* II, no. 10 (October 1975): 224–26. https://nsarchive.gwu.edu/sites/default/files/documents/5301807/National-Security-Agency-Cryptolog-Vol-2-No-10.pdf.

———. "TEABALL: Some Personal Observations of SIGINT at War." *Cryptology Quarterly* (Winter 1991): 92–97. Accessed September 20, 2023. https://www.nsa.gov/.portals/75/documents/news-features/declassified-documents/cryptologic-quarterly/teaball.pdf.

Pratt, John C. "Air Tactics against NVN Air Ground Defenses December 1966 – 1 November 1968 (U)." Contemporary Historical Evaluation of Combat Operations (CHECO) Report, August 30, 1969. HQ PACAF, Directorate, Tactical Evaluation, CHECO Division.

"Route Packs in Vietnam." Air Force History and Museums Program. Accessed May 10, 2023. https://www.afhistoryandmuseums.af.mil/About-Us/Fact-Sheets/Article/639570/route-packs-in-vietnam/.

Schick, George B. Jr. "Command History of USS WORDEN 01 January 1972 to 31 December 1972." OpNav Report 5750-11. Naval History and Heritage Command. Accessed August 15, 2023. https://www.history.navy.mil/content/dam/nhhc/research/archives/command-operation-reports/vietnam/Worden%201972.pdf.

Schlight, John. "Rules of Engagement (U) 1 January 1966 – 1 November 1969." Contemporary Historical Evaluation of Combat Operations (CHECO) Report, August 31, 1969. HQ PACAF, Directorate, Tactical Evaluation, CHECO Division. Accessed July 24, 2024. https://apps.dtic.mil/sti/pdfs/ADA586305.pdf.

Seltzer, Joseph. Memorandum for the United States Intelligence Board Subject: SNIE 10-65: Communist Military Capabilities and Near-term Intentions in Laos and South Vietnam, February 1, 1965. Central Intelligence Agency Collection, box 7, folder 111. The Vietnam Center and Sam Johnson Vietnam Archive, Texas Tech University. Accessed December 28, 2022. https://vva.vietnam.ttu.edu/images.php?img=/images/041/04107111001.pdf.

Stephenson, Henry E., ed. "FAAWTRACEN San Diego Trains Air Intercept Controllers for Pacific Fleet." *Navy Training Bulletin*, Winter 1969–70, 30–31.

U.S. Air Force. "4080 Strategic Wing." Accessed February 28, 2023. https://usafunithistory.com/PDF/4000/4080%20STRATEGIC%20WG.pdf.

———. "Characteristics Summary – Reconnaissance RB-47H." Accessed February 16, 2023. https://www.avialogs.com/aircraft-b/boeing/item/3633-2751rb-47hstratojetcharacteristicssummary-october1963.

———. HQ PACAF. "Rules of Engagement (U) - 1 January 1966 - 1 November 1969." Contemporary Historical Evaluation of Combat Operations (CHECO) Report, August 31, 1969. HQ PACAF, Directorate, Tactical Evaluation, CHECO Division. Accessed July 24, 2024. https://apps.dtic.mil/sti/tr/pdf/ADA602256.pdf.

———. "Major General Kenneth C. Dempster." Accessed January 30, 2023. https://www.af.mil/About-Us/Biographies/Display/Article/107265/major-general-kenneth-c-dempster/.

———. "Major General Walter B. Putnam." Accessed March 15, 2023. https://www.af.mil/About-Us/Biographies/Display/Article/105882/major-general-walter-b-putnam/.

U.S. Air Force, Historical Report Division. "1972 Operations Linebacker I." Accessed August 16, 2023. https://www.afhistory.af.mil/FAQs/Fact-Sheets/Article/458990/1972-operation-linebacker-i/.

U.S. Air Force Security Service. "Growth of USAFSS ACRP in Terms of Tasking and Mission Aircraft and Historical Resume of Hostile Reaction to U.S. Reconnaissance Activities [highly redacted declassification] Governmentattic. Accessed May 27, 2023. https://www.governmentattic.org/13docs/SpecHistStudyGrowthUSAFSS-ACRP1966.pdf.

U.S. Army. "S-125 Neva/Pechora (SA-3 Goa) Russian 6x6 Surface-to-Air Missile System." ODIN. Accessed September 11, 2023. https://odin.tradoc.army.mil/WEG/Asset/S-125_Neva::Pechora_(SA-3_Goa)_Russian_6x6_Surface-to-Air_Missile_System.

———. "Secondary Aircraft Control System." *Military Review* XLVI, no. 9 (September 1966): 101.

———. "'Seek Dawn' Aircraft Control System." *Military Review* XLVIII, no. 10 (October 1968): 97.

U.S. Army, Center of Military History. *History of Strategic Air and Ballistic Missile Defense Volume II: 1956–1972*. Washington, DC: Center of Military History, 2009.

U.S. Department of Defense, Vietnam War Commemoration. "North Vietnamese Easter Offensive, March 30, 1972." The United States of America Vietnam War Commemoration. Accessed July 14, 2023. https://www.vietnamwar50th.com/1972-.

U.S. House of Representatives. *Department of Defense Appropriations for 1962, Hearings before the Subcommittee of the Committee on Appropriations House of Representatives*, 87th Cong., 1st Sess., Part 3. Washington, DC: Government Printing Office, 1962.

———. *Department of Defense Appropriations for 1968, Hearings before a Subcommittee of the Committee on Appropriations House of Representatives*, 90th Cong. 1st Sess. Part 1, Military Personnel. Washington, DC: Government Printing Office, 1967.

———. *Hearings on Military Posture and H.R. 3818 and H.R. 8687 . . . Before the Committee on Armed Services House of Representative*, 92nd Cong., 1st Sess., Part 2. Washington, DC: Government Printing Office, 1971.

U.S. Intelligence Advisory Committee, National Intelligence Estimate *11-5-57*. "Soviet Capabilities and Probable Programs in the Guided Missile Field." TOP SECRET, 12 March 1957. National Security Archive. Accessed January 23, 2023. https://nsarchive.gwu.edu/document/19840-national-security-archive-doc-14-intelligence.

U.S. Navy, Bureau of Personnel. "On Guard in the Pacific." *All Hands* no. 572 (September 1964): 3–5.

U.S. Navy, CinCPacFlt. "Analysis Staff Study 8-68: An Analysis of SA-2 Missile Activity in North Vietnam from July 1965 through March 1968." October 31, 1968. FPO San Francisco: Commander in Chief United States Pacific Fleet, 1968.

U.S. Navy, Naval Air Command. "Standard Aircraft Characteristics: Navy Model EA-1F Aircraft." American Aviation Historical Society. Accessed July 28, 2023. https://www.aahs-online.org/images/Navy_SAC/EA-1F.pdf.

U.S. Navy, Naval Air Force, U.S. Pacific Fleet. "Squadron History." Electronic-Attack-Squadron-VAQ-132. Accessed July 25, 2023. https://www.airpac.navy.mil/Organization/Electronic-Attack-Squadron-VAQ-132/About-Us/History/.

U.S. Navy, Naval History and Heritage Command. "*Chicago* III (CA-136)." Naval History and Heritage Command. Accessed July 22, 2023. https://www.history.navy.mil/research/histories/ship-histories/danfs/c/chicago-iii.html.

———. "H-Gram 070: The Easter Offensive-Vietnam 1971(1)." Accessed April 27, 2022. https://www.history.navy.mil/about-us/leadership/director/directors-corner/h-grams/h-gram-070.html.

———. "USS *Maddox* (DD-731)." Accessed December 21, 2022. https://www.history.navy.mil/browse-by-topic/ships/modern-ships/uss-maddox.html.

U.S. Senate. *Department of Defense Appropriations for 1968, Hearings before a Subcommittee of the Committee on Appropriations United States Senate*, 90th Cong. 1st Sess. on H.R. 10738, Part 2. Washington, DC: Government Printing Office, 1968.

———. *Fiscal Year 1974 Authorization for Military Procurement, Research and Development, Construction Authorization for the Safeguard ABM and Active Duty and Selected*

Reserve Strengths, Hearings before the Committee on Armed Services United States Senate, 93rd Cong., 1st Sess. on S 1263 Part 6 Tactical Air Power. Washington, DC: Government Printing Office, 1973.

INTERVIEWS

Robert P. Breault [Wild Weasel IB pilot], interview at his home in Tucson, AZ, July 20, 2024.

VIDEOS AND PODCASTS

"A-4E Skyhawk CP-741 Toss/Loft Bombing Tutorial." DCS World. Accessed March 24, 2023. https://www.youtube.com/watch?v=4GMLCfMhDPc.

Nowell, Larry. *Sea Control 315*. "Fighter Control over Vietnam with Master Chief Larry Nowell (ret.), Pt. 1." CIMSEC, February, 3, 2022. https://cimsec.org/sea-control-315-fighter-control-over-vietnam-with-master-chief-larry-nowell-ret/.

"Using the CP-741 Bombing Computer in the A-4E." Reddit. Accessed March 24, 2023. https://www.reddit.com/r/hoggit/comments/jj0r0g/using_the_cp741_bombing_computer_in_the_a4e/.

INDEX

Abrams, Creighton, 150, 153
Aerojet Mk 26 rocket motor, 103
Aerospace Historian, 12
After Burner: Naval Aviation in the Vietnam War, 57
Air Defense Command (U.S.), 118
Air Defense Simulator #1, 76
Air Intercept Controller, 165–67
Air Proving Ground, Eglin AFB, Wild Weasel evaluation, 92
air strikes and bombings, Ai Moi warehouse, 198; Bac Giang transshipment center, 196, 199; Bac Mai airfield, 201–2; Chann Hoa barracks, 10–11; Chap Le barracks, 8, 10–11; Doc Noi RR yard, 205; Doc Noi SA-2 assembly site, 207; Dong Hoi barracks, 8–9 Gia Lam RR yard, 201, 205; Haiphong petroleum storage area, 203; Hanoi Int'l Radio Station, 185, 187, 194, 196; Hanoi RR repair shop, 185; Hanoi/Haiphong, 136; Hoa Lac airfield, 184; Kep airfield, 184, 185; Kep highway bridge, 65–67; Kinh No vehicle repair shop, 185, 187; Lang Truc airfield, 187; Me Xa highway bridge, 66; Ngyen Khe oil storage, 51–52; Phuc Yen airfield, 184; Thanh Hoa Bridge, 115; Vinh POL, 154; Vinh RR yard, 154; Yen Vien RR yard, 185, 187, 196, 198, 199
Air University Review, 69
Air Warfare Center, 127
Airborne Communication Program, 112
aircraft:
 A-H: Flaming Dart II, 10
 A-3D, 18
 A-3D-2: conversion of, 25–26
 A-4: Pierce Arrow, 3; receives APR-27, 57
 A-4C: MiG engagement, 115
 A-4E: AGM-45 combat debut, 86; and Project Shoe Horn, 44–45; Flaming Dart, 8,9; Iron Hand strikes, 64–66; receives APR-23, 55; strike SAM sites, 154; struck by SA-2s, 42, 60; use of ALQ-51, 64
 A-6A, 34; conversion to A-6B, 104; Haiphong strike, 136; Iron Hand missions, 64–65, 154, 187
 A-6B, 104
 A-7, 203; downed by SA-2, 194; Iron Hand missions, 187

AD-5N, 24
B-52: cell formation, 178; elephant walk, 187; first missions north of DMZ, 170; jamming SA-2s, 158, 193; loss rate, 202, 204, 207; nuclear mission, 177; Operation Freedom Porch Bravo, 136; Operation Freedom Train, 154; psychological effect of, 176; SAM threat to, 182; sorties flown, 208; tonnage dropped, 208
B-52D: attacks on Hanoi, 203–5; Big Belly conversion, 177–78; bomb load compared to B-52G, 200; brought down by SA-2, 194; damaged by SA-2, 182–83; difference with G-52G, 177, 179; ECM malfunctions, 183; first attacks on Hanoi, 184; Phase V ECM, 180–81
B-52G: difference with 52-D, 177–79; ECM effectiveness, 195; ECM suite, 188; E/F jammer failure,198; first attacks on Hanoi, 184; flight deck layout, 179; importance to SAC, 178, 195
B-57: Rolling Thunder, 12
B-66B Brown Cradle, 18–20; arrives in theater, 34, 66
Big Eye: targets below 8,000 ft, 156
Big Look, 140
C-130B-II, 112
C-135, 131
College Eye: IFF upgrade, 130; replacement of, 131; targets below 8,000 ft, 156
Combat Apple, 133, 162
Combat Lightning, 163, 163
Combat Martin, 135
Combat Tree, 168–69
CV-2 transport, 8
Deep Sea, 140

Disco EC121K, 163–64; targets below 8,000 ft, 156
E-1B, 141
EA-1F, 24–25; first mission, 28
EA-3B: Deep Sea 140; tasked to excite Fan Song radars, 63
EA-6A: communication jamming, 147n; ECM equipment, 34–35; EW support in Linebacker II, 174; TACOS, 26; use of ALQ-55, 146, 147
EA-6B, 214; characteristics, 171; ECM suite, 172; EW support in Linebacker II, 174; stand-off jamming mission, 173–74
EAK-3B, and ALQ-92, 134
EB-66: ALQ-59, 134; replacement for, 174; SIGINT mission, 123; stand-off jamming, 183; Wild Weasel support, 78–79
EB-66B, 23, 66
EB-66C, 23; attack on SA-2 sites 6&7, 41; jamming sorties, 151; QRC-160-1 evaluation, 50
EB-66E, 151; Linebacker II jamming missions, 187
EC-121D, AC system, 123; and GPA-122, 130; Big Safari feasibility study, 125; crew, 119; electronic equipment, 118–19; IFF operator, 121; missions, 119–22; radios, 122
EC-121K, 127
EC-121M, 118–19; and NTDA, 140; ARCP mission, 123; Brigand equipped, 64; design 17; ECM equipment, 17; Rolling Thunder, 16–17
EC-121P: modified for Sea Trap, 125–26
EC-121T, 147; call sign "Disco," 156; radio limitations, 156
EC-135, command and control, 122

INDEX

EC-21D, River Top, 126
EF-105F, 94
EF-10B: attack on SA-2 sites 6&7, 41; ECM configuration, 32; jamming missions, 34; lack of refueling, 32–33; limited jamming capability, 32; ordered to Vietnam, 30; origins, 30; rate of climb, 33
EF-111A, 214
EF-4C, 174
EKA-3B: disrupts MiG intercept 146–47; EW support in Linebacker II, 174n; Tanker Countermeasures/strike, 26
Electric Spads, 24
F3D-1, 30, AGM-45 test bed, 84
F3D-2Q, 46
F-4: combat loses 158; forms chaff corridor, 154; lost to MiGs, 150; MiG CAP armament, 112
F-4B: APR-25s installed, 89–90; as pathfinder with APR-24, 56; receives APR-27, 57
F-4C: APR-24, 55–57; chaff laying, 183, 187–88, 203; first loss to SA-2, 33, 40; MiG CAP, 33; Strikes SA-2 sites, 41; Wild Weasel IV, 174
F-4D: Combat Tree, 168–69; guided by LORAN, 206; Linebacker I MiG CAP, 168; weapon systems operator, 168
F-8, flack suppression, 115; Flaming Dart, 8; Maddox Incident, 4; receives APR-27, 57; MiG engagement, 115
F-100: battle damage of, 15; designed for, 15; Flaming Dart II, 10; lock stabilizer modification, 15; ordnance capacities, 15; QRC-160, 28; Rolling Thunder, 12; Thunderstick II modification, 15

F-100D, 74; MiG engagements, 115; modified for MAX-DIX, 92–93
F-100F: first use of AGM-45, 90; flying characteristics, 74; Gold Fire I, 69; selected for Wild Weasel, 43,73; use of IR-133, 77; Wild Weasel configuration, 90; Wild Weasel evaluation, 92; Wild Weasel operations, 78–82
F-101F, stores carried, 102
F-104: use during attack on SA-2 sites 6&7, 41
F-105, 33; losses, 118
F-105D: ALQ-71, 101; ECM pod formation, 51; evaluates QRC-160-1, 50; Iron Hand missions, 41–42, 63, 65–66, 183; MiG engagement, 115
F-105F: AGM-78, 105; Combat Martin, 135–36; QCR-128, 135; selected for Wild Weasel role, 94; upgraded ECM, 108
F-105G, 109; Iron Hand missions, 171; stores diagram, 110; Wild Weasel modifications, 170–71
F-111, 174; Linebacker II, 187; supports B-52s, 203, 204, 207
FR-8G, 147
HH-43, 12
KC-130, radio relay aircraft, 122
KC-135A, 164
Lockheed L-1049, 17, 118
MiG-15, 114
MiG-17, 133; 921st Fighter Regiment, 114; attacks by, 115; Chinese use of, 123–24; coordinated with MiG-21s, 134; jamming of, 147; supplied to North Vietnam, 111; number of, 149
MiG-19, number of, 149

MiG-21, 133; attack on F-105D, 51; GCI control, 134; jamming of, 147; number of, 149; tactics, 157
MiG-23, armament, 115
O-1F, 8
RB-47 Silver King, 6–7
RB-47H: ELINT survey of North Vietnam, 6; Operation Long Arm, 61, 63, 67; United Effort ELINT mission, 68
RB-66, 18, 25; anti-aircraft jamming, 32
RB-66C, 18–23, 33
RB-66C Big Sail, 18; electronic warfare suite, 22
RC-135M, 131–32; EWO stations, 133
RC-135M Combat Apple, 162
RF-8: discovers first SAM site, 27; receives APR-27, 57
RF-101, first mission, 28; receive QRC-160-1, 29; Rolling Thunder, 12
Rivet Card, 131–32
Rivet Gym, 130
Rivet Top, 126–30; electronics, 156
Sea Trap, 125–27
U-2: danger from SA-2, 27; ECM, 27; Olympic Torch, 161, 162, 163; photographs Moscow SAM sites, 36; SIGINT platform, 161
UH-1B, 8
Wild Weasel I: AGM-45 first use, 90; configuration, 90; evaluation of, 92; name origin, 73
Wild Weasel III, 94, 174; AGM-78 modification, 104; AIM-9 requirements, 102; deployment, 95, 97; losses, 97–98; training, 95
Wild Weasel IV, 174
Wild Weasel IV-A, 175

aircraft, unmanned:
  Lightening Bug, 62
  Ryan Model 147A, 62
  Ryan Model 147B, 61
  Ryan Model 147D, 61–63
  Ryan Model 147E, 61, 65, 68
aircraft lost to SAMs, 33
aircraft tracking system Seek Dawn, 139
AJB-3, 86–87
Alfa Strikes, 40
Alvarez, Everett, 3
America (CV 66), 171
Anderson, Andrew B. Jr., 184
Anderson, Robert Jr., 27
Angle Gating, 89
Anti-SA-2 symposium, 47
Applied Technology Inc. (ATI), 70, 75
ATI. *See* Applied Technology Inc.
Axe, David, 63
Ayer, William E., 70
Azimuth-Elevation System (AZ-EL), 94–95

B-52 Project Office, 70
Babcock, Elizabeth, 84–85
Back Up Interception Control system (SAC), 140
Baird, Robert V., 12
Baucom, Donald R., 15
Bendix Corporation, 69, 92, 104
Biery, Bud M. II "Bud," 87
Big Eye missions, 123
Big Eye, 121; missions, 123; project, 127; Task Force, 118, 122, 125, 127
Big Look, 64, 123
Big Safari program, 125
Black Friday, 60
blue bells ringing, 33
bomb damage assessment system, for F-105F, 108; logic, 109
Bombing Interlude, 1, 2
Bombing systems. *See* AJB-3

INDEX   297

Boslaugh, David, 144
Brees, Anton D., 29, 30
Brigand, 17–18, 64
Brown, Charles "Charley," 194
Brown, Donald H. Jr., 42
Brown, Steven, 185
Bull, Norman S., 17
Bull's Eye system, 159–60
Bunker Ramo Corporation, 35
Burch, Robert, 111
Burns, John J., 173

Cagle, Malcolm W., 42
Cahill, Malcolm W., 60, 63
Cahill, William, 6
Carter, Tom, 146
CBU-24 cluster bombs, 79
CBU-52 bomblet, 171
Central Intelligence Agency (CIA), 9; data on Fan Song radar, 61, ELINT missions, 77; Office of ELINT (OEL), 68; Intelligence Division, 68; Moscow SAM sites, 36
Chaff: B-52 use off, 199; corridors, 154, 155; degrades North Vietnamese radar, 154; M-129 bomb, 188; SAC forbids, 191
chaff dispensers: ALE-2, 26; ALE-24, 180; ALE-25, 180; ALE-29, 58; ALE-32, 35; ALE-38, 155; ALE-41, 35; ALE-48, 187
Chapman, Edward, 70–73, 75
Chicago (CG 11), 141, 165; NTDS, 143–44
CINCPAC. *See* Commander-in-Chief Pacific
CINCPAC loss study, 13
College Eye, missions, 123; project, 132, task force, 147
Colonel Computer, 135
Combat Apple, missions, 132, 135; Information collected, 163

Combat Information Center (CIC), 140
Combat Lightning, 122, 147; Seek Dawn, 139–40; SIGNT intercepts, 127
combat operations:
  Constant Guard, 170
  Flaming Dart, 2, 8
  Flaming Dart II, 9–10
  Freedom Porch Bravo, 136
  Freedom Train, 154, 156
  Linebacker I, 1–2, 133, 148–58; named by Nixon, 154
  Linebacker II, 1–2, 174, 177–208; cost in airmen, 210; first bombing strikes, 184; purpose, 176; sorties conducted, 209
  Long Arm, 61, 62
  Old Bar, 63
  Pierce Arrow, 1, 2, 3–4, 111
  Deep, 150, 151
  Queen Bee, 112
  Rolling Thunder, 1–2, 10–12, 111, 212; end of, 148
  Silver Dawn, 112
Combat Tree. *See* IFF systems; APX-80
COMINT, 132; Y service, 158
Commander-in-Chief Pacific, 5, 8, 13, 146, 147, 150
Connell, John P., 43
Constellation (CVA 64), 3, 57
Coral Sea (CVA 43), 8, 10, 27
Corder, John A., 53
Cordes, Harry, 200
Corman, Otis W., 31
Cuban Missile Crisis, 119, 126, 212
Curatola, John M., 210

Davies, Peter, 64, 79
Davis, Larry, 70, 76, 170, 175
Demilitarized Zone (DMZ), 148, 153
Dempster, Kenneth: decides to buy Vector IV & WR-300, 93; establishes task force, 43, 47; orders

F-105F for Wild Weasel III, 94; requests test of RAWH devices, 92; Vector IV demonstration, 70; WR-300 briefing, 75
Dempster Task Force, 43, 71
DIANE (Digital Integrated Attack and Navigation Equipment, 64–65
Dictionary of American Naval Aviation Squadrons, 86
Donovan, John E. "Jack," 76, 80
Doolittle, James H., 14
Doyle. William C., 75
Duẩn, Lê, 207

E-248 airborne radar feed, 140
Easter Offensive, 153
ECM equipment: ALQ-100, 173, ALR-28 display, 26, installations, 46; pods, 173–74; QRC-380/ALQ-155, 170
Eighth Air Force, 177; B-52 tactics, 180; Linebacker II briefing, 184; Linebacker II planning, 181–82, 205; questions B-52G ECM effectiveness, 195
Electronic Warfare Officers: identifying and jamming Fan Song radars, 190–91; procedure for locating radars, 21; roles in RB-47H, 7; selected for Wild Weasels, 76
Electronic Warfare Test Division, 47
ELINT, 132; aircraft, 17; high gain antenna, 126; receiver, ALQ-28, 17; tracks Spon Rest radar, 148
Enterprise (CVN 65), 172

Fan Song radar, 33; attempts to locate, 83; CIA intelligence on, 77; computer guidance, 191; early jamming of, 34; hardware vans, 39; operator's response to AGM-45, 98–100; side pointing technique, 197; Wild Weasel tactics against, 81

Fan Song B radar: modes and computer processing, 37–38
Fan Song surrogates: Flint Stone, 44; SADS-1, 76–77, 204
Ferguson, Walter, 193
Fighting in the Electromagnetic Spectrum, 212
Fino, Steve, 156
flak trap, 34
flare dispenser: ALE-20, 180
Fobar, Roscoe, 33
Foreman, Merlin L., 152–53
Fowler, Robert M., 58
Franklin D. Roosevelt (CV 42), 56
Frankum, Ronald B., 6
Fubini, Eugene, 68

Gallotta, Albert A. Jr., 173
Galveston (CLG 3), 152
Garbera, Cecil, 64
Gaylor, Noel, 158, 202
General Dynamics Corporation, 103
General Electric Company, 28
Giant Stride program, 180–81
Gold Fire I exercise, 69
Graham, Gordon M., 13
Green Door syndrome, 163, 213
Grier, Peter, 177
Grimes, Alton B., 42
Grisby, John, 71–72, 76
Ground Control Interception, 115; working with SAM sites, 150
Grumman Aircraft Engineering Company, 103
Gulf of Tonkin Resolution, 1, 2, 6, 7

half-angle correction mode, 38
Hampton, Dan, 41, 73
Hancock (CVA 19), 115: Flaming Dart, 8, 10
Hanoi International Radio Station, 196
Hannah, Craig, 14
Hanyok, Robert J., 5

Harris, Hunter Jr., 127
Harry Diamond Laboratory, 76
Hartzel, Curt, 95
Haugen, Ingwald, 47
Hays, Philip, 152
HBR Systems Division Singer Corporation, 67–68
Head, William P., 176, 177
Herman, Michael, 213
Herrick, John, 3–5
History of US Electronic Warfare, 17, 71
Ho Chi Minh Trail, 106–7, 150
Hobson, Chris, 114
Hognose Silent Warrior, 133
Holcombe, Kenneth B., 121
Hold, William H., 97
Horn (DLG 30), 166
Houser, William D., 173

IFF, 123; and Big Safari, 125; North Vietnamese use of, 177; Red Crown use, 145
IFF systems: APX-49, 119, 126; APX-80, 168–69; GPA-122, 130; QRC-248, 126–27, 129, 168; SRO-2 (USSR), 126
Improved Shrike. *See* missiles: antiradiation, AGM-45
Independence (CA 62), 45, 64
Iron Hand missions, AGM-45 warhead for, 91; and electronically guarded corridors, 100; composition of, 171; failures of, 60, 195; first success, 64; formations, 82; Haiphong area, 203; operational name, 42
Ironhorse, 113, 139, 147; MiG plots, 163

Jagiello, Leonard T. "Lee," 84
jammers:
 barrage, ALT-16, 180
 communication, ALT-13, 188; ALQ-55, 146; ALQ-59, 134; ALQ-92, 26; ALQ-92, 134–35; QRC-128, 135
 continuous wave, ALT-22, 180
 deception, ALQ-41, 26, 35; ALQ-51, 26, 44–45, 64; ALQ-51A, 45–46
 high band, ALT-32H, 180
 low band, ALT-32L, 180
 noise, ALT-6B, 35, 180, 188, 191; ALT-17, 32; ALT-19, 32; ALT-28, 188, 191; Hallicrafters QRC, 20
 radar, ALT-2, 24–25, 32; ALT-27, 26
 slow sweep, ALT-B, 20
 tactical systems, ALQ-99, 173–74
jamming: beacon, 55; communication, 147; down link, 55; of SA 2 guidance, 158; synchronized noise, 48; systems integrated receiver, 172; types, 47–48
jamming pods:
 ALQ-31B, 35
 ALQ-71, 108; down link jamming, 55; effectiveness, 53; F-105D installation issues, 52; F-105 standard equipment, 101; jamming onboard receivers, 102; tactical use, 53
 ALQ-87, 55
 ALQ-101, 108
 QRC-72, 54
 QRC-160-1, combat evaluation, 50; dimensions, 47; evaluation, 48, 50; new versions, 54; on RF-101, 28; problems with, 29–30; redesignated ALQ-71, 52; use against North Vietnam, 54
 QRC-160-4, 54
 QRC-160-8, 54
 QRC-160-A, 212; wide-scale use, 100
jet engines, Westinghouse J34-WE-36, 30
Johnson, Calvin R., 183, 299; breach of protocol, 202
Johnson, Harold, 73
Johnson, James T., 92

Johnson, Lyndon B., 34, 79–80, 130; announce bombing halt, 106; authorizes Flaming Dart II, 9; begins Rolling Thunder, 10; Gulf of Tonkin Resolution, 1, 2, 6; orders Pierce Arrow, 3, 6; selects targets for Flaming Dart, 8–9

Joint Chiefs of Staff, 150, 176; instructions for round-the-clock bombing, 184; Project Left Hook authorized, 61; proposes bombing campaign, 11; reacts to F-4C loss, 40, 60; restrictions on surface-to-air-missiles, 144–45; strike permissions, 206; target selection, 184

Keegan, George J., 158
Keirn, Richard P., 33, 40
Kerzon, Warren, 105–7
Kilmek, Robert A. Jr., 97
King, Ernest J., 141–42
Kissenger, Henry, 207
Kitty Hawk (CVA 63), 104
Kopp, Carlo, 38
Korean War, 14; COMINT, 158–59, 161; ECM, 31
Kosygin, Alexei, 9
Kulla, Vern, 115

Lacouture, John, 25
Lake, Julius S., 25, 89; suggests ALQ-51 installation, 44–45
Lamb, Allen T., 80
Lang, Delmar, 160–62
Larson, Doyle E., 163
Lawson, Cliff, 84
LeMay, Cutis E., 13–14
Levy, Irwin J. "Piere," 71–72, 89
Lexington (CV 16), 64
Lightning Bug War Over North Vietnam, 68
Ling-Temco-Vought's Electrosystems Division, 125–27

Little Ears, 58
Long Arm drone. *See* aircraft, Ryan Model 147E
Long Beach (CGN 9), 145–46, 147
long range navigation system. *See* LORAN-D
look-through, 35
Loral Corporation, 97
LORAN-D (ARN-92), 206
Lovelady, David, 114–15
Low Blow, 201
Luyen, Dao Dinh, 114

M-61 cannon, 79
MacDonald, Thomas, 31
Maddox (DD 793), 3–5
Maddox Incident, 1, 2, 3–6, 140
Magnusson, James, 115
Mahan (DLG 11), 165–66
Maifeld, Hank, 127–28
Mare, Howard P. "Hap," 93
Marine Tactical Data System, 113
Maut, Nguyen X., 101
MAX-DIX, 92–93
Maxon Corporation, 92; wide band receiver, 108
McCarthy, James, 182, 195, 200
McCarthy, John, 211
McClellan Air Force Base, 118
McDonnel-Douglas Corporation, 175
McGuigan, William M., 58
McIntire, Scott W. "Scottie," 105, 107
McMakn, Charles, 17
McNally, Irvin, 142–43
McNamara, Robert, and Maddox Incident, 5; cancels SAM strike, 27; requests analysis of losses, 13
Melpar Electronics Corporation, 55
Menard, David, 79
Meyer, John C., 176, 177, 182, 200, 202
Michel, Marshall, L., 121, 155, 164, 180–81, 190, 192, 194; analysis of SA-2

effectiveness, 208; and importance of after-action feedback, 198; explains LORAN, 206; on North Vietnamese GCI, 134; on success of North Vietnamese defenses, 199; recounts discussion at SAC, 200; SAC's poor planning, 199; Steffen interview, 194
Midway (CV 41), 42
MiG CAP, 33, 105, 118; Linebacker I missions, 168
MiG threat, 17, 129, 133, 150; Bull's Eye map, 160; classification problems affecting, 158; dealing with, 121; F-4 tactics against, 162; first appearance, 24; negative kill ratio, 158; River Top success against, 129
missiles, air-to air:
  AIM-7 Sparrow, 112, 156, 169, 175; AGM-45 development, 84–85
  AIM-9 Sidewinder, 101, 112; and AGM-45 development, 84–85
  Atoll (USSR), 158, 159
missiles, air-to-ground: Zuni, 115
missiles, antiradiation:
  AGM-45 Shrike, 83, 85–92, 175, 202, 210; Air Force request for, 89; combat debut, 86; distinctive flight profile, 197–98; first Air Force use, 90; hits USS *Worden*, 137; in Wild Weasel IV, 175; importance to Weasels, 98–99; kill rate, 98; number fired, 195; on EA-6A, 35; origin, 85; seeker head, 103; tactical impact, 100–101; test period, 88; supplied to Air Force, 47; variations produced, 91
  AGM-45A-3 Shrike, 89
  AGM-78A Standard ARM, 101, 136, 151, 210; and ER-42 receiver, 103; flight tests, 105; number fired, 195; origins, 103; typical use, 105–96; unsuited for F-4C, 175

  AGM-78B Standard ARM, 108–9; BDA Logic, 109; for F-105G, 170; SDU-6/B target indicator, 108
Corvus, 84–85
RGM-8H Talos, 151–53
missiles, surface to air:
  Guideline (NATO code name). *See* SA-2
  RIM-2 Terrier, 144
  RIM-8 Talos, 145, 146
  RIM-66 Standard, 103
  S-25 Berku (USSR), 36
  S-75 Dina (USSR). *See* SA-2
  S-125 Neva/Pecora (USSR). *See* SA-3
  SA-1 V-300 (USSR), 37
  SA-2, 37; aircraft destroyed by, 33, 208; battalion equipment, 39; chaff countermeasure against, 155; ECM against, 48, 98, 157–58; efforts to detect, 43; failure to down aircraft, 154; flight characteristics, 87, 191; guidance, 38, 193; number of, 55, 148, 186, 210; origins, 37; reduced effectiveness, 157; Star of David pattern, 27; strikes against, 43, supply issues, 40, 185, 194; system modifications, 157, 204–5; tactics, 46, 107–8, 157, 197; warhead, 33, 19, 196; warm up time, 157
  SA-3, 201
  Wasserfall (German), 36
Miz, John, 201
Momyer, William A., 50
Monkey Mountain, 112–13, 129, 138, 139
Moore, Dwight, 186
Moore, Joseph H., 27, 41
Moorer, Thomas H., 150
Moran, William J., 83

Nalty, Bernard, 33, 55, 121, 147
Nash, John, 58

National Security Agency, 126, 135; ARCP program, 112; Project Left Hook, 61; R8 office, 138
Naval Air Systems Command, 35, 103
Naval Air Work Facility, Alameda, 25
Naval Ordnance Test Station China Lake, 83, 86; AGM-45 display, 88; AGM-45 range tables, 87; improvements on AGM-45, 90
Naval Security Group, 3, 165–66; Maddox Incident, 4
Naval Tactical Data System, 140–44; vs Teaball, 164–65
Nazzaro, Joseph J., 11
Nederveen, Gilles, 161, 163
Nellis Air Force Base, 82, 95
Nguyen Hue Campaign, 153
Nicols, John, 24
Nixon, Richard M.: and Hanoi Int'l Radio Station, 196; and Linebacker II, 184–209; halts bombing, 207; initiates Operation Freedom Train, 154; orders B-52 bombing campaign, 176, 177; orders bombing to continue, 203, 207; orders renewed bombing, 170
North Vietnam Air Defense Command: Battalions: 57th Battalion, 201; 59th Battalion, 191; 77th Missile Battalion, 193, 197; 89th Battalion, 23; 93rd Battalion, 198
Control: Air Citation Center, 117; Air Weapons Control Staff, 117–18; headquarters, 117–18; headquarters monitoring battle, 190
missileers, 46
radar companies: 45th Radar Company, 188–90
regiments: 63rd Missile Regiment, 66; 161st Regiment, 191; 236th Air Defense Regiment, 39, 41, 157; 274th Missile Regiment, 23

North Vietnam Air Defense System: assets, 111
early warning system: coverage, 117; warning radars, 116; Rolling Thunder, 15
GCI: effective range, 156; communication links, 161–62; controllers, 26, 130, 133, 134; system, 133; use of, 115–16, 118
IFF susceptibly discovered, 170
tactics: analyzes, 197; bolsters Hanoi defenses, 197; failure to down B-52s, 197; for defeating Wild Weasels, 197–98; "How to Fight the B-52," 182; response to Rolling Thunder, 23; studies B-52 jamming, 182
North Vietnamese air force: 921st Fighter Regiment, 114–15, 133; 923rd Fighter Regiment, 133; Jet order of battle, 149; MiG loses, 158
North Vietnamese antiaircraft, 41, 100, 111; early estimate of, 12; PUAZO-5 director, 16; S-60 anti-aircraft gun, 16
Nowell, Larry, 165–66
NTDS. *See* Naval Tactical Data System

O'Brian, J. T., 32
Oh Shit Light, 76
Oklahoma City (CLG 5), 152
Oriskany (CVA 34), 65
Ostrozny, Norbert J., 183
Outlaw, Edward C., 27

Pacific Air Force Joint Intelligence Center, 111
Parks, Hays, 206
Paterson, Pat, 5
Pathet Lao, 151
Peacetime Aerial Reconnaissance Program, 211
Pemberton, Gene, 98

INDEX   303

People's Republic of China, 111
PIRAZ. *See* Positive Identification Radar Advisory Zone
Pitchord, John J., 79
Point Yankee, 136
Porter, Melvin F., 106, 154, 155
Positive Identification Radar Advisory Zone, 141, 143, 144, 145; value of SPS-48E radar, 164
Powers, Gary, 27
Powers, Trent R., 65
Price, Alfred, 17, 25, 41, 52, 53, 66; estimate of SA-2 launched, 208n
Project Hammock, 112, 123, 138
Project Lamplight, 142–43
Project Left Hook, 61
Project Problem Child, 48
Project Quick Look, 125
Project Rivet Rambler, 180
Project Rivet Top, 127
Project Seek Dawn, 139–40
Project Shoe Horn, 44–45, 47, 58
Project Teaball, 133, 162; effect on kill ratio, 163; operations center, 163; system delays, 164
Project United Effort, 63, 65, 68
Project Vampyrus, 50, 52
Putnam, Benjamin R., 76
PVO-Starny, 156

Queer Whale, 26
Quick Reaction Capability (QRC) program, 29

radar direction finders: APA-69A, 24, 32; APD-4, 7, 18–19
radar homing and warning receivers: ALR-18, 180
ALR-31, 97, 170
ALR-42, 172
APR-17, 7
APR-20, 190

APR-23, 55–56, 64, 65, 66; with AGM-45, 86–87
APR-24, 55–57, 89, 212; display, 90
APR-25, 20, 89, 108, 170, 175, 180; B-52D Phase V, 180; F-105D installation, 94, 97; service in F-4Bs, 89–90; Wild Weasel IV-A, 175
APR-26, 20, 109, 174; F-105D installation, 94, 97; Wild Weasel IV-A, 175; with AGM-78, 105
APR-27, 173
APR-30, 57
APR-32, 26
APR-35, 170
APR-136, 108, 170; F-105D installation, 94
APR-137, 108, 170
APS-107, 92–93
APS-107B, 104
CMR-312, 58
DPN61, 92
IR-133, 71–75, 77, 79, 81; in F-105D, 94
QRC-153-2, 69
QRC-327A, 104
Vector IV, 70–74, 79, 81; F-100D installation, 92; variations produced, 93
WR300, 75, 93
radar homing and warning systems: CIA System XII, 27–28, 70; CIA System XVII, 67; CIA System XV, 28
radar pulse analyzers: ALA-3, 32; ALA-74, 7
radar receivers: ALQ-53, 34–35; ALQ-76, 35; ALQ-86, 35; APL-5B, 24; APR-4, 18–19; APR-9, 17, 18
radar receivers, panoramic: ALR-8, 32; ALR-30, 26; APR-1, 17; APR-6, 17; APR-35, 108; ARR-13, 24; ER-142, 108; IR-133, 59, 171–75
radars, air search: SPS-48E, 164–65

radars, airborne: APG-54, 87; APS-20, 17, 126
radars, airborne interception: APQ-35, 30
radars, fire control: B-200 (USSR), 37; Fire CAN (SON-9 USSR), 16, 25, 32, 34, 50, 55, 56, 66, 90, 99, 100; Whiff (SON-4 USSR), 16, 25, 28
radars, guided missile fire control: Low Blow (SNR-125 USSR), 201
radars, height finding: APS-45, 119, 121; Side Net (PRV-11 USSR), 201
radars, search and acquisition: Flat Face (P-15 USSR), 77, 116, 189; Spoon Rest (P-12 USSR), 37, 39, 77, 116–17, 148, 188–89, 191
radars, search and early warning: A-100 (USSR), 37; APS-95, 119; Back Net (P-80 USSR), 77; Bar Lock (P-35/37 USSR), 77, 151–51; Knife Rest (P-10 USSR), 26, 77; Tall King (P-14 USSR), 189; SPS-48E, 164–65
radars, tail warning, APS-54, 70
radars, target acquisition and guidance: RSNA-75 (USSR). *See* Fan Song
radars, Target Identification and Acquisition System, APS-117, 88
radio delay system, ARC-89, 122
Ranger (CVA 61), 10, 26
RCA, 55
Reboli, William P. Jr., 121
Red Crown, 140–41, 147, 163; advantages over Teaball, 164; CIC, 164–65; duties, 144; radar coverage, 164
Reddel, Carl, 130
Republic Aviation, 108
Reynolds, R.C., 100
RHAW, 94, 95; equipment summary, 59. *See also* radar, homing and warning
Rissi, Donald, 191, 193
Roberge, Francis D., 42
Roberts, Thomas S., 121

Robinson, Bob, 75
Rochester, Nathan, 142
Rock, Edward T., 97
Rocket Launchers: LAU-3, 79; Lau-77, 104
Rome Air Development Center, 204
Route Packages, 40, 50, 105, 148, 150, 151; map, 96
Rucker, Edward A. "Count," 84
rules of engagement, 169
Russell, Duane J. "Jack," 85
Russell, Glenn A., 199
Russian air defense specialists, 40
Ryan Aeronautical, 65
Ryan, John D., 109, 158

Sacramento Air Logistics Center, 125, 136
SAGE air defense computers, 140, 142
SAM sites: difficulty in locating, 60–61; discovery of, 30; number of, 148
Sames, Monroe E., 98–99
Sapp, Danny, 88
Sather, Richard C., 3
Schreader, George, 123, 132
Schuster, Carl, 68
Schweinfurt raid (WW II), 208
Sacramento Air Material Area, 92
Sea Trap program, 125–27
SEAFOR. *See* Southeast Asia Operational Requirement
"Second Generation Weaponry in SEA," 106
SEE SAM: ALR-31, 97; QRC-327A, 104
Selmanovitz, Hyman "Mart," 135
Senate Ad Hoc Subcommittee on Tactical Air Power, 173
Sensabaugh, Gerald R. Jr., 28
Sharp, Ulysses G., 41, 42, 57, 211; and Maddox Incident, 5; requests multi-service collaboration, 63
Sherwood, John, 57

Side Net, 116
SIDs, 88
SIGINT, 112, 132, 138: and Maddox Incident, 3, 5; assets, 213; Red Crown, 140
Signal Processors, AS-400, 17
SIOP (Single Integrated Operations Plan), 177
Sjolund, David, 202
SKQ-1 telemetry station, 152
Smart, Rudy, 188
Smeteck, Ronald T., 163n
Smith, Bernard "Barney," 84
Snake Eye, 66
Southeast Asia Operational Requirement, 125
Southworth, Harrison B., 65
Soviet National Air Defense Force, 156
Spravleniye (half-angle correction) mode, 38
Stalin, Joseph V., 36
Stanford University System Engineering Laboratory, 70
Steffen, Bob, 194
Stockdale, James B., 4–5
Strategic Air Command, 191; and B-52G ECM, 195; and loss of B-52Gs, 199; assesses bombing campaign, 200; assessment of North Vietnam defenses, 198; B-52 tactics, 180; befuddled by losses, 202; domination of AF strategy, 13; Linebacker II planning, 181–82; permission to maneuver, 195; Project Left Hook, 61; transfer planning to Eighth Air Force, 205
Sullivan, Glenn, 202
Supplemental Radio space, 166–67
SupRad. *See* Supplemental Radio space
SUU-30 submunition dispenser, 171
Swanhausser, Robert, 65
Systems Engineering Group Wright-Paterson, 54

TACC. *See* Tactical Air Control Center
TACOS, 26
Tactical Air Command: lack of assets, 14; limits R&D testing, 28; mentality, 29; need for training, 14; nuclear strike role, 14
Tactical Air Control Center, 113–14, 127; NS, 139, 158
tactical ELINT, 21
Tactical Jamming System, 173n
Tactical Warfare Center, Eglin AF, 108; Wild Weasel evaluation, 92
Tactics and Techniques of Electronic Warfare, 121
Tan Son Nhut Air Base, 118
Target Identification and Acquisition System (TIAS), 88; TPS-118, 104
Task Force Ozark, 69
Task Force Sue, 69
Teaball. *See* Project Teaball
Telford, Joseph W., 92–93
Thang, Nguyen, 193
The 11 Days of Christmas, 208, 210
The Hunter Killers, 41, 73
Thomas, George G., 136–37
Thomas, Guy, 166
Thomas, Robert, 193
TIAS. *See* Target Identification and Acquisition System
Ticonderoga (CVA 14), 3–5
Tillman, Barret, 24
Topeka (CLG 8), 141
Treokh Tochek (three-point) mode, 38
Trier, Robert D., 79
Tripoli (LHA 7), 137
Turner Joy (DD 951), 4–5

U.S. Air Force:
2nd Air Division, 32, 34, 41, 78, 112
Aviation Systems Division, 28–29
Security Service, 130; 6908th SS, 161, 163; 6924th SS, 112, 138, 147; at Nakhon Phanom, 162;

IFF interrogation, 126; Ironhorse established, 139; Korean War COMINT, 158

Seventh Air Force, 98, 112; and Linebacker II, 184; discontinues beacon jammer, 158; headquarters, 113; MiG losses, 134

squadrons, fighter: 13th TFS, 95; 67th TFS, 12; 354th TFS, 115; 355 TFS, 97–98; 561st TFS, 170; 6234th TFS, 78

squadrons, tactical electronic warfare: 42nd TEW, 151

squadrons, training: 9th TRS, 18; 15th TRS, 29

squadrons, strategic reconnaissance: 82nd SRS, 132

squadrons, Wild Weasel: 17th WWS, 170

wings, bomb: 17th BW, 19; 96th BW, 183

wings, airborne early warning: 552nd AEW, 118, 125

wings, fighter: 335th TFW, 50, 65; 27th TFW, 74, 76; 338th TFW, 98; 357th TFW, 104; 366th TFW, 170; 388th TFW, 53, 78, 132, 136

wings, strategic reconnaissance, 55 SRW, 6; 4080th SW, 62

U.S. Marine Corps squadrons VMC-3, 30; VMCJ-1, 30–31, 146; VMCJ-2, 34

U.S. Military Advisory Command (MACV), 7–8

U.S. Navy Squadrons VA-164, 65

U.S. Navy: Seventh Fleet, 113, 140, 184

U.S. Navy aviation squadrons:
 airborne early warning: VAW-13, 24, 26
 airborne reconnaissance: VQ-1, 16–17
 attack: VA-23, 86; VA-74, 65; VA-75, 64, 104
 air test and development: VX-5, 44
 electronic attack: VAQ-132, 171–72
 fighter: VF-14, 56–57; VF-142, 57; VF-142, 57
 light photographic: VFP-63, 27

Van Nederveen, Giles, 23
Van Staavern, Jacob, 8–9, 63
Van Dinh, The, 193
Vershinin, Kostantin A., 9
Vietnam People's Air Force, 113, 148
visual identification requirements, 156
Vito, Carmine, 36
Vittoria, Andrews N. Jr., 190
Vogt, John W., 158, 161, 162, 164, 182
Voland, Bud, 29

Wainright (DLG 28), 143
Weapons Evaluation TESTS (WEXVAL), 19–20
Westinghouse Electric Corporations, 55, 108
Whatley, Douglas, E. "Doug," 93
Wild Weasels, 214; AGM-45 use, 98, 107; AGM-78 use, 151; destruction first SAM site, 80; F-105G, 170; first combat loss, 79; first mission, 79; Nellis school for, 104; North Vietnam tactics to defeat, 197; objectives, 78; rules of engagement, 79–80; shortcomings, 81, 88; tactics, 81;
Wildeboore, Edward, 191
Willard, Gary A. Jr., 78–79
Williamson, William B. "Willie," 73, 92
Wilson, Hal "Red," 194
Worden (DLG 18), 136–37

Y Service, 158
Yankee Station, 136
Young, James Jr., 98

Zalogz, Steven, 13, 156–57
Zucker, Eugene M., 13

ABOUT THE AUTHOR

Thomas Wildenberg is an award-winning scholar with specialization in aviation, aviators, and technical innovation in the military. He is the author of many books, including *The Origins of Aegis: Eli T. Reich, Wayne Meyer, and the Creation of a Revolutionary Naval Weapons System* (2024) and *Fighting in the Electromagnetic Spectrum: U.S. Navy and Marine Corps Electronic Warfare Aircraft, Operations, and Equipment* (2023), both with the Naval Institute Press.

The Naval Institute Press is the book-publishing arm of the U.S. Naval Institute, a private, nonprofit, membership society for sea service professionals and others who share an interest in naval and maritime affairs. Established in 1873 at the U.S. Naval Academy in Annapolis, Maryland, where its offices remain today, the Naval Institute has members worldwide.

Members of the Naval Institute support the education programs of the society and receive the influential monthly magazine *Proceedings* or the colorful bimonthly magazine *Naval History* and discounts on fine nautical prints and on ship and aircraft photos. They also have access to the transcripts of the Institute's Oral History Program and get discounted admission to any of the Institute-sponsored seminars offered around the country.

The Naval Institute's book-publishing program, begun in 1898 with basic guides to naval practices, has broadened its scope to include books of more general interest. Now the Naval Institute Press publishes about seventy titles each year, ranging from how-to books on boating and navigation to battle histories, biographies, ship and aircraft guides, and novels. Institute members receive significant discounts on the Press' more than eight hundred books in print.

Full-time students are eligible for special half-price membership rates. Life memberships are also available.

For more information about Naval Institute Press books that are currently available, visit www.usni.org/press/books. To learn about joining the U.S. Naval Institute, please write to:

Member Services
U.S. Naval Institute
291 Wood Road
Annapolis, MD 21402-5034
Telephone: (800) 233-8764
Fax: (410) 571-1703
Web address: www.usni.org

www.ingramcontent.com/pod-product-compliance
Ingram Content Group UK Ltd.
Pitfield, Milton Keynes, MK11 3LW, UK
UKHW012017021225
465666UK00003B/3